Prospects and Utilization of Tropical Plantation Trees

Prospects and Utilization of Tropical Plantation Trees

Edited by
Kang Chiang Liew

CRC Press
Taylor & Francis Group
Boca Raton London New York

CRC Press is an imprint of the
Taylor & Francis Group, an **Informa** business

CRC Press
Taylor & Francis Group
6000 Broken Sound Parkway NW, Suite 300
Boca Raton, FL 33487-2742

First issued in paperback 2021

© 2020 by Taylor & Francis Group, LLC
CRC Press is an imprint of Taylor & Francis Group, an Informa business

No claim to original U.S. Government works

ISBN-13: 978-1-138-33689-6 (hbk)
ISBN-13: 978-1-03-208625-5 (pbk)

Visit the Taylor & Francis Web site at
http://www.taylorandfrancis.com

and the CRC Press Web site at
http://www.crcpress.com

Contents

Preface

This book relates to the key areas for tropical plantation trees in terms of their prospects and utilization. In order to increase the breadth of this work, it encompasses the topics of growth performance, nursery practices, soil properties, planting stock production, raw material cellulose, anatomy, pulping and papermaking, fiber modification and the properties of wood composites.

The first six chapters address the upstream process of different tree species used in plantations and would be beneficial to students and workers in the field who wish to have some knowledge of the upstream process, especially for tropical tree species. Numerous references are attached for further information.

The remaining six chapters deal with the properties and utilization of plantation trees. The information provided on this downstream process allows readers to stay abreast of the different wood materials being converted into products. This book provides a background and framework for improvements in the tree plantation industry, emphasizing its prospects and utilization.

Kang Chiang Liew
Universiti Malaysia Sabah

About the Editor

Kang Chiang Liew graduated in the year 2002 with a Doctor of Philosophy in Wood Science from Universiti Putra Malaysia. He is currently an associate professor at Universiti Malaysia Sabah with expertise in pulp and paper, conventional and hybrid composites and other lignocellulosic-based utilization.

Contributors

Nurlaila Fauriza Abdul Rahman
Faculty of Science and Natural
　Resources
Forestry Complex
Universiti Malaysia Sabah
Sabah, Malaysia

Shirley Marylinda Bakansing
Faculty of Science and Natural
　Resources
Forestry Complex
Universiti Malaysia Sabah
Sabah, Malaysia

Michelle Boyou
Faculty of Science and Natural
　Resources
Forestry Complex
Universiti Malaysia Sabah
Sabah, Malaysia

Eunice Wan Ni Chong
Faculty of Science
Universiti Sains Malaysia
Penang, Malaysia

Hui Ching Chong
Faculty of Science and Natural
　Resources
Forestry Complex
Universiti Malaysia Sabah
Sabah, Malaysia

Nor Hayati Daud
Faculty of Science and Natural
　Resources
Forestry Complex
Universiti Malaysia Sabah
Sabah, Malaysia

Rahila Dahlan David
Faculty of Science and Natural
　Resources
Forestry Complex
Universiti Malaysia Sabah
Sabah, Malaysia

Melissa Sharmah Gilbert
Faculty of Science and Natural
　Resources
Forestry Complex
Universiti Malaysia Sabah
Sabah, Malaysia

Affendy Hassan
Faculty of Science and Natural
　Resources
Forestry Complex
Universiti Malaysia Sabah
Sabah, Malaysia

Crispin Kitingan
Sabah Forest Development Authority
Sabah, Malaysia

Julius Kodoh
Faculty of Science and Natural
　Resources
Forestry Complex
Universiti Malaysia Sabah
Sabah, Malaysia

Kang Chiang Liew
Faculty of Science and Natural
　Resources
Forestry Complex
Universiti Malaysia Sabah
Sabah, Malaysia

Mandy Maid
Faculty of Science and Natural
    Resources
Forestry Complex
Universiti Malaysia Sabah
Sabah, Malaysia

Rhema D. Maripa
Faculty of Science and Natural
    Resources
Forestry Complex
Universiti Malaysia Sabah
Sabah, Malaysia

Su Xin Ng
Faculty of Science and Natural
    Resources
Forestry Complex
Universiti Malaysia Sabah
Sabah, Malaysia

Mohd Hamami Sahri
Faculty of Science and Natural
    Resources
Forestry Complex
Universiti Malaysia Sabah
Sabah, Malaysia

Grace Singan
Faculty of Science and Natural
    Resources
Forestry Complex
Universiti Malaysia Sabah
Sabah, Malaysia

Yu Feng Tan
Faculty of Science and Natural
    Resources
Forestry Complex
Universiti Malaysia Sabah
Sabah, Malaysia

Bryant Jia Ming Wong
Sapulut Forest Development Sdn. Bhd.
Sabah, Malaysia

Norman Shew Yam Wong
Sapulut Forest Development Sdn. Bhd.
Sabah, Malaysia

Hardawati Yahya
Faculty of Science and Natural
    Resources
Forestry Complex
Universiti Malaysia Sabah
Sabah, Malaysia

1 Tropical Tree Plantation and the Economic Value Chain

Bryant Jia Ming Wong, Michelle Boyou and Norman Shew Yam Wong

CONTENTS

1.1 INTRODUCTION

Land use in the early stages of the 21st century has been defined by a more enlightened perspective on the mix of social, environmental, and economic factors that must be taken into account to ensure a sustainable way of life. Climate change and income inequality have seen large focus on a global scale. In Asia, and particularly in Southeast Asia, these effects are even more pronounced given the remaining levels of forest cover, combined with nations still mired in nation building and populations that are undergoing urbanization with massive exposure to the forces of globalization.

Demand for wood and wood products is growing in line with a rising global population, and this is expected to triple from current rates by 2050 (Midgley and Arnold, 2017). Supplying this demand provides massive economic and social opportunities for Southeast Asian states, particularly in the areas of industrial production and rural employment. On the other hand, this also creates threats to already diminishing

1

natural forest cover as pressure is put on forests that are unsuitable for large-scale industrial demand in order to satisfy the needs of the state and its stakeholders.

Much of the world's tropical rainforest is located in the Amazon Basin in South America, while the Congo Basin and Southeast Asia have the second and third largest areas of tropical rainforest, respectively. Tropical rainforest can also be found in the Caribbean islands, Central America, India, South Pacific, Madagascar, West and East Africa, Central America and Mexico, and parts of South America outside of the Amazon. The distribution of the world's tropical rainforest is scattered along the equator of the Earth, as can be seen in Figure 1.1. Asia has the highest percentage of forest plantation with 62% of the total area, followed by Europe with 17%, North and Central America with 9%, South America with 6%, Africa with 4%, and Oceania with 2% (Figure 1.2).

As time goes by, tropical rainforests are depleting at a worrying rate. Man has cleared vast areas of the forest for building materials and firewood and to utilize the cleared land for local animals and planting crops. The large-scale clearing of the tropical rainforests in Amazonia was principally to make way for cattle ranches, and in Asia, the tropical forests were cleared for palm oil and rubber plantations (Raj and

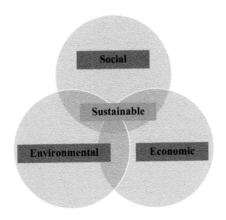

FIGURE 1.1   Sustainability and triple bottom line. (From Carle et al., 2002.)

Distribution of forest plantation area by region.

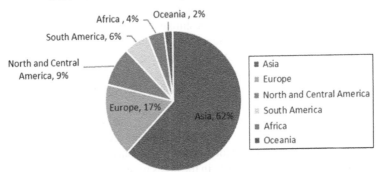

FIGURE 1.2   Distribution of forest plantation area by region.

Lal, 2013). Due to the notable growth of the world's demand for wood, the international market for wood is undergoing swift changes and has put a lot of pressure on the world's remaining natural forests.

Industrial timber planting on medium to large scales is key to addressing these opportunities and threats. A rudimentary SWOT analysis quickly demonstrates its suitability as a land use solution in response to economic, environmental, and social needs (Table 1.1).

Enters et al. (2004) lay out a process of integrating enabling and encouraging policies through direct and indirect incentives, based on studies of the Asian context as well as the history of timber planting internationally. Byron (2001) identifies key factors in the success of smallholder forestry. The studies and the strategies highlighted by the analysis above clearly indicate the state's major role in guiding a complex mix of public and private stakeholders.

Knowledgeable landowners and managers are essential in establishing productive plantations. Management is complicated by long payoff horizons as well as numerous

---

TABLE 1.1

SWOT Analysis on Land-Use Solutions in Response to Economic, Environmental, and Social Needs

Strengths
1. 300% more productive compared to natural forest yield per hectare.
2. Consistent, uniform supply of species and size in boles.
3. More efficient as a carbon sink.

Opportunities
1. Development of local supporting and downstream industries.
2. Rural development in and around planted areas.
3. Timber producing forest planting more attractive compared to cash crop planting.

S/O Strategies
1. (S1, S2, O1): Target economies of scale and associated benefits from consistent, predictable supply.
2. (S1, O2): Institute policies that enable and encourage smallholder timber planting, market access, etc.
3. (S1, S2, S3, O3): Institute policies that encourage timber planting in order to benefit from better ecosystem services.

S/T Strategies
1. (S1, S2, T1): Leverage higher efficiency of timber plantations in order to reduce forest clearing for industrial purposes.

Weaknesses
1. Less biodiversity compared to natural forests.
2. Requires large amount of initial investment with payoff horizons >5 years.
3. Management intensive.

Threats
1. Increased clearing of natural forest land to satisfy demand.

W/O Strategies
1. (W2, W3, O1, O2): Enable long-term planning by providing a stable business environment.
2. (W3, O2): Ensure that education and training are available and aligned with stakeholder needs.
3. (W1, O3): Ensure that planting plans and regimes take a holistic view towards sustainability.

W/T Strategies
1. (W1, T1): Produce and implement clear, precise land-use plans in order to ensure sustainability of forests and plantations.

other factors, including shortcomings in supporting industries and unpredictable markets, as well as the variety of options that are available. Thorough knowledge of the value chain of timber plantations will go a long way to ensure that correct decision making is exercised from day 0 of the planting regime.

1.2 HISTORY OF PLANTATION TIMBER IN MALAYSIA

The history of plantation forestry has been focused on expanding plantations' estates for wood production rather than to achieve a wider set of values and outcomes. Over the years, there has been a continuous change in plantation forestry, which occurred in response to several factors, which including the ever-growing global demand for forest products, market and technological forces in favor of plantation wood, the diminishing supply from natural forests, policy decisions designed to protect forests from being harvested, the declining competitiveness of other land uses, and the adoption of policies to promote plantation forests. The mix of public and private, priced and unpriced, ecosystem goods and services associated with plantation forests, and the spatial and temporal variability associated with many of them, means that developing policies to enhance their provision is particularly challenging (Kanowski, 2010).

Policy planning can be viewed from two perspectives: (i) vertically, as different levels of scale – international, regional, on-site, and operational – and (ii), in relation to time – the progress and implementation of a project, weekly work programme, or national policy as they develop (Evans and Turnbull, 2004). The emergence of 'new generation' approaches to environmental and sustainability policy is particularly relevant to the provision of ecosystem goods and services from plantation forests (Kanowski, 2010).

In Malaysia, forest plantations are not new in concept or in practice. The awareness of forest plantations has arisen from and is influenced by from fear of timber deficits. From Table 1.2, we can see the history of plantation forestry in the country, especially in Peninsular Malaysia. However, in Sabah and Sarawak, the situation is more straightforward and less illustrative. Figure 1.3 shows a *Acacia mangium* plantation located in Sabah.

When we talk about forest plantations in Malaysia, we have to look at three different areas, and these should be treated separately. Most of the prospects for and constraints of plantations took place first in Peninsular Malaysia. The States of Sabah and Sawarak were in a position to pick the best options suited to their specific needs following the trials and the experiences gained in Peninsular Malaysia. Table 1.3 shows a summary of the current status of forest plantation development in Malaysia in terms of area location, species, and ownership.

Timber plantations in Malaysia have not had the same success as in other countries due to a host of different factors. Using Byron's four keys as a guide quickly outlines the situation:

1. Clear tree ownership: Different situations in different states. Even if tree ownership is not in dispute, there are collections by forest departments once they are sold or transported, creating an extra expense for the grower compared to that of other crops.

TABLE 1.2
Summary of the Most Notable Events in the History of Forestry in Peninsular Malaysia

Year	Events
1877	Rubber (*Hevea brasiliensis*) planted in Kuala Kangsar.
1884–1900	Small trials of exotics started.
1900–1913	Regular plantations of gutta percha (*Palaquim gutta*) and rubber (*Hevea brasiliensis*); line planting of chengal (*Neobalanocarpus heimii*) in forest reserves; experimental planting on abandoned mining land.
1927–1941	Forest Research Institute set up in Kepong and experimental plantations in lowlands started; plantation experiments in Cameron Highlands (ca. 1,500 m asl); teak planted in Langkawi Island.
1945–1950	Experimental teak plantations in north-west Malaya; plantings in forest clearings resulting from disturbances during the war.
1952	FAO *Eucalyptus* study tour in Australia and extensive species trials with *Eucalyptus* spp.
1954–1958	Species trials with *Pinus* spp. with potential pulp value initiated; experimental plantations started on tin tailings; taungya system tried using *Gmelina arborea* in tobacco farms; line plantings of kapur (*Dryobalanops aromatica*) established in Kanching.
1959–1962	Large-scale experimental planting with *Pinus caribaea* and *P. insularis* in the lowlands. *Pinus* spp. from Central America and *Populus* spp. from Kenya also tested; experimental plantings in shifting cultivation areas; line planting and small scale plantings of secondary growth of *Dryobalanops aromatica, Eusideroxylon zwageri, Flindersia brayleyana, Fragraea fragrans, Khaya* spp., *Pentaspadon officinalis,* and *Shorea macrophylla.*
1963–1965	Bigger trials of *Pinus* spp. conducted in Selangor.
1966–1970	Under UNDP assistance, pilot plantations of quick growing industrial tree species initiated, mainly for production of pulp. Plantations of pine expanded in Selangor, Johore, Pahang, Negeri Sembilan and Kedah; Shorea and *Dryobalanops* spp. planted under the Taungya system in Negeri Sembilan. Jelutong (*Dyera costulata*) plantations expanded in Sungei Buloh F.R.
1971–1976	Mixed plantations of *Pinus* and *Araucaria* tested on poor soils in Bahau; enrichment planting using indigenous species become an important forestry practice.
1981–1992	The Compensatory Forestry Plantation Project through Asia Development Bank (Asia Development Bank) loan initiated. Quick growing tropical hardwoods like *Acacia mangium, Gmelina arborea,* and *Paraserianthes falcataria* chosen for producing general utility timber. Compensatory plantations come under review and the planting for sawlog production put on hold. Planting for pulp production continues.
1992–1996	Planting of teak begins earnestly even in wetter sites; Sentang (*Azadirachta excelsa*) also gains importance as a plantation species.

Source: FAO (2002)

FIGURE 1.3   *Acacia mangium* plantation located at Sabah.

2. Reliable markets: Relatively robust industry in Peninsular Malaysia, limited investment and modernization in Sabah.
3. Sympathetic legal and regulatory framework: Framework requires revision to enable and encourage planting of quality timber and to guide infrastructure decisions.
4. Robust technical package: Severely lacking.

The benefits of a thriving timber planting industry are varied and many. Wood processors can have access to a consistent, predictable, uniform supply of input material that will benefit them in creating economies of scale. Nursery and tissue-culture industries can develop in order to supply these timber plantations. More attention, resources, research, and innovation will be invested in beneficial tropical forest management, an area that must be addressed if sustainable goals are to be reached. Employment in forest plantations requires skilled and unskilled labor in rural areas as well as labor from highly trained researchers and managers, making them ideal in addressing employment concerns for Sabah and other developing areas in the 21st century.

Timber planting as an industry is underdeveloped in the state. Although the legal framework to establish medium to large-scale plantation areas was created through the Sabah Forest Management License Agreement (SFMLA) in 1997, supporting industries both upstream (timber nurseries, research and development) and downstream (primary/secondary/tertiary processors and their relevant suppliers) Integrated Tropical Plantation (ITP) entities themselves have not developed to a point of sustainable economic impact.

TABLE 1.3

Summary of the Current Status of Forest Plantation in Malaysia

State	Area	Location	Species	Ownership
Peninsular Malaysia	Divided into three principal types: 1. Specialty species (teak) 2. Pine plantations for producing pulp 3. General utility timber plantations (*Acacia mangium*)	In the 1950s, the teak plantations were initially tried out on an experimental basis in northern states of Perlis and Kedah. Then, in the 1960s, it expanded into larger-scale plantations. Finally, in the 1980s, teak planting expanded from Perlis to Johore.	Compensatory plantations, *Pinus* spp., teak and Sentang.	Forest Department, private sectors, and small holders.
Sabah	Divided into two types: 1. Industrial pulpwood production 2. High-quality timber species planted in reforestation schemes in degraded forests	In the 1990s, commercial plantations grow rapidly. The Forestry Department planted some but they are mainly trial plots in the Research Stations at Sibuga, Gum Gum, Kolapis, Segaliud Lokan, Telupid, Sosopodon, and Sook.	*Acacia mangium, Paraserianthes falcataria, Gmelina arborea, Eucalyptus deglupta, Tectona grandis, Eucalpytus* spp., *Pinus caribea,* and other spp. (*Dipterocarpus, Swietenia, Pterocarpus, Araucaria, Pinus*)	Forestry Department and private sectors.
Sarawak	1. Indigenous species of the Engkabang group for illipe nut production 2. Durian plantation 3. Indigenous hardwood species plantations	No major plantings were done after the initial trials until 1965, when interest in a 'Reforestation Research Programme' was initiated. This was to test out fast-growing exotic tree species, especially conifers, for reforesting land. In the early 1970s, the Forest Department began experimenting with some of the fast-growing exotic tropical hardwoods. The pace quickened, and during the period from 1991 to 1995, about 7,500 ha were planted.	*Acacia mangium. A. auriculiformis, Alstonia,* and *Dryobalanops* spp., *Anthocephalus cadamba, Anthocephalus chinensis, Araucaria hunsteinii, A. cunninghamii, Artocarpus* spp., *Azadirachta excelsa, Calamus* spp., *Ceiba pentandra, Dipterocarpus* spp., *Dryobalanops* spp., *Durio zibethinus, Eucalyptus* spp., *Gmelina arborea, Hevea brasiliensis, Paraserianthes falcataria, Parkia leucocephala, Pinus caribaea, Shorea* spp., *Shorea parvifolia,* and *Swietenia macrophylla.*	Forestry Department and private sectors.

Governments play a vital role in the development of plantation forests over time. This could be through direct public investment or involvement, for example, as in Australia, China, India, New Zealand, or the United Kingdom; or by facilitating private-sector investment, through financial policy and mechanisms such as facilitating access to financial policy and mechanisms as done, for example, in Brazil, Chile, Indonesia, Thailand, or South Africa (Kanowski, 2010). The United Nation's International Tropical Timber Agreement (1984) and the Convention on Biological Diversity (1992) are examples of such agreements that may influence national forest planning (Evans and Turnbull, 2004). At a national level, forest plans must be linked to plans for other sectors, and there should be a clear linkage between local and regional plans and the national forest and national development plans that address broader issues of economic and social development. National and sub-national policy regimes more likely to enable positive environmental and sustainability outcomes, as can be seen in Figure 1.4.

The Malaysian Timber Certification Council (MTCC) was established in Malaysia in 1998, and it is an independent organization set up to develop and operate the Malaysian Timber Certification Scheme (MTCS) in order to provide independent assessments of forest-management practices in Malaysia and also to meet the

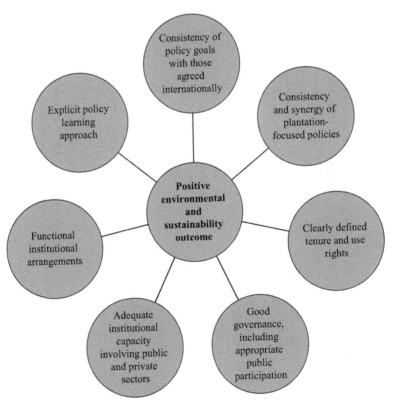

FIGURE 1.4 Characterization of national and sub-national policy regimes that enable positive environmental and sustainability outcomes.

demand for certified timber products. Certification is a market-linked tool to promote and encourage sustainable forest management and to provide an assurance to buyers or end consumers that the timber products they buy come from sustainably managed forests. The Malaysian Criteria, Indicators, Activities and Standards of Performance for Forest Management Certification (MC and I) is the standard that is used for assessing forest-management practices in the Permanent Forest Estate (PFE) at forest management unit (FMU) level for the purpose of certification. Certification is a market-linked tool to promote and encourage sustainable forest management and to provide an assurance to buyers or end consumers that the timber products they buy comes from a sustainably managed forests. Several factors affect the timber trade when timber certification is implemented, which include power, social, economic, and environmental effects (Figure 1.5).

Enters et al. (2004) emphasize the importance of consistent policy as the fundamental starting point for a state to implement and maximize the benefits that can be gained from a strong timber-planting industry. Whether or not it is recognized as such, the success of the timber industry in Peninsular Malaysia is an indicator of the value that can be generated from well-implemented timber plantations.

The most important step in understanding the value chain of industrial timber plantations is to be aware that timber plantations are not the first part of the value chain for finished wood products. Timber plantations must be treated as an industry in and of themselves in order to fully realize their value. Timber plantations require the availability of strong suppliers, supporting industries, and supportive policies that are essential to any industry in order to succeed. Suppliers include research, nursery, and tissue-culture laboratories able to provide appropriate planting materials as seeds or seedlings, chemical manufacturers able to formulate suitable fertilizers and/or insecticides, and logistics and equipment providers necessary to accommodate an industry with heavy dependence on physical infrastructure. Further requirements and opportunities are numerous, with remote sensing and information technology (IT), the Internet of Things (IoT), and data science all able to play a big part in a successful future for the industry. Failure to recognize these requirements will result in double the opportunities lost: employment and business opportunities for a wide spectrum of individuals and companies, and the failure of an industry that could have become a great provider of raw materials for other development opportunities.

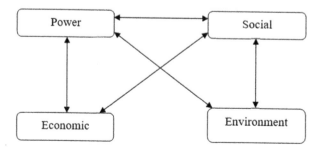

FIGURE 1.5   Effect types of timber certification on tropical timber trade.

1.3   VALUE PROPOSITION OF PLANTATION TIMBER

The value of plantation timber goes above and beyond its higher efficiency in production versus land area and much shortened rotation time of fast-growing species. Well-managed plantations are able to produce consistent, uniform supplies of raw material that enable producers of wood products to pursue economies of scale in a way that the unpredictable supplies of material types in the natural forest do not.

This also means that more benefits can be reaped per dollar spent in contribution towards establishing a productive value chain for plantation wood. Money spent in research and development in production and the products of the raw material will garner much higher returns per unit cost. Contributions from the state through money and non-monetary benefits are much easier to justify the larger the potential market for locally produced raw materials.

In tropical countries without a supply of plantation timber, economies of scale are either impossible to achieve due to the variety of species encountered in the natural tropical forest, or come at a high environmental cost, with large-scale clear-felling being necessary in order to produce a consistent supply of certain types of timber. These costs manifest themselves in the economic evaluations of manufacturers in different ways.

Most types of machinery have to be adjusted to specific species in order to achieve the necessary efficiency, for example, rotary peelers for veneer with specific blade angles, or kiln drying protocols for different wood types. Switching costs of this nature can significantly reduce the efficiency of processing facilities, both in time lost to these processes as well as reduction in recovery from raw material to finished product.

Another major cost that is often not accounted for properly is the logistical cost for manufacturers. Mills that are only able to accept a small number of species types will accrue higher logistical costs the larger the number of suppliers they must turn to in order to sufficiently supply their production capabilities. Logistical costs become cumulatively lower the larger the load that is transported per trip from supplier to buyer. In order to achieve these economies of scale, the manufacturer may have to resort to waiting for a sufficient supply of specific wood types to build up in a certain number of supplies (which may have an adverse effect on freshness if dealing in round logs, which in turn will have a very adverse effect on the mill's recovery rate). They may attempt to purchase specific wood types from multiple suppliers, which may not be acceptable to most suppliers, as most buyers are mainly interested in similar log types, leaving them to deal with a high volume of wood type that may not have a market in that situation.

On a macro scale, not having a supply of raw material from which economies of scale can be derived can have devastating opportunity costs that are impossible to numerate. As mentioned, economies of scale also result in higher returns from each dollar spent on research and development. If there is no potential for this to be achieved, there may not be any dollars at all spent on research and development, since there is no way for the spender to project sufficient returns for their initial outlay. An example of this type of supply-driven encouragement of market development is Peninsular Malaysia's furniture industry. Malaysia had furniture exports of RM

7.9 billion in the 6-month period from January to August 2018. The industry is still mainly driven by the export of rubberwood furniture, which only developed because of an excess supply of rubberwood timber due to the vast planting of rubber trees for latex production and the subsequent supply of aging trees as well as the salvaging of trees from clearing caused by the declining market for latex. This consistent, uniform supply of a singular type of wood meant that great potential returns were to be made, having been harnessed correctly.

1.4 IMPLEMENTATIONS OF TIMBER PLANTATIONS

Numerous economic forces have contributed to reduced forest cover in tropical forests. The immense clearing of the tropical rainforests in Amazonia was principally to make way for cattle ranches. In Asia, tropical forests were cleared for palm oil and rubber plantations (Raj and Lal, 2013). However, as the global population increases, so does the demand for wood products, leading to increased pressure on rapidly depleting forest lands to supply this growing need. Tropical tree plantations are key in responding to these conflicting forces, as sustainable, efficient sources of in-demand, valuable raw material.

Given the myriad demands placed on tropical forest lands in terms of competing land use, the demand for wood and non-wood forest products, as well as ecosystem services on both local and global scales, it is imperative that the state and all other stakeholders take into account more inclusive methods of evaluating value contributions than purely financial measures when making broad landscape-scale decisions. This is the case even though tree plantations have proven to be more lucrative in terms of their overall contribution to production than palm oil in cases where the government has put in place an ecosystem suitable for them to exist.

The Global Forest Resources Assessment (FRA) programme has been collecting statistical information regarding the classes of 'natural resources' and 'plantations'. In 2005, FRA introduced two additional forest classes: 'modified natural forests' and 'semi-natural forests'. The new classification, which includes primary, modified, semi-natural, and plantation forests, is based on the degrees of human intervention and method of regeneration. Table 1.4 presents these categories with their brief descriptions.

Development of tropical plantations can be traced back to the 16th and 17th centuries with the expansion of European influence by colonial powers, whereby the exploitative timber export trade was encouraged and resulted in the damaging of natural forests (Evans, 2009). Before 1900, plantation activities included the introduction and testing of exotic species, especially teak and eucalypts, and the introduction of taungya and irrigated plantations. Evans also mentioned that the initiation of government agencies, employment of trained foresters, and definition of forest policies and legislation provided an institutional framework on which extensive forest plantations of the 20th century could be based.

Nambiar (2018) notes that plantation forests are renewable natural resources primarily managed for growing wood for a range of purposes. In addition to being a source of a key raw material, two important roles of plantation forests in environmental terms are to provide ecosystem services which includes carbon absorption

TABLE 1.4

Forest Classifications or Categories According to Global Forest Resources Assessment (FRA)

Classification			Description
Primary			Forest of native species, where there are no clearly visible indications of human activities and the ecological processes are not significantly disturbed.
Modified natural			Forest of naturally regenerated native species where there are clearly visible indications of human activities.
Semi-natural			Assisted natural regeneration through silvicultural practices for intensive management.
Planted forests subgroup	Natural		Forest of native species, established through planting, seeding, coppice.
	Plantation	Productive	Forest of introduced species and, in some cases, native species, established through planting or seeding mainly for production of wood or non-wood forest products.
		Protective	Forest of native or introduced species, established through planting or seeding mainly for provision of services.

and landscape restoration. He also stated that the share wood supply from planted forests is set to increase from the present 50% to 75% by 2050, and the per capita consumption of wood is also on the rise in highly populous countries where economies are growing rapidly. The supply of plantation resources is increasing at a yearly rate of about 2.5 million hectares, largely in sub-tropical and tropical environments in Asia and South America and to a much lesser extent in Africa. Most plantations in the tropics are to provide wood for fiber-based industries and their significant economic, environmental, and social impact.

In most countries, forest plantations represent a way to supply the ever-increasing demand for wood on a sustainable basis. However, there a several factors that affect and become a constraint in plantation forestry. These factors include ecology, land, site selection, species selection, inadequate supply of elite planting material, labor and mechanization, and economic factors, as can be seen in Table 1.5.

Based on the information provided by the Center for International Forestry Research – CIFOR (2001), the elements within the overall definition of 'plantation' are distinguished by purpose and followed by nature which includes stand structure and composition. The elements which fall outside (or half outside) a 'plantation' are described mainly by their nature and the means of establishment by which they are managed. Table 1.6 shows the types of plantations that are distinguished not only by their different purpose, but also by their spatial scale, management intensity, structure, and ownership (Bauhus et al., 2010).

It has been emphasized that well managed tree-planting in the tropics engages all parts of the triple bottom line. Planting for firewood, for other village needs, in agroforestry development, and for protection, which reduce soil erosion, control

TABLE 1.5

Constraints in Plantation Forestry According to Raj and Lal (2013)

Factors	Description
Ecology	The establishment of forest plantation involves extensive alteration of the ecosystem, particularly when heavy equipment is employed. The complex closed nutrient cycle in tropical rainforests is disrupted for a long time, which can lead to a reduction in productivity unless ameliorative measures are undertaken. Monocultures further destabilize the system and require heavy use of fertilizers and pesticides. Many slow-growing species grown under fast plantations are of poorer quality. There is the problem of species-site matching for the heterogenous area of large plantations, and the danger of forest fire may also increase in exotic species plantations.
Land	For a forest plantation investment to be commercially viable, a large area is required. The size of the piece of land required will vary with the objective of the plantation. If the timber is for sawmilling and furniture manufacturing, then a smaller area would suffice. If the objective is to establish a chip or pulp and paper mill, then larger areas are required for raising plantations. It is always preferable have a single contiguous piece of land, one that is close to basic amenities and has an accessible road system and is within an economic range of a processing mill or market.
Site Selection	It is important to know about the local site conditions and silviculture in order to decide which species is to be raised under what conditions. It is not advisable to raise some susceptible tree species in unsuitable soil and climatic conditions. Susceptible species are at huge risk of pest and disease attack and failure to produce the desired yield. It is also advisable to have knowledge of the plant and root growth of a particular tree species under the site conditions. Soil and climatic conditions are important factors in the success of a plantation programme.
Species Selection	For plantation forests, although indigenous species are available, there is a greater preference for the selection of exotic species. This is because there is generally a lack of adequate knowledge on the propagation and silvicultural management of indigenous species. There is a plentiful supply of seeds of exotic species, exotic species are easy to handle, and exotics are fast growing and high yielding.
Inadequate supply of quality planting material	High levels of productivity are achieved when the genetic and physiological potential of the species is well matched with management practices which promote rapid growth. Valuable improvements can be made in important properties such as stem form and wood density through selection and breeding. One major constraint that is currently perceived is the shortage of good planting material for the various plantation programs. Quality seeds and trees that have been selected and reproduced by vegetative methods are inadequate to meet current and projected needs.
Labor and mechanization	Although the labor requirement in forest plantations is less than in agriculture, it still has to compete for labor in an expanding economy. One option to alleviate the labor shortage is increased mechanization. Machines developed in countries like Finland and Canada, for example, are environmentally friendly and highly flexible in their operation in forest plantations.
Economic factors	High initial capital investment to establish forest plantations is required. The long period that elapses between initial planting efforts and harvesting is another important concern, and there is thus corresponding concern for interest being carried until harvesting period. The high biological and economic risks involved in forest plantations are major limiting factors. Unattractive and inappropriate incentives provided by the Government for forest plantation investments results in a lack of interest on the part of private sectors in commercial ventures in forest development.

TABLE 1.6
Typology of Planted Forests According to CIFOR (2001)

Typology	Description
Industrial plantation	Intensively managed forest stands established to provide material for sale locally or outside the immediate region, by planting or/and seeding the process of afforestation or reforestation. Individual stands or compartments are usually with even age class and regular spacing and: • Of introduced species (all planted stands) and/or • Of one or two indigenous species • Usually either large scale or contributing to one of a few large-scale industrial enterprises in the landscape.
Home and farm plantations	Managed forest, established for subsistence or local sale by planting or/ and seeding in the process of afforestation or reforestation with even age class and regular spacing. Usually small scale and selling, if at all, in a dispensed market.
Environmental plantation	Managed forest stand, established primarily to provide environmental stabilization or amenity value, by planting or/and seeding in the process of afforestation or reforestation, usually with even age class and regular spacing.
Managed secondary forest with planting	Managed forest, where forest composition and productivity are maintained through additional planting or/and seeding.
Managed secondary without planting	Managed forest, where forest composition and productivity are maintained through natural regeneration processes, which can include the use of seed trees.
Restored natural/ secondary forest	Restored forest, through either planting or/and seeding or through natural regeneration processes, where restoration aims to create a species mix and ecology approaching that of the original natural forest.

water runoff, combat desertification, and provide shelter and shade, play vital roles. Table 1.7 summarizes the various types of plantation development. Four main types of tree-planting projects can be identified, and each of these types plays an important role in contributing to the development of plantation forests.

Tree plantations can be considered as an alternative way to restore forest landscapes, at least in the short term and especially on very badly degraded soils. Montagnini states that tree species chosen for a plantation in the context of forest restoration can provide benefits in the form of tree products such as timber, fuelwood, and leaf mulches, and in terms of their ecological effects, for example, nutrient recycling, or attracting birds and other wildlife to the landscape.

There are several factors that influence the choice of a tree species for plantations, depending on whether both productive and ecological advantages can be achieved in the same system. Usually, the preferred choice for restoration would be natural regeneration within the context of a forest landscape. Planting would only be a second option in cases where natural regeneration cannot proceed due to obstacles such as poor soil condition, long distances to seed sources, isolation, or invasion by aggressive grasses. There should be a balance of socioeconomic goals in terms of productivity, for instance, and biodiversity objectives for restoration, within a landscape context. Table 1.8 shows factors that influences the species choice for plantations.

TABLE 1.7
Types of Tree-Planting Projects

Types of Plantation Development	Description
Planting by private companies	Almost entirely carried out by commercial enterprises (includes all plantation development in Brazil and much of it in countries such as Indonesia, Philippines, South Africa, Swaziland, and Thailand). In some countries, governments initially established a critical mass of plantations, but many are now ceding ownership and responsibility for plantation establishment and management to the private sector. Some government incentive is provided to assist with the costs of plantation establishment.
Specific development projects	Often for creating an industrial resource, but organizations funding projects are as much concerned with aiding investment and development in developing countries as with the direct commercial profitability of the operation.
National afforestation programs	Often forms a part of a nation's planned national expenditure and invariably includes tree-planting for non-industrial purposes.
Small-scale tree-planting projects	Promoted by private companies and non-governmental organizations (NGOs).

Once an area has been identified as an industrial planting zone, however, it is imperative that policy is in line with creating an economic environment that is conducive to long-term success. Industrial timber plantations are the most efficient and effective option for land use to accommodate the economic demands of the people and the state, and, with effective land-use policy, are necessary to the survival of other, environmental-focused land use of tropical forests.

1.5 VALUE CHAIN OF TIMBER PLANTATIONS AND CONSIDERATIONS

The most important step in understanding the value of industrial timber plantations is to be aware that timber plantations are not the first part of the value chain for finished wood products. Timber plantations must be treated as an industry in and of itself in order to fully utilize their value. Timber plantations require the availability of strong suppliers, supporting industries and supportive policies that are essential to any industry in order to succeed. Suppliers include research, nurseries and tissue culture laboratories able to provide appropriate planting materials as seeds or seedlings, chemical manufacturers able to formulate suitable fertilizers and/or insecticides, as well as logistics and equipment providers necessary to accommodate an industry with heavy dependence on physical infrastructure. Further requirements and opportunities are numerous, with remote sensing and IT, IoT and data science all able to play a big part in a successful future. Failure to recognize these requirements will result in double the opportunities lost: employment and business opportunities for a wide spectrum of individuals and companies, and the failure of an industry that can become a great provider or raw materials for other development opportunities.

TABLE 1.8

Factors Influence Species Choice for Plantations (Montagnini, n.d.)

Factor	Description
Target Ecosystem Productivity and Biodiversity	Fast-growing, native pioneer species with high productivity are recommended for the initial stages of restoration of degraded lands. These species help in facilitating the environment for later successional, longer-lived species whose end products are more valuable. This means that the timber is of better quality.
Saving Endangered Local Species	Fast-growing exotic species such as eucalypts, acacias, or pines should be used only when there are no available seeds of native species or where the environmental conditions are just too harsh for any native species to survive. Exotic species predominate both in industrial and rural development plantations worldwide; however, native trees are more appropriate than exotics for the following reasons: a. Better adapted to local environmental conditions b. Seeds are more available c. Farmers usually familiar with species and their uses d. Helps to preserve genetic diversity and serves as habitat for the local fauna The following are the disadvantages of the use of native species: a. Uncertainty regarding growth rates and adaptability to soil conditions b. General lack of guidelines for management c. Large variability in performance and lack of genetic improvement d. Seeds of native tree species often not commercially available and have to be collected e. High incidence of pests and diseases
End use of products	Half of forest plantations are for industrial uses such as timber and fiber, and one quarter are for non-industrial uses such as for home or farm construction, local consumption of fuelwood and charcoal, and poles. Among the non-specified uses are small-scale fuelwood plantations and plantations for wood to dry tobacco. Species choices reflect the end use of each plantation while considering the purpose of forest restoration.

The phrase *value chain* refers to a set of interdependent economic activities and to a group of vertically linked economic agents; depending on the scope of the study the focus of the analysis can be on the activities or on the agents (Bellu, 2013). Bellu also stated that a value chain starts with the production of a primary commodity, ends with the consumption of the final product, and includes all the economic activities undertaken between these phases, such as processing, delivery, wholesaling, and retailing. Applying this definition to timber plantations is troublesome because in this case, timber, a primary commodity, is itself an output of a group of economic decisions and processes. Figure 1.6 illustrates the simple value chain for a timber plantation.

However, taking this as the value chain will still lead to a loss of value for the planter due to the nature of marketing timber and timber products. Pricing of round logs is determined by a tiered pricing model where the price per volume unit increases in steps according to diameter classes, as shown in Figure 1.7.

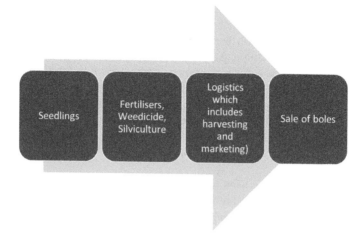

FIGURE 1.6   Value chain without processing for a timber plantation.

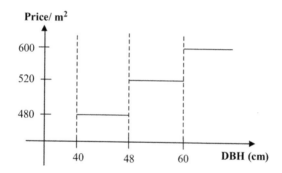

FIGURE 1.7   Tiered pricing graph.

Given these variables, it is important that managers are aware of the mix of characteristics that they are attempting to plant for and how this will affect the returns they will achieve. By understanding thoroughly the value chain and how the decisions made will influence these variables, managers can mitigate, or at least minimize, the market risk that they are exposed to.

This is due to differing recovery rates when the round log is processed further, with larger diameter logs typically leading to higher recovery rates and better quality product. The manager, then, must find the best way to maximize value from two inversely related variables: Do they plan to grow a large volume of small-diameter logs, or do they grow a smaller number of higher value large-diameter logs?

Managers thus face a series of decisions with numerous tradeoffs from the time the decision has been made to plant industrial timber. Making these decisions incorrectly can have catastrophic effects on the realizable value of the plantation, with the worst-case scenario somewhere along the lines of wasting 8 years of time and investment capital over tens of thousands of productive land only to find that they now have a product that has no realizable value.

It is imperative, therefore, that the manager takes a broader approach when analyzing the value chain of timber plantation to account for the differences in value for different processors. This is further complicated because different processes will require different characteristics and qualities of raw material.

The success of tropical timber plantation establishment in Southeast Asia has been mixed. Sustainable and thriving timber plantations can bring numerous benefits: less deforestation in pursuit of timber raw materials, more efficient land use, rural development, and more efficient carbon offset (Evans and Turnbull, 2004). These goals can all be achieved while creating a consistent, profitable supplier of raw materials to industries that can further benefit the development of a state.

While these benefits have been well understood by most stakeholders involved with Southeast Asian tropical forests, tropical plantations have thus far typically fallen into two categories: large-scale plantations for feed stock for pulp and paper mills, and small to medium-scale establishments focused around smallholder planting. Other common types are small to medium-size privately held plantations of high-value, slow-growth species such as teak or agarwood. For various reasons, most of these regimes are neither of the appropriate scale or type necessary to maximize land use for the benefit of the state and its stakeholders. Figure 1.8 shows land clearing to make room for a teak plantation located in Sabah, while Figure 1.9 shows a small patch of forest left in the middle of land-clearing.

On the other hand, a lot of damage can be done in assuming that a one-size-fits-all approach to tropical plantations is the better option. Land use and, in particular, forest use is a dynamic matter that involves stakeholders ranging from entire states that count on raw materials for development, to sun bears and hornbills, to butterflies that are endemic to particular forest types not found anywhere else on Earth. Any approach to maximizing utilization of these lands must take account of all the benefits and compromises that will impact the long-term sustainability of the state and its forests.

FIGURE 1.8 Land clearing to make room for teak plantation.

FIGURE 1.9 Small patch of forest left in the middle of land clearing.

The principal cause of these inefficiencies can be boiled down to improper decision making. Due to the scale of both time (fast-growing species take between 5 and 12 years to reach economical rotational age) and land use (industrial plantations ranging from 1,000 hectares to 100,000 hectares and beyond), the effects of misinformed or misguided decisions made during day 1 of a plantation regime can be exacerbated in multiples over periods of time during which the entire industrial landscape can have shifted numerous times.

In order to realize the maximum efficiencies and benefits of sustainable tropical timber plantations, consideration must be given to the entire value chain, from seed to finished product. This chapter will attempt to illustrate why this is so and hopefully be a useful guide for how this could be achieved. The value of a tropical timber plantation consists of two variables: the economic value that it produces, versus the land that is being used to support it. These are negatively related values when determining the overall value, with the aim of a plantation being to maximize economic value while minimizing the area used to produce.

With this in mind, the overall economic value to the state must now be properly considered. The only way to properly determine this value is to consider the entirety of the value chain, from planting material to finished product. Writing on the importance of raw material use in a state's economy, Jacobs (2000) suggests that focusing on import stretching, the maximization of value attained from the materials being brought into a state, is key to the goals of increasing employment and increasing value generated from those employed. She further suggests that a state's naturally occurring raw materials be considered a form of natural import that the state should treat with the same care and focus. Timber plantations can be a key part of a well-performing state by maximizing the production of such 'naturally occurring' timber raw materials that have the potential to supply diverse industries ranging from commodity products (pulp products, veneer products, sawn lumber) to specialized items (architectural elements, high-end furniture) and beyond (wood-based plastics, mass engineered timber, etc.). This diversity of use, and the industrial and economic potential derived from it, cannot be matched by other raw materials and must be accounted for when making land-use decisions.

The consequences of improper planning are potentially astronomical. If the output of the plantation is not of the type desired by downstream processors, the planter may find that he or she has an inventory of excess stock that is costly to maintain and getting costlier the more time passes after the optimal rotational age of their stock has been reached. The liquidity of raw wood materials can further be affected enormously by factors such as the current market supply, the makeup of downstream industries, logistics within and without the plantation area, and even the political landscape with regard to international trade.

These uncertainties can be minimized if the entire value chain is considered from the beginning of the planting regime. The appropriate type (species, size, physical properties) of timber can be targeted in order to match what the market is demanding or is able to accept. The costs of producing these types of materials can be calculated against the profit that can be realized from them. Further considerations can be accounted for, such as the cost versus the benefit of a planter starting their own processing facilities, moving further downstream in the value chain.

Given the economic goal of the state to maximize employment and the value derived from it, timber planting should be considered second only to food in terms of its importance as a major land use for agriculture lands. Understanding the complete value chain illustrates this in multiple ways and also draws attention to the overarching conditions that can lead to successful establishment.

1.5.1 LAND USE

Land selection will be the first decision that a manager must make when choosing to undertake timber planting. Land-use planning or land resource management is a key issue in the development of rural areas. The United Nations Conference on Environment and Development (UNCED) recognized that issue and in 1992 proposed an integrated approach to the planning and management of land resources involving environmental, social, and economic issues and the active participation of local communities. The three criteria that must be considered are a) land capability, b) land use, and c) land tenure and planting mechanism (Evans and Turnbull, 2004). Going into the specifics of these points is beyond the scope of this chapter, so focus will be given to how considering the value chain can affect decision making in this early part of the planting process.

Land capability values are related to the geographic properties of the land available. A site's climate, existing vegetation, previous use, and current conditions have a tremendous effect on its viability as a timber plantation. These factors will have a bearing on the choices available after determining where to plant, which in turn will affect all other decisions made throughout the value chain.

Site selection will immediately determine what species of trees are suitable for planting. Differences in weather, soil conditions, and site elevation can cause great variance in the growing performance of different species of tree. The immediate implication of this is that the market, and thus the future value, of the planter's production of timber is already determined at this early stage. The planter must determine what product can be made from the logs produced by their plantation, whether or not existing timber users accept of the types of wood being produced if it is not already an established material, and other points of suitability, such as wood properties at rotation age and the size of round logs – all values that will determine the final price offered for the material.

These factors in turn will come into consideration in determining the management prescriptions of the plantation. Planting with the purpose of producing larger diameter, defect-free boles for use in high-value sawn-timber products or veneers will require a higher degree of silviculture treatment than would be necessary for a plantation with a focus on volume of production. The decision must be made whether to engage in these higher costs with the goal of maximizing value per volume of wood versus lower returns on a lower cost to produce. Again, these final calculations will have already been largely determined by the species suitable for the site and the current market for wood both domestically and internationally.

Land tenure must be considered if it in any way affects the options available to the timber planter. This can take place when the owner of the land and its manager are different parties. An example of this is the disallowance of cash crop and agricultural planting in industrial timber-planting areas, as seen in Sabah's Sustainable Forest Management License Agreement (SFMLA), whereby public forests are licensed to private managers to be developed on behalf of the state. This precludes any form of agroforestry or interplanting of cash crops as seen in regimes in other parts of the world.

Another instance where land tenure can affect options is when the planting area covers multiple parcels belonging to different owners. Planting regimes must be agreed by multiple parties who would have to be willing to go along with the arrangement for a minimum of one harvest rotation, and more if economies of scale are to be approached. This is especially complicated due to the nature of rural land tenure in the tropics, where boundaries can be indistinct and legal arrangements challenging to enforce.

Considering these factors with regard to the entire value chain can make or break the financial viability of a timber plantation. These will also greatly affect financial projections and budgets, which will be necessary if the manager has to source funding from external sources through banks or other forms of capital investment.

1.5.2 Selection of Planting Materials

Once the planting area has been selected, the planter must then choose what to plant. As mentioned above, utmost consideration must be given to the type of species that will be perform well on the selected site. The next consideration must be regarding the availability of the selected planting material, either through internal production through a nursery or tissue-culture generation, or general availability on the open market. Finally, it is essential to consider the final use of the raw material produced.

An extra consideration when choosing species to suit a site is to consider the option of planting multiple species that can be maintained in a system of crop rotation. This should be evaluated due to the advanced soil depletion that occurs when a fast-growing crop is planted in monocrop fashion over any extended area. Further granulation into selection can be evaluated, such as the planting of different species at different slopes and elevations due to performance differences, or whether multiple species need to be planted for a complete product line, for example, for production of both face/back and core quality veneers.

It is important to ascertain the availability of the chosen planting material. Given the variety of species in tropical forests, there are also vast differences in flowering, seed types, and germination protocols, some being more costly, and thus more expensive, than others. Tissue culture material can be also considered if available. Whatever material is chosen, the planter must take into account the overhead nursery costs and requirements even if the transition period from reception to planting is kept to a minimum.

Wood is an extremely versatile material. It is likely that whatever the quality of wood produced on the plantation, there will be a market for it. However, the planter must consider what price they are able to find on the market and whether this will be worthwhile for their planting regime. Wood that is only usable for pulp production will fetch considerably lower prices than wood for timber or veneer products. Uses of specific wood types will be heavily influenced by research and development into its uses. This is where coordination between the planter, the wider market, and research institutions is paramount. Given the benefits to the state should wider timber planting be successful, this coordination should be seen as benefitting the wider public and should be addressed accordingly.

These two factors must be considered in concert when selecting the species. The more resources spent in silviculture treatments such as pruning and weed control, the more the volume produced and the better the quality of the volume. Again, this must be balanced against the price the market is willing to pay based on wider conditions that the planter must be aware of.

1.5.3 Deciding the Planting Regime and Planting Distance

When calculating the performance of a plantation, a typical measure used is to calculate the volume of wood produced per hectare. When taking a value-chain approach to decision making, however, attempts should be made to calculate the profit per hectare to a reasonable degree of certainty. Typically, prices for round logs follow a step price increase at discrete points depending on the market they are sold in. The price per volume of round logs increases, in some cases quite drastically, within different diameter classes.

Wood quality, taken here as recoverability and suitability from round log to final product, is directly related to the size in diameter of bole produced and is influenced by both the degree of silviculture treatments and the spacing of the planting regime, which will directly influence the metric of volume per hectare. Again, the planter faces a decision in balancing the loss in volume versus the higher profits paid for higher quality boles. This will be made much clearer if they consider the entire value chain instead of ending value calculations at round log volume.

Besides the aspects of the planting methodology that will be determined by the climate and soil type of the area, there are several other decisions for the manager to make at this point. Land preparation costs and practices will differ heavily depending

TABLE 1.9

Mean DBH (cm) Differences in Kenangan Manis and Gum Gum Plantation, Sabah (Mannan, 2015)

	Mean DBH (cm)	
Species	Kenangan Manis (Year 6)	Gum Gum Plantation
Laran	24.6 (5 m × 6 m)	12.6 (2.44 m × 2.44 m, Year 6)
Binuang	25.5 (5 m × 7 m)	14.7 (2.44 m × 4.88 m, Year 5)

on the existing condition of the site. Sites still covered with vegetation will cost more to clear; however, these costs may be offset by the sale of salvage logs obtained from preparation procedures. Previously cleared, arid areas will cost more in soil regeneration procedures, depending on the relative soil health and type.

If the area is still covered with vegetation, some form of land clearing will need to be done in order to prepare the land for planting. Clearfelling in tropical forests for planting is widely considered unacceptable from the environmental perspectives of conservation and soil erosion (Evans and Turnbull, 2004). Evaluating the performance of clearfelled plantations in tropical areas, there is also an economic case against this practice, as performance is especially poor in following cycles. The initial costs of clearfelling may be lower due to the easier nature of work compared to a less intensive method and the higher volume of salvage logs available for sale. This will need to be balanced against the costs of restoring acceptable soil quality in the future as well as other environmental effects, such as pest and diseases or fire that may affect a vast monocrop plot in a tropical climate.

Once the land has been prepared, the planter will have to make a decision on the distance allowed in between each planting point. This will have direct implications for the overall value calculation of the plantation, as there is a tradeoff between the number of trees planted per hectare and the size of each tree, which, in turn, affects the final price of the bole.

In Sabah, planting of red laran (*Neolamarckia cadamba*) and binuang (*Octomeles sumatrana*) in two different sites with smaller planting distances has produced trees with much smaller mean diameters at similar ages, as can be seen in Table 1.9.

The immense implications of variances in planting spacing are obvious and again illustrate the importance of considering the entire value chain before operations begin. A small number of sawgrade logs may be more valuable than a large number of fiber-grade logs, depending on the context of the plantation.

1.5.4 SILVICULTURE AND PLANTING MAINTENANCE

Silviculture is the manipulation of forest structure and dynamics at the stand level, with the specific aim of producing certain goods or services (Bauhus et al., 2010). The degree of maintenance performed in the plantation will have impacts on the quality of the final bole. There are numerous methods of performing these actions, and costs can vary greatly. Proper weeding is necessary to ensure the tree grows to its full potential size. Pruning aims to ensure a bole has minimal defects such as knots or compressed wood. Fertilization may be necessary for initial establishment in less fertile areas. Pest and animal control will be required if fauna are affecting the saplings.

Weeding can be done manually or mechanically depending on available equipment or labor. Fertilizer costs usually end up as the highest or second-highest variable cost per planting point. Pest and animal mitigation options and their costs are determined largely by external factors such as government regulations on wildlife matters and pesticides, or costs of materials for tree guard type controls.

Fertilizing, besides being one of the most expensive inputs the planter will have to account for, also needs to be studied for its effects on tree growth and log quality. Most forest use of fertilizer is to correct known deficiencies, for example, deficiency of micronutrients such as boron in much of tropics and deficiency of macronutrients such as phosphorus in impoverished forests (Raj and Lal, 2013). While fertilizing may cause an increase in growth rate and thus shorter rotation lengths, research must be done into the effects this may or may not have on the wood properties of the log at that rotational age. The increase in growth rate may have an adverse effect on the density of the wood produced, which can be detrimental for the workability or even suitability in production of the final product. This is especially a concern for fast-growing, lower density species such as *Albizia falcataria* or *Neolamarckia macrophylla/cadamba*, and maybe less so for *Acacia mangium* or Eucalyptus. Again, these are variables that the planter must coordinate, and it is difficult to see the longer-term effects without considering all the elements in the value chain.

1.6 MARKETING CONSIDERATIONS

Most of the marketing factors relevant for accessing markets for tropical timber products are similar to those of timber products in general. There are several factors to be considered: products, distribution channels, promotion, and price.

These considerations require some assumption of the future realizable selling price of the planted timber materials as well as the costs that will be incurred in the process of selling. The selling price will be determined by a wide range of factors, from the quality of roads in and around the plantation area to geopolitical influences on foreign markets.

1.6.1 MARKET FACTORS

The domestic price of the timber will primarily be determined by the processing capacities of downstream industries. The planter should already have been aware of the industrial landscape prior to establishment. The ideal scenario for the planter would be a robust, diverse industry that would quickly be able to realize the economic potential of the homogenous, consistent supply being produced. In reality, if the species of timber produced is different to what is widely used, the planter may have to take initiatives to introduce potential customers during the gestation period of the plantation.

The challenges associated with attracting potential customers are considerable. Different wood types can respond very differently to different methods of sawing, drying, peeling, gluing, or other types of processing. They may also have a range of properties depending on variances in their age or planting area. Ideally, this type of data would already be available widely and from a reliable source. This is where the state can play an important role in the success of wide-scale timber plantations. Studied guidance from them would be invaluable to planters and processors across the entire value chain and would greatly accelerate the benefits mentioned in the value proposition section.

1.6.2 LOGISTICAL COSTS

Logistical costs are a complex enough factor to require their own section for discussion. It is the underlying factor no matter what market the timber is being sold to, and it can sometimes be major enough to play a large part in the feasibility of selling to certain markets.

The costs of transporting logs involves many inefficiencies. Due to the step-pricing nature of different sizes of logs, the monetary value of the amount of material transported per truck can vary greatly, given that physical space is constrained but the unit cost of material is variable. A truck or container carrying 30 tons of small-size logs is much less efficient in value terms than a truck or container carrying 30 tons of large-size logs. Maximizing value then becomes a matter of optimizing a FIFO (first in, first out) sequence – necessary given the natural deterioration of logs – with the added complication of choosing optimal timing and lot sizes when considering what and when to transport. This, however, mainly applies to salvage logging operations, as an established plantation should provide a standardized inventory that can be transported much more efficiently.

If a plantation is made up of many small holders, there is again the added complication of consolidating raw material from various sources. There is also the consideration of different lots becoming available at different times, necessitating more decision making on the part of buyers and sellers.

As mentioned, a lot of these decisions are made much simpler if material is coming from an established plantation. Logs of similar size and age are available consistently, making transport, handling, and documentation all much easier to process.

Logistics can also be made much more efficient the more completely the wood is able to be processed on site. Sawn timber recovery of 50% close to site means that the direct costs of logistics can be halved if sawn timber is being transported over the same distance compared to logs. Further processing would bring these costs down even further. On-site processing is also more sustainable because less fuel/energy is being consumed per unit of product produced.

The initial costs related to plantation establishment are significant enough that the planter may need to process the timber material before sale in order for the plantation to be financially viable. In large-scale tropical plantations, it is suggested that the options available should be decided in relation to the size of the planted area (Figure 1.10).

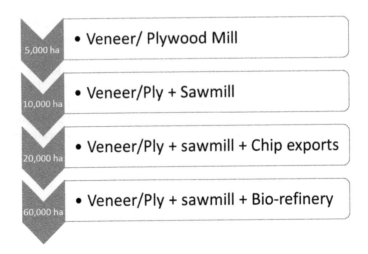

FIGURE 1.10 Hierarchy of plantation size for downstream industries.

The larger the scale of the processing, the higher the investment cost and the more material needed for the facilities to be viable. Large-scale tropical timber planting for pulp or woodchip feed stocks are the most inefficient forms a plantation regime can take. Arguments typically used to justify these regimes include that they are those with the fastest timeline for realizable profits and that fast-growing tropical species are not suitable for sawn-timber feed stock or high-value veneer feed stock. It bears repeating that when determining the economic potential of a plantation regime, consideration must be given to the entire value chain with respect to the land that is being used to support it.

1.7 CONCLUSIONS

Economics of the 21st century will be defined by the balancing of triple bottom line factors while supplying the needs of a rising population that is increasingly fractured by the divisions in income and the rural–urban divide. New advances in wood utilization have the potential to revolutionize everything from construction (cross-laminated timber, glulam), automotive manufacturing, plastics, and clothing to the way we power the manufacture of these goods through biofuel. All of these possibilities start from having access to consistent, uniform, sustainable supplies of timber raw materials. Industrial timber plantations in the tropics are the most efficient method we have of creating this supply in a sustainable manner. We need to change the way we are satisfying our growing needs, and tropical timber plantations are key to sustainability.

REFERENCES

Bauhus J., Meer P. V. D. and Kanninen M. 2010. *Ecosystem Goods and Services from Plantation Forests—Plantation Forests: Global Perspectives*. Earthscan, London.

Bellu L. F. 2013. *Value Chain Analysis for Policy Making: Methodological Guidelines and Country Cases for a Quantitative Approach*. Food and Agriculture Organization of the United Nations, FAO.

Byron R. N. 2001. Keys to Smallholder Forestry. *Forests, Trees and Livelihoods*, 11(4): 279–294.

Carle J., Vuorinen P. and Lungo A. D. 2002. Status and Trends in Global Forest Plantation Development. *The Journal of Forest Products*, 52(7): 1–13.

Center for International Forestry Research. 2001. *Typology of Planted Forests*. CIFOR Info brief, Centre for International Forestry Research (CIFOR). Retrieved from http://www. cifor.org/publications/pdf_files/typology/john-typology.pdf 10 December 2018.

Enters T., Brown C. L. and Durst P. B. 2004. Chapter 13: *What Does It Take? Incentives and Their Impact on Plantation Development*. Asia-Pacific Forestry Commission. Retrieved from http://www.fao.org/docrep/007/ae535e/ae535e0f.htm#bm15.1 on 30th December 2018.

Evans J. 2009. The History of Tree Planting and Planted Forests. *Planted Forests: Uses, Impacts and Sustainability*. FAO.

Evans J. and Turnbull J. 2004. *Plantation Forestry in the Tropics*. 3rd Edition. Oxford University Press, Oxford.

Food and Agriculture Organization. 2002. Forest Plantations Working Papers—Case Study of the Tropical Forest Plantations in Malaysia. Retrieved from http://www.fao.org/docrep/005/y7209e/y7209e00.htm#Contents on 16 December 2018.

Forest Resources Assessment. 2005. Progress Towards Sustainable Forest Management. Food and Agriculture Organization of the United Nations (FAO) Forestry Paper 147, Rome. Retrieved from http://www.fao.org/docrep/008/a0400e/a0400e00.htm 2 December 2018.

Jacobs J. 2000. The Nature of Economies. Retrieved from https://www.penguinrandomhouse.com/books/86060/the-nature-of-economies-by-jane-jacobs/9780375702433/ on 15th December 2018.

Kanowski P. 2010. Policies to Enhance the Provision of Ecosystem Goods and Services from Plantation Forests. In: Jurgen Bauhus, Peter J. van der Meer, Markku Kanninen (eds.), *Ecosystem Goods and Services from Plantation Forests*, Earthscan Publications Ltd, London, pp. 171–204.

Mannan S. 2015. The Potential of Woodlots in Sabah as a Supplementary Income in Land Development. Retrieved from http://www.forest.sabah.gov.my/images/pdf/publications/AR2015/03.pdf on 10th December 2018.

Midgley S. J., Stevens P. R. and Arnold R. J. 2017. Hidden Assets: Asia's Smallholder Wood Resources, and Their Contribution to Supply Chains of Commercial Wood. *Australian Forestry*, 80(1).

Montagnini F. n.d. Selecting Tree Species for Plantation. Retrieved from http://www.bf.uni-lj.si/fileadmin/groups/2716/downloads/%C4%8Clanki_vaje/Mansurian_forest_restoration_38_tree_species_plantation.pdf on 12th December 2018.

Nambiar E. K. S. (Editor) 2018. Site Management and Productivity in Tropical Plantation Forests. Retrieved from https://www.researchgate.net/publication/282915323_Site_Management_and_productivity_in_tropical_plantation_forests on 10th December 2018.

Raj A. J. and Lal S. B. 2013. *Forestry Principles and Application: Plantation Forestry*. Scientific Publishers, Jodhpur, India.

2 Managing Planting Materials and Planting Stock Production of Tropical Tree Species

Mandy Maid, Crispin Kitingan and Julius Kodoh

CONTENTS

2.1 INTRODUCTION

With the global trend of increased demand for fiber, paper, and wood products, there is an ever-urgent need for reliable and sustainably produced raw materials. Timber supply from developed countries (e.g. United States, Russia, Canada) and from natural forest harvest in Southeast Asia (SEA) have shown declining trend since the

1990s. In Thailand, depleted natural forest resources and enforced logging bans have necessitated the import of timber import to fill local demand gaps (Intongkaew and Liu, 2017). Australia is the exception as its timber plantations are gradually replacing the declining supply from natural forests (Brand, 2015). Over-harvesting of timber in SEA's natural forest means that it may take many decades before it reaches commercial harvestable levels. Therefore, timber plantations are expected to become the primary source of timber supply (Intongkaew and Liu, 2017).

However, global demand for low-cost timber products is expected to increase much more than the potential supply from forest plantations (WWF, 2018). Thus, it is imperative that the forestry sector strives to increase productivity by improving all aspects of production, from improved seed production and management strategies (e.g. silviculture, nutrition, and pest management) (Brand, 2015) to effective harvesting of existing and declining forested areas. Productive planted forests are usually established using exotic species as monocultures under intensive management practices for timber, but they are often associated with negative social, environmental, and governance issues (Brotto et al., 2016). On the other hand, planted forests are also accounted for in climate-change mitigation and adaptation efforts such as the Clean Development Mechanisms (CDM) and Reducing Emissions from Deforestation and Forest Degradation (REDD+) schemes (Stanton et al., 2010). Fast-growing tree plantations cover approximately 54.3 M ha globally, with the United States, China, and Brazil having the largest areas of planted forest at 5 M ha each (Indufor, 2012).

This chapter will review the management and production of planting materials or planting stock with special attention to tropical species. The production of reliable and good-quality forest reproductive materials (FRMs) will be imperative for the various types of planting programs, that is, for agroforestry, social forestry, enrichment, rehabilitation, and industrial tree plantation (ITP). Various tree-planting programs in the tropical region were conducted in varied vegetation, soil, and climatic conditions (i.e. rainforest, subtropics, dry or arid) and disturbance levels under trial conditions that resulted in failures or relatively little success, especially during the earliest attempts. Reasonable progress was achieved in producing high-quality FRMs of native (Aminah, 1991; Aminah et al., 1995; Anon, 1998; Itoh et al., 2002; Sheikh Ibrahim, 2006a; Ajik and Kimjus, 2013; Castellanos-Castro and Bonfil, 2013) and introduced tree species (Darus et al., 1989; Gohet al., 2014; and Goh and Monteuuis, 2016) for numerous planting programs.

The perceived benefits of producing high-quality planting materials were not always realized after the commencement of the planting programs due to the complex interface between research and application (Ng, 1996). Ill-suited fast-growing exotic species such as *Pinus* spp., *Eucalyptus camaldulensis*, *E. deglupta*, *Falcataria moluccana* (syn. *Albizia falcataria*), *Gmelina arborea*, *Muesopsis emenii*, and *Acacia mangium* were introduced for plantation establishment in West Malaysia in the hopes of producing sawn timber. These species failed to meet the plantation objectives due to different and varied reasons. Pines failed to produce seeds, the trees were weak and spindly foxtailed, *M. emenii* had poor wood working properties, and *F. moluccana* had very poor form (Ng, 1996). *Eucalytpus* and *Acacia* species initially showed very impressive growth but were later badly affected by leaf disease and fungal heart rot disease (Selvaraj and Muhammad, 1980). Elsewhere, forest plantations failed to establish because of early mortality due to poor species-to-site matching (Sheikh Ibrahim, 2006a), weed encroachment and discontinued maintenance (Francis, 1998), and planting on waste land beyond the regeneration capabilities of the species (Appanah and Weinland, 1993). Knowledge of the site is important; for instance, in areas where water tables can markedly rise and fall during the year (Perumal et al., 2017), this will limit tree roots' access to water or drown the roots. Sites with hard pans or heavily compacted soils should be avoided unless they can be cultivated using machinery. Soils that are rocky, are heavy in clay, have low productivity levels, and have high erosion risks must be excluded from conversion into plantations (Lee et al., 2008).

Land designated for forest plantations are more often on waste land or marginal land that is severely degraded, poorly stocked, and do not have any natural regeneration potential (Lee et al., 2008), such as grasslands, ex-shifting cultivation areas, or remnants of other land use such as ex-tin-mining soils and other problem soils (sandy, saline soils, peat). Some of the steps taken to rehabilitate these lands were to plant hardy exotic species such as *Acacia* spp. or other leguminous species due to their nitrogen-fixing ability to improve soil conditions (Fisher, 1995). After some time, planted *Acacia* trees were then used as nurse trees for reforestation with indigenous trees with some success (Norisada et al., 2005; Daisuke et al., 2013). However, afforestation on Beris soil (raised sea beaches) has not been successful (Appanah and Weinland, 1993).

Following the failure to establish plantations using exotic tree species, some tropical countries have turned to enrichment planting or reestablishing forest using native species (Bhat et al., 2001; Ng, 1996; Subiakto et al., 2016). In some circumstances, exotic tree species were not well-received compared to multipurpose trees by local communities that are involved in agroforestry and social forestry tree-planting programs (Bhat et al., 2001; Tchoundjeu et al., 2002; Tsobeng et al., 2016). Since tree-planting activity is not only for ITP but also encompasses conservation and multipurpose usage, continuous research and development covering as many tropical tree species as possible is required to secure the supply of planting materials or FRMs. However, knowledge on the propagation of most native species of tropical forest is still scarce (Bonfi and Trejo, 2010; Castellano-Castro and Bonfil, 2013) despite increasing anthropogenic threats to existing forests. Although local communities preferred native species, reforestation programs with native species in the upland tropical forest of the Philippines have often failed (Chechina and Hamann, 2015). It was seen that greater social preference or higher value was accorded to late-successional tree species compared to early-successional tree species, and the least for tree plantation species. But late-successional tree species cannot survive on degraded, open-field, and abandoned farmland (Uhl et al., 1988). Therefore, identifying useful pioneer or early-successional tree species such as *Agathis philippinensis*, *Ficus septica*, *Shorea contorta*, and *Casuarina equisetifolia* could lead to more success (Chechina and Hamann, 2015). More recent research comparing the outcomes of using a multistoried enrichment planting approach using native species in secondary forest (98%), *A. mangium* plantation (78%), and selectively logged over forest (76%) in West Malaysia have shown promising survival and growth a year after planting (Parsada, 2013). At Sempadi FR Sarawak, line-planted native species *Shorea macrophylla* showed high survival rates, good growth, and adaptability despite recurring inundation in an undisturbed area (Perumal et al., 2017).

There are specific conditions under which exotic-tree plantation has been relatively successful. In Thailand, ITP companies actively engaged farmers to plant forest trees on farm lands through contract farming of *Eucalyptus* tree species, which has provided a reliable source of income due to the strong market demand for *Eucalyptus* wood. Farmers involved in contract farming have better access to new knowledge provided by the contracting company (Boulay et al., 2013). Similarly, tree-growing on privatized individual household farm lands has had strong support from the government of Vietnam, which rapidly enhanced the socio-economic status of local farmers and transformed the country's landscape to dominant farm-based plantations (Sandewall et al., 2010). In the east Malaysian state of Sabah, degraded grassland has been successfully rehabilitated using *Acacia mangium* since the 1970s by statutory state agencies Sabah Softwoods Sdn. Bhd. (SSSB) and the Sabah Forest Development Authority (SAFODA). Forest plantation development in Sabah was relatively successful due to the support and active involvement of state government–linked organizations that were dedicated to forest plantation establishment. Large tracts of degraded state land were set aside for forest plantation establishment, and the first industrial paper mill was established in the 1980s by Sabah Forest Industries Sdn. Bhd. (SFI). Industrial tree plantation has continued to expand with additional investment from the private sector since the 1980s. By 2006, there were approximately 205,000 ha of total planted area with trees consisting of *A. mangium, Falcataria moluccana, Gmelina arborea, Eucalyptus* spp., and rubber plantations (Lee et al., 2008).

Commonly cited problems related to the production of FRMs included a lack of basic information on seed germination, seedling survival, and seedling growth performance for native species (Yang Lu et al., 2016); a shortage of planting material because of unreliable flower and fruit production, especially for Dipterocarp species; low-quality planting material; poorly developed tending and silvicultural treatment post-planting; a lack of trained and skilled forest workers (Sheikh Ibrahim, 2006a); and the rapid infestation of insect and fungal diseases (Grippin et al., 2018). Vegetative propagation methods have shown feasibility, but large-scale production of FRMs requires advanced technology and efficient infrastructure (Adjers et al., 1998), and many species (34%) cannot be cultivated using this propagation method (Itoh et al., 2002).

2.2 CHOICE OF SPECIES

The question of which tree species to choose for any tree-planting program is influenced by past experience with success or failures. The following are important points that should be taken into consideration for selecting tree species for planting-stock production.

1. The purpose of tree planting program

    The purpose of the tree-planting program, whether for ITP, social forestry, ecosystem rehabilitation, enrichment planting, or protection forests (e.g. watershed, wildlife), will heavily affect the choice of tree species.

Highly valued tree species for social forestry (Chechina and Hamann, 2015) may include those that can provide fruit, medicine, or resins or those that can provide additional income. A different set of species will be more suited for a planting program in a protected watershed area (e.g. *Cedrela odorata, Swietenia macrophylla, S. mahagoni, Azadirachta indica, Acacia spp., Hibiscus elatus, Tabebuia pentaphylla, Cordia gerascanthus, Cupressus lusitanica, Guaiacum officinale, Pinus caribaea* var. *honduren-sis* in Jamaica) (Camirand, 2002). Hardy species are required on problem soils (e.g. sandy, saline soils, Figure 2.1).

2. Native versus exotic species

Exotic tree species (*A. mangium, A. aulacocarpa, A. crassicarpa, Azadirachta excelsa, Eucalyptus* spp., *Falcataria moluccana, Gmelina arborea, Hevea brasiliensis, Swietenia macrophylla, Tectona grandis*) consisting of fast-growing and high-value exotic and fast-growing native species (*Octomeles sumatrana, Neolamarckia cadamba*) have been selected for ITP in Sabah, Malaysia (Lee et al., 2008) in consideration of site suitability and secured land tenure. Some exotic tree species that had a longer history of domestication and planted as ITPs in the tropics, especially tropical *Acacias, Eucalyptus*, and *Pinus*, were preferred initially. The availability of more information on native tree species, covering topics from seed and vegetative plant production to silviculture, harvesting, and end-product processing, have gradually changed tree-species preferences.

3. Species-to-site matching

Tree species trials are expensive but necessary to determine the response of potential species over different soil, climate, altitudinal, and vegetation zones. These abiotic and biotic factors are also associated with temperature and the amount of rainfall, which can determine the suitability of tree species (Camirand, 2002). Aspects related to soil concern parent material, the presence of waterlogging, salinity, pH conditions (Camirand, 2002), moisture availability, rooting conditions, nutrient availability, and high-risk sites (Lee et al., 2008). For example, a species trial was conducted to select nine out of 37 suitable framework native tree species to complement the natural

FIGURE 2.1 Attempt to establish *Acacia* stand on sandy soil (left) and saline soil (right) in Hue, Vietnam.

regeneration of forest ecosystems and to support biodiversity on degraded sites (Elliott et al., 2003). In a vulnerable forest ecosystem such as the peatlands of Kalimantan, four out of 21 tree species were suitable for the site but faced obstacles because of the risk of forest fire and a fluctuating water table (Lampela et al., 2017). At such sites, site preparation (e.g. weeding, mounding, and fertilizer) can enhance seedling growth and survival in the field (Lampela et al., 2018).

4. The length of forest rotation and nature of the products

The products from forest plantation are either timber or non-timber forest products (e.g. resin, oil, traditional medicine from plant parts). The length of forest rotation is determined by the period required to obtain timber of the desired volume and quality, and by the natural maturation of the tree species (e.g. short, medium, or long-generation). Tree species grown for fuelwood, pulp, and paper products (e.g. *Acacias*, *Eucalyptus*) commonly have a short rotation length compared to those grown for high-quality sawn timber and veneers (e.g. *Tectona grandis*). Multiple-use (Siti Maimunah et al., 2018) and high-market-value (Chechina and Hamann, 2015) tree species are preferred for community forestry.

5. Access to market

Tree species are more likely to succeed as a product of forestry when planters can benefit from access to local and export markets (Boulay et al., 2013) and when the species are economically sustainable (Intongkaew and Liu, 2017).

2.3 SEED PROCUREMENT, HANDLING, AND STORAGE

2.3.1 THE SOURCE AND SUPPLY OF SEEDS

Due to the high inputs and investments involved, planted forests are expected to be productive and to yield good-quality products. Even before planting takes place, planters have the daunting task of securing the supply of seeds, correctly handling the seeds, and maintaining seed viability during storage. Most logged-over natural forest stands have an insufficient number of seed trees, and these have poor characteristics, which in turn produce poor seed crops. Earlier established seed stands and seed orchards were too young to produce sufficient seeds (Evans, 1982). Knowledge on the timing of seed collection; methods of seed extraction and storage; control of seed pests; and seed testing, grading, and certification had limited the use of many tree species (Evans, 1982) for replanting activities. Other seed-related problems include difficulties in collection, short viability, dormancy, and other handling difficulties (Schmidt, 2000).

Prior to the establishment of *ex situ* germplasm collection areas (Figure 2.2), seeds, wildings, and plant parts (e.g. marcots) were collected from well-stocked natural forests. Effective seed collections are only possible with sufficient knowledge of the flowering and fruiting seasons of the different target tree species (Sheikh Ibrahim, 2006a). For example, the phenology of dipterocarps differs in seasonal and aseasonal regions. Some dipterocarps species flower annually with varying intensities

FIGURE 2.2 Process of establishing *ex situ* germplasm collection for *Acacia mangium* and *A. auriculiformis* at (Universiti Kebangsaan Malaysia, Bangi): (a) site selection, (b) land preparation and installation of facilities, (c) tree planting and labelling, (d) protection and fertilizing, (e) new establishment, (f) mature stand.

in seasonal regions but tend to flower gregariously at intermittent years in aseasonal regions (Ashton et al., 1988; LaFrankie and Chan, 1991), which is also known as flowering in a 'mast' year. Thus, the planning of seed collection and field planting using dipterocarp species must essentially be conducted during mass fruiting years (Sheikh Ibrahim, 2006a). Furthermore, dipterocarp seeds are affected by pre- and post-dispersal predation by insects (e.g. moth larvae) and long-tailed macaques, while some portions are affected by non-viability six months after fruit fall (Chong et al., 2016). Seeds collected in years of light flowering and seeding will contain a higher portion of non-viable seeds, and thus are genetically less representative of the sampled tree population. These seeds will produce genetically inferior plants (Guarino et al., 1995).

Generally, there are four types of seed sources – seed orchard (SO), seed production area (SPA), seed stand, and seed trees. The establishment and management of the different types of seed sources is described in much detail in other publication (Granhof, 1991; Krishnapillay, 2002; Finkeldey, 2005; Lee et al., 2008). Seeds from SPA, seed trees, and seed stands are not recommended for operational planting, and the origins of the trees or populations are usually not known when they are introduced to a location. The quality of the mother tree is not yet proven, and the seed quality is usually moderate. The precondition for producing improved seed is the identification of seed sources, which should be taken from stands of good provenance (i.e. where the species grows naturally) or land races (i.e. where the species has naturalized after introduction) (Krishnapillay, 2002). The SPA can be established from natural populations or plantation stands. Remnant populations and SPA of two key restoration species in south-eastern Australia, *Acacia montana* and *Dodonaea viscosa* subsp. *cuneate* had good levels of genetic diversity but had significant levels of inbreeding. The SPA was also biased towards two remnant source populations. New germplasm from other remnant populations is necessary to broaden the genetic base of the seed crop (Broadhurst et al., 2017). A provenance resource stand is a type of SPA that is founded concurrently with a provenance trial, which is the earliest stage of a breeding program. The provenance resource stand consists of a combination of progenies from various seed trees of a single population of superior provenance. Upon maturity, the stand is useful for the production of seeds and can be converted to SPA. The provenance resource stand is not subjected to statistical design and is less costly to establish than a provenance field trial (Finkeldey, 2005).

The seed orchard (SO) is a means of producing a large quantity of genetically improved seed which is established from known-origin, genetically selected seeds from plus trees (Lee et al., 2008) and to realize genetic gains (Granhof, 1991). A seed orchard is a plantation of genetically superior trees (clones or progenies) which have been isolated to decrease pollination from outside sources, and intensively managed to produce frequent, abundant, easily harvested seed crops (Zobel, 1958). Seed orchards are identified by generation (e.g. first, second, or advanced generation), depending on the number of improvement cycles. First-generation orchards originate from plus trees selected in natural stands or in unimproved plantations (Granhof, 1991). Seed orchards can be categorized based on the objectives of seed production (Barner, 1975) and function (Granhof, 1991). The objectives of SO establishment are (Barner, 1975):

1. Provenance-hybrid orchards, in which trees of two provenances of distinctly different origin are selected to produce a special hybrid effect.
2. Species-hybrid orchards, in which trees from two species are selected to produce a beneficial hybrid effect (e.g. hybrid vigor). For example, single *A. auriculiformis* and *A. mangium* clones have been planted in alternate rows to produce *Acacia* hybrids (Wickneswari, 1989).
3. Advanced orchards, referring to the mass production of material already subjected to advance or specific breeding methods, for example, back-crossing.

The functions of seed orchards are (Granhof, 1991):

1. Production seed orchards, which mainly function to mass-produce seeds. The size is correlated to the desired quantity of seed, and the life span depends on the natural ageing of the tree species.
2. Breeding seed orchards are part of long-term breeding program which involves using several breeding populations to obtain additional genetic variation. The breeding population can be established as a progeny trial or plantation with many families, family control, and large plots on isolated and good sites for flowering and seed production.

Seed orchards (Figure 2.3) can be established using improved seeds raised from selected parents through natural or controlled pollination, called *seedling seed orchards* (SSOs); or using selected clonal materials propagated through vegetative means (grafts, cuttings, or plantlets from tissue culture), called *clonal seed orchards* (CSOs). SSOs result in progenies with a broader genetic base but lower selection differences (Krishnapillay, 2002). SSOs are planted at normal plantation spacing, with the identities of families maintained to allow for roguing (genetic thinning) within and among families based on their phenotypic performance (Granhof, 1991). The disadvantages of SSOs are the conduct of early selection when juvenile-mature correlation may still be low (Toda, 1974), and that flowering and seed productivity will have to wait for the age of maturity (Granhof, 1991). SSO are preferred for tree species with early reproductive maturity (e.g. eucalypts, acacia, tropical pines), when large number of trees (phenotypes) are selected (>100), and when species show unsuitable clonal characteristics (e.g. deformation of roots) (Barrett, 1985).

FIGURE 2.3 *Acacia mangium* seedling seed orchard in Sabah.

Extensive seedling seed orchards (ESSOs) (Nickles and Spidy, 1984) are SOs consisting of a balanced mixture of seeds from a minimum of 60 good parents of verified excellent combining ability, while maintaining family identity at seed collection and nursery to ensure balanced seedling lots with equal representation from all families. However, in the field, the seed lot identity is given up, and successive thinning is for silvicultural purposes. Therefore, ESSOs are not suitable as breeding orchards but are useful seed production orchards (Grandhof, 1991). Findings by Harwood et al. (2004) demonstrated the need to minimize the impact of selfing or inbreeding in operational seed production (e.g. *Acacia mangium* SOs) because it causes significant reduced growth and vigor. Similar observations of decreased vigor were found in the inbred loblolly pine (*Pinus taeda* L.) parental group, although one parent group had positive growth response and thus presents an opportunity for advanced-generation breeding among related individuals (Ford et al., 2015).

In a CSO, the ramets (replicate from a clone) are planted at wide spacing to encourage crown development for maximum seed production. The identity of each ramet is judiciously retained through labels and records (Granhof, 1991). A CSO is only suitable for species that are easily propagated through vegetative methods, e.g. *Eucalytpus* and *Tectona grandis,* and only for organizations that have sufficient financial (e.g. for research and development) and technical resources to manage a sustainable clonal program. The choice of clonal or seed production program can greatly affect plant unit costs (Griffin, 2014).

Therefore, seed production requires a set of skills that includes but is not limited to silviculture, tree improvement, and plant propagation. The establishment of SOs entails similar preparation to a forest plantation setup, including the selection of seedlings or clones, determination of a good planting site, fulfillment of the species-specific requirement and management requirement, protection, isolation from pollen contamination, and secure land tenure. This is followed by the field establishment procedure, which includes design and spacing; site preparation and planting; and silvicultural management, such as weeding, roguing, water, fertilizer, and hormone applications. Record keeping of the various activities and of seed production is essential (Krishnapillay, 2002; Granhof, 1991). After roguing, about 50%–75% of the original number of clones or families are eliminated after sequential evaluations, and thus initial establishment requires a high number of clones or families (Granhof, 1991). The forest department and the private forestry sector of many tropical countries have made considerable progress by establishing *in situ* and *ex situ* germplasm collection areas, provenance-cum-progeny trials, SPAs, and SOs. However, these are often limited to a number of native (e.g. *Shorea leprosula*) (Naeim et al., 2014) and exotic species with high commercial value.

In Scandinavia and Finland, plantations established using improved forest trees have resulted in gains in objective traits and positive economic impacts that were beneficial for society. Tree breeding which uses phenotypically superior plus trees showed an increase in volume of 10%–25%, and bare land value associated with genetically improved trees gave a better return on investment and a shorter rotation period compared to unimproved forests (Jansson et al., 2017).

2.3.2 SEED PROCUREMENT

The planning and preparation involved in seed collection from woody trees of the tropical region have been described in great detail by Guarino et al. (1995), Schmidt (2000), and Krishnapillay (2002). Fruit and seed collection (Figure 2.4) in the tropics is arduous, time-consuming, and costly, and it can pose a safety hazard for seed collectors. Therefore, careful consideration needs to be given with regards to appropriate sampling approaches, and whether to collect and maintain seeds of individual trees or carry out bulk provenance collection. The recommendations for sampling trees varies according to purpose: for provenance or progeny trials, 10–20 individuals from each provenance or family; for *ex-situ* conservation, 25–50 individuals per population or more; and for improvement programs and seed orchards, more than 200 individuals per population for a genetically broad program. The individual trees should be distanced from each other according to distance of normal seed dispersal, and usually at a minimum of 100–200 m between sampled individuals (Guarino et al., 1995). This is to avoid sampling closely related individuals, for example, from the same mother tree or set of parents. Seed collection involving separating seeds from individual trees is necessary for genetic study and tree-breeding programs. Bulk provenance collection is usually sufficient for first-stage sampling for provenance trials; however, collectors must be certain of the identity of the species and collect equivalent quantities from individual trees, and the bulk seed should be thoroughly mixed (Guarino et al., 1995).

Due to difficulties in seed collection in the natural forest, simple methods are normally favored: collecting from the ground, using elevated platforms (Lee et al., 2008), using sling shots, and using bow and arrow. The services of trained tree climbers, who cut off the fruiting branches, are used for tree species that have early maturity (< 3 years old), and at sites easily accessible using vehicles and equipment. Felling individual trees for seed collection should not be practiced as genetic resources will be eroded. Seeds collected should be fresh, with a moisture content of more than 15%, to achieve a high germination rate (Schmidt, 2000), but this will vary with species and fruit maturity. For successful collection of dipterocarp seeds in an aseasonal region like West Malaysia, the joint Forest Department–ITTO project recommended limiting seed collection to SPAs to ensure economic quantities of seed materials. Other essential recommendations were to have adequate trained tree

FIGURE 2.4   Ripened *Acacia mangium* pods (left) and pod collection at Mantanau, Sabah.

climbers and climbing equipment, carry out regular monitoring of the dipterocarp flowering and fruiting season, ensure well-maintained SPA through silvicultural treatments, use the best planting material from selected mother trees, and have an adequate budget for the production of planting stocks (Sheikh Ibrahim, 2006a). The different seed collection methods, their applicability, advantages, and limitations are described in great detail by Schmidt (2000).

2.3.3   SEED HANDLING AND DISPATCH

While obtaining fruits and seeds in the field, the material should be protected from the elements and labelled with information such as the species, seed lot number, geographic location, name of seed source, weight of seed, date of collection, and the collector's name (Willan, 1985). In the tropics, seeds are lightly and loosely packed in material with good aeration such as gunny sacks, bamboo baskets, open-mesh baskets, and cotton bags. When plastic bags are used, the tops are left open and they are punctured all over (Krishnapillay, 2002). Delicate dipterocarp seeds are preferably stored in moistened cloth bags placed in sheds with temperatures maintained at 20–25°C (Lee et al., 2008).

Seed collection areas are sometimes located some distance from collection and storage centers. Therefore, to maintain the viability of seeds, temporary fruit and seed processing and storage areas must be prepared in the field, with good aeration and sheltered from rain and pests. In the field, seeds can be separated from ripened fruit and air-dried in sheds, with daily turning of the loose-piled seed to reduce moisture accumulation. Seeds with pulp must be checked and selected for good condition and the pulp removed quickly. Germinated recalcitrant seeds during the collection period must be separated and placed in containers or baskets lined with moist vermiculite or moist paper. A Mobile Seed-Seedling Chamber (MSSC) has a storage unit furnished with sensors that can maintain an environment suitable for the storage and transportation of tropical seeds (Marzalina, 1998). The MSSC equipped with lighting can maintain a cool temperature (20±5°C) and high humidity (80–95%) and reduce the respiration of seeds. The survival rate of recalcitrant seeds was improved from 25% survival under conventional transportation to 70% survival using an MSSC. The MSSC improved the condition of intermediate type seeds but not orthodox seeds (Marzalina, 1998).

2.3.4   SEED CONDITIONING

Seeds harvested from trees must be processed immediately at collection centers to retain their freshness and viability. *Seed conditioning* is the process of producing high-quality seeds and involves the elimination of unwanted materials such as trash, leaves, weeds, other crop seeds, insects, and inert matter (chaff, stems, and stones). The purposes of seed cleaning are to remove all contaminants from the intended seed; decrease the loss of good seeds; improve seed quality by removing non-viable seeds (decayed, damaged, and empty seeds); and produce consistent seed quality. Ultimately, it is to acquire pure live seed percentage or the maximum percentage of pure crop seed with maximum germination potential (Copeland and McDonald, 1995).

In the tropics, the seeds of many native species are processed by hand due to variation in fruit and seed characteristics (Figure 2.5). Non-seed materials attached to the fruit and seeds, such as the pulp, seed pods, wings, and fruit coats, need to be removed. (Krishnapillay, 2002). Large tropical seeds (e.g. *Potoxylon melagangai*) are easier to clean manually. A small amount of very fine seeds can be sieved using mesh of different sizes to separate the chaff (e.g. *Neolamarckia cadamba, Octomeles sumatrana*). Seeds contained in fleshy fruits require de-pulping by alternately soaking in water and mild abrasion. For large quantities of seeds, the pulp can be removed using a modified coffee de-pulper or a cement mixer containing rubber

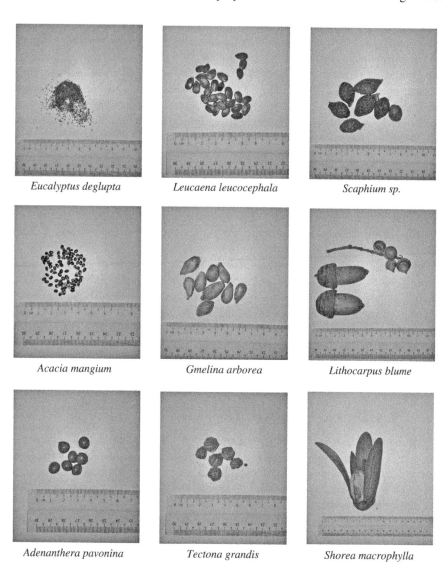

FIGURE 2.5 Tree seed types.

balls to rub the pulp off the seeds. Fine seeds can be rubbed lightly against fine wire mesh, soaked in water to allow the seeds to sink. Cleaned seed can be dried on layers of paper and turned frequently to ensure even drying of each seed batch. Pods and cones are spread on screens or canvas in the sun or in sheds, are turned frequently for even drying, and should be protected from rain, animals, and insect pests. Seeds detached from their pods during drying must be removed to avoid excessive exposure to direct sunlight. Hard pods must be mechanically or manually pried open to release the seeds, which are then air-dried. Dipterocarp seeds must be removed from the wings (Krishnapillay, 2002).

After conditioning, seeds must be graded, because seed size affects seedling emergence, uniform growth, and seed quality. Seed grading can be conducted manually by screening and sieving (separation by thickness and diameter); using an indented centrifugal cylinder (separation by length); liquid flotation and blowing, fanning, and winnowing (separation by specific gravity); and frictional cleaning (separation by surface texture) (Krishnapillay, 2002). Seeds can also be cleaned and graded simultaneously using various types of seed-conditioning equipment. Seed cleaning usually starts with the use of a pre-cleaning machine (e.g. scalper, debearder, or huller-scarifier) that removes the different types of non-seed material (inert matter, awns, beards, lints, hulls, or pods) attached to the seeds. Air-screen machines or fanning mills use airflow and perforated metal or wire screens to separate seeds base on size, specific gravity, and resistance to airflow. The specific gravity separator can be used to separate undesirable seeds and inert contaminants that are similar in size, shape, and seed-coat characteristics. This machine can remove damaged, empty, and off-color seeds that have decreased specific gravity, and heavy particles such as mud balls, soil particles, and small stones.

Other seed-conditioning machines separate or grade seeds based on seed dimension (e.g. length, width, and thickness separators or graders), seed texture, color, electrostatic properties, vibration, and affinity for liquids of seeds. The seeds maybe treated with chemical substances to reduce, control, or repel seedborne, soilborne, or airborne organisms. Seed treated with highly toxic substances should be labelled appropriately (with a skull and crossbones) and a precautionary statement, or, if treated with substances not labelled as highly toxic, they must bear a label with appropriate precautionary statement (Copeland and McDonald, 1995).

2.3.5   SEED CLASSIFICATION AND STORAGE

Seeds are a natural means and form for storing, moving, and collecting plant genes and germplasm (Appanah and Turnbull, 1998). The purpose of seed storage is to maintain seed viability over a prolonged period of time and to delay seed deterioration. Thus, good-quality seed can be supplied for planting programs that stretch over an extended period of time. Seed storage is particularly important in areas with seasonal climates and short planting seasons (Schmidt, 2000), and for tree species that only yield good seed crops once every few years (Wang, 1975). Some tree species can flower and fruit throughout the year, or flower copiously on an annual basis. But collecting surplus seed for the next year's planting need is more cost efficient. Seed kept in storage act as a safeguard between the period between demand and production of seeds, which can be supplied to tree nurseries or potential clients (Schmidt, 2000). Sudden circumstances such as a high mortality of tree seedlings due to pests and diseases or other damage can occur in the nursery or the field, thus necessitating additional seedlings.

Seed storage is closely related to the planning of seed collection but limited by the technical and physiological storage potential of seeds. The two main seed storage classifications are *orthodox* seeds and *recalcitrant* seeds. Orthodox seeds can be desiccated to a low moisture content ($\leq$5%) and stored for several years at -20°C with limited loss of viability. Recalcitrant seeds are not amenable to desiccation (12%–30%) without losing germination ability and cannot be stored for long periods of time (Roberts, 1973; Schmidt, 2000). Recalcitrant species retain a high moisture content at maturity (30%–50%) (Schmidt, 2000), which can lead to microbial contamination and germination during storage (Chin and Roberts, 1980). Recalcitrant seeds are also prone to chilling damage and cannot be stored below 16°C. In ambient conditions, the variable humidity levels can cause the death of recalcitrant seeds (Krishnapillay, 2002).

There is a relatively continuous range of seed-storage physiology in between the extremely recalcitrant to the orthodox seed classification (Farrant et al., 1986). A third indefinite category termed *intermediate* was described as orthodox with limited desiccation ability (OLDA) because of its partial tolerance of drying and sensitivity to low temperatures (Ellis et al., 1990); one example is the seed of *Swietenia macrophylla*. The physiological differences between intermediate seed and orthodox seed could be very little but they have different practical handling difficulties (Appanah and Turnbull, 1998). Other minor categories within the main seed categories are sometimes referred to as *sub-orthodox* and *sub-recalcitrant* seeds (Schmidt, 2000).

Orthodox and intermediate seeds can be kept in airtight vessels under conditions of 4%–8% moisture content, 55%–65% relative humidity, and -18–2°C temperature. Under these conditions, seed respiration is decreased, seed aging delayed, and seed viability extended (Harrington, 1972), which also prevents fungal activity during storage. Orthodox seeds are usually kept with silica gel that absorbs moisture and are stored in air-conditioned rooms or ultra-low-temperature containment units. Dried lump wood charcoal can also be used in place of silica gel for short-term storage in nurseries (Elliott et al., 2013).

Tropical recalcitrant seeds stored in environments with 23%–55% moisture content, 55%–75% relative humidity, temperatures of 15–25°C (Marzalina, 1995), and good aeration can maintain 50% seed viability during a year of storage. 1kg of recalcitrant seed requires soaking in 1% fungicide/3L water for 15–20 minutes and should then be air-dried on paper to the lowest seed moisture content for the species. Then, the seeds can be sealed in plastic bags and stored at 15°C. It is possible to store germinated seeds in controlled conditions that delay their growth, but this requires a growth chamber. When stored at 16°C, 90% relative humidity, 4 hours photoperiod, and 400 lux by fluorescent light intensity, the germinated recalcitrant seeds can be maintained for 3–18 months (Krishnapillay, 2002).

The only viable method of long-term seed storage is at ultra-low temperatures or cryopreservation in liquid nitrogen (−196°C). At very low temperatures, cell division and metabolic processes cease. Prior to storage, moisture content or free water within seeds must be reduced because it can cause ice formation and death of the seeds. Cryopreservation requires very little space and maintenance and protects the seeds from contamination (Krishnapillay, 2002).

2.3.6   SEED QUALITY, TESTING, AND DOCUMENTATION

The quality of seeds largely depends on the genetic quality of seed sources (the origin and the genetic base), and is later influenced by the successive procedures of seed collection, conditioning, handling, and storage, as discussed in the previous sections of this chapter. The seed quality will in turn affect the seedling quality (Jaenicke, 1999). Good-quality seeds are mature; healthy; true to type; pure; without inert matter and contamination; viable with good germination; uniform in texture, structure, and appearance; without damage; and pest and disease free (Munjuga et al., 2013). Seed viability and quality are ensured through periodic seed testing and documentation throughout the storage period. Seed-technology and seed-testing handbooks and manuals meticulously cover the various aspects of seed quality and testing (Copeland and McDonald, 1995; Willan, 1985).

Seed testing is the discipline of assessing seed quality and reveals the overall value of the seed for its intended purpose. The seed lot potential can only be known when its genetic components, mechanical components, and viability have been determined. The information derived from seed testing is useful to the seed producer as it can provide quality assurance for their seeds, allows the seed purchaser to identify whether losses in the field are associated with poor-quality seeds (Copeland and McDonald, 1995), and allows a monetary value to be attached to the quality of seeds. Most seed lots are a heterogenous mix of seeds due to differences in seed characteristics such as seed weight, maturity, lodging, and disease occurrence. The seeds are also harvested from different trees and locations and lack consistency in terms of harvesting, storage, and conditioning. Thus, a representative sample of the seed lot must be obtained. Firstly, a submitted sample drawn from bulk seed lots is sent to the laboratory. At the laboratory, the submitted sample is further divided into small working samples for testing (Copeland and McDonald, 1995). Low germination rates and low seedling vigor in germination tests (Figure 2.6) reflect poor seed quality, which could be the result of the inherent genetic properties of certain provenances, immature seeds, damaged, or improper seed handling and storage (Krishnapillay, 2002).

2.3.7   QUANTITY OF SEEDS AND SEEDLINGS

Tree seeds and seedlings are very costly to collect or purchase. Estimating the amount required for an annual planting program will ensure that short-term difficulties in obtaining seeds or seedlings will not cause unnecessary delay of a project. Whether small or large amounts of seeds are collected, it makes more economic sense to collect seeds for future requirements due to the cost and efforts involved. The quantity of seed required depends on the density of seedlings planted per hectare, number of seeds per kilogram, seed viability of percentage of germination, anticipated seedling

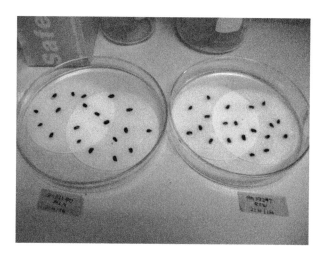

FIGURE 2.6 Germination test of *Acacia mangium* seeds in the laboratory.

survival rate in the nursery, and anticipated seedling survival rate in the field (Lee et al., 2008). Seed requirements can be estimated as follows (Evans, 1982):

$$\text{Weight of seed required} = \frac{\text{Ha} \times \text{N}}{\text{S} \times \text{PS}} \qquad (2.1)$$

Where,
Ha = size of annual planting program for each species
N = intended number of seedlings to be planted per hectare
S = estimate of survival after planting
PS = estimated number of seedlings that can be planted per kilogram of seeds

The estimated number of seedlings that can be planted (PS) can be determined according to Evans (1982). Values for N, G, PU are results from seed testing, and M is estimated from past records in the nursery.

$$PS = N \times G \times PU \times (1 - M) \qquad (2.2)$$

Where,
N = number of seeds per kilogram
G = germination percentage
PU = seed purity percent
M = estimate of seedling mortality during nursery life

Alternatively, the quantity of seeds can be estimated as follows (Lee et al., 2008):

$$\text{Quantity of seeds required} = \frac{L}{PS \times FS \times NS \times G \times SK}(kg) \qquad (2.3)$$

Where,

   L = land area
   PS = planting space
   FS = field survival rate
   NS = survival rate in the nursery
   G = percentage of germination
   SK = number of seeds per kilogram

The number of seedlings required for field planting depends on the size of the planting area, spacing and planting design, expected seedling survival during transportation, and expected seedling survival in the field (Lee et al., 2008).

2.3.8 Dormancy and Seed Pre-Treatment

Seed dormancy is a genetically inherited trait, and its extent is modified by the environment during seed development. Wild or recently domesticated plants tend to show *dormancy*. Dormancy is a state in which seeds are prevented from germinating under favorable environmental conditions. Seed dormancy can be categorized into *primary* and *secondary* dormancy. Primary dormancy can be caused by exogenous factors or can be due to the physical properties of the seed coat which prevents the external requirements for seed germination to penetrate the seed coat (e.g. water, light, and heat). Therefore, the seed coat of dormant seeds is impermeable to the external factors because of the condition of hard seeds and the presence of mechanical restrictions (Copeland and McDonald, 1995).

Endogenous dormancy is the most prevalent dormancy found in seeds and is due to the inherent properties of the seed. Seeds may possess an excess of inhibitor, and this may be caused by the environmental conditions wherein the seeds developed, such as day length, moisture status, position of the seed in the fruit or inflorescence, age of the mother plant, temperature during seed maturation, and maturation level of seeds. Endogenous dormancy can only be released when physiological changes take place in the seed such as undeveloped embryo dormancy, response to growth regulators, changes in temperature, exposure to light, and endogenous rhythms (Copeland and McDonald, 1995).

Under natural conditions, exogenous dormancy is broken down by freezing-thawing of the soil, ingestion by animals, microorganism activities, forest fires, natural soil acidity (Copeland and McDonald, 1995), and repeated soaking-drying of the seeds. In the laboratory, mechanical and chemical scarification or the removal of the seed coat can be carried out. Mechanical scarification involves grinding seeds with abrasives, heating, chilling, drastic temperature shifts, immersion in boiling water, nicking the hard seed coat, or exposure to certain radio frequencies to permit entry of water and gasses. Chemicals that are used to degrade the seed coat include sulfuric acid, sodium hypochlorite, hydrogen peroxide, and seed-coat enzymes (e.g. cellulose, pectinase). Solvents such as alcohol and acetone can be used to dissolve water-insoluble compounds around the seeds (Copeland and McDonald, 1995).

Leguminous trees such as *Calliandra calothyrsus*, *Gliricidia sepium*, *Leucaena trichandra*, *Sesbania sesban*, *Faidherbia albida*, and *Acacias* have hard seed coats.

These seeds can be treated by soaking them in boiling water and leaving them overnight until the seed coats are softened. Alternatively, very hard seed coats can be nicked using cutters and then the seeds soaked in water over night to allow water permeability. Species with softer seed coats (e.g. *Dalbergia, Gmelina, Tephrosia*) can be soaked in water without heat treatment or nicking of the seed coat (Mbora et al., 2008; Munjuga et al., 2013). Species like *Tectona grandis* (teak) need a longer period of stratification to release dormancy, involving 6–12 weeks of alternating cool moist-aerated conditions followed with warm-moist conditions in sand (Hall, 2003). Publications by Bhat (2001) and Krishnapillay (2002) provide a list of native tree species and their seed pre-treatment requirements for their respective countries.

2.4 FORESTRY NURSERIES

2.4.1 PLANNING AND ESTABLISHMENT OF FOREST NURSERIES

Although direct seeding is practiced in reforestation to some extent, seedlings have not yet passed the vital early stages of development and are subjected to the harsh conditions in the field (Smith et al., 1997). Germinated seedlings of most trees cannot survive competition with existing plants, and sufficient care and maintenance cannot be given to seedlings in the field for economic reasons. The need to establish a nursery or to purchase seedlings depends on seedling availability, cost, and the remoteness of the planting site. An annual requirement of 20,000 seedlings warrants the establishment of small nursery. (Evans, 1982). Therefore, forestry nursery is essential for producing healthy, vigorous and good-quality planting stock for forestry programs in the tropics.

Forest nurseries usually produce seedlings for specific purposes, and thus tree seedlings are of the desired species, specific size, and hardiness; are adequate in number; and should be ready at the right time (e.g. in the wet season in the tropics) (Evans, 1982). The most important consideration in establishing a forest nursery is market analysis, which encompasses the evaluation of profit margins compared to alternative investment of the capital, so that the investment can be economically justified. The extent and nature of forestry operations and the demand for planting stock are codependent. Establishment of forest nurseries must complement the demand for tree-planting material with adequate quality control measures, create employment opportunity for local society, and involve the capacity building of local expertise in operating forest nurseries (Krakowski, n.d.).

Many publications have addressed the planning, establishment, and management of formal plant or tree nurseries for different target audiences, such as local communities or personnel involved in planting stock production and tree planting as small holders, industrial sectors, or government agencies (Elliott et al., 2013; Evans, 1982; Hall, 2003; Krishnapillay, 2002; Mbora et al., 2008.; Munjuga et al., 2013; Mason, 2004; Smith et al., 1997). The establishment of the plant nursery must consider other factors related to the location, the size of the nursery, design and construction, and facilities and resources (Figure 2.7). The location must be easily accessible; protected from biotic (e.g. pest and diseases) and abiotic factors (e.g. frost, strong winds, floods, bushfires); have easy access to water and available

FIGURE 2.7   Tree nursery facilities.

planting material (e.g. soil, sand); ideally be on a gentle slope to allow for drainage; have access to available labor; and be secure. The soil at the nursery should ideally be deep, have good structure, and be easily pulverized; shallow soils with very hard sandstone, sandy soil, or clayey soil should be avoided (Munjuga et al., 2013). The size of the nursery will be determined by the availability of land, the type of planting stock, the expected production capacity, water availability, and land ownership (Mbora et al., 2008). A forest restoration program of 1 hectare per year will require 3,100 trees, which will need a nursery space of approximately 80 m² (Elliott et al., 2013).

The facilities and resources (Figure 2.7) require basic nursery tools, chemicals and their storage area, fencing and walls, internal paths, water tanks and distribution systems, a stock plant area (e.g. for cuttings), a seed bed or propagation area, a potting-up and working area (with pot, soil, machines, benches, and storage), a soil and compost shed, a production area, a growing-on area, a hardening off-area, shelter and amenities for workers (e.g. a wash room), and an administration area (Hall, 2003; Mason, 2004; Elliott et al., 2013). The availability of an affordable source of energy (Krakowski, n.d.) is also essential for undisrupted production.

2.4.2 TYPES OF FOREST NURSERIES

Forest nurseries can be established as permanent, temporary, or extension nurseries (Evans, 1982). A permanent forest nursery produces seedlings in large quantities over a long period of time, and it also supplies to temporary nurseries adjacent to planting sites. There are very expensive to establish and maintain but are well equipped and easily accessible (Appanah and Weinland, 1993). The location of a permanent nursery is carefully selected and usually close to the staff living quarters.

A centralized nursery operation permits better planning, maintenance, and supervision; good record keeping; and stock control, which allows better prediction of production and costs. It is highly invested and equipped with modern technologies for cultivation and irrigation and for pest, disease, and weed control, which can ensure high seedling production, survival, and quality. The nursery operation is much efficient and so reduces the unit cost per seedling (Evans, 1982). However, long-term flexibility is diminished with the amount of investments made for nursery development (Krakowski, n.d.).

A temporary or 'flying' nursery is established for a short duration (<5 years) to meet a specific and usually small-scale need of a planting program (Evans, 1982). In most instances in the tropics, tree-planting activities are carried out as enrichment planting or small-scale reforestation programs in small blocks of forested units. Temporary sowing beds are made in small forest clearings or under existing planted trees, and make use of locally available fruiting and naturally regenerated wildings (Appanah and Weinland, 1993). A temporary nursery is usually located in remote areas near planting sites (Hall, 2003), on flat areas with workable soils, good water supply, easy access to mother trees, and staff lodgings. The advantage of using a temporary nursery is to provide a transit site located nearer to the planting site so as to protect seedlings and sun-sensitive tree species like the dipterocarps (Appanah and Weinland, 1993).

Transportation to very remote planting sites represent a risk during the wet season because road conditions can be harsh. Thus, in the absence of temporary nursery, seedling supply to the planting site is hindered. The establishment of small nurseries can aid in isolating disease and preventing other types of damage. Although the capital investment of a temporary nursery is low, the cost of raising individual plants is higher due to the small scale of the operation and the low number of plants produced. The site usually lacks permanent facilities and is subject to low levels of supervision and control, and it is thus at risk of plant loss. The tree species selected should ideally be those that do not require special attention (Appanah and Weinland, 1993; Evans, 1982). In Sabah, temporary nurseries are also set up by ITPs because of the remoteness of the planting site, as well as the limited number of workers and inadequate transportation to make the frequent trips involved in transporting workers and seedlings. Field workers normally stay in temporary camps near the nursery during the planting season.

The third type is the extension nursery, which rarely forms part of a regular plantation program but is established to provide plants of many species that are useful to the local community for amenities, firewood, fodder, posts, and poles (Evans, 1982). Many forest reserves (FRs) are located adjacent to villages, and some communities are found within protected forest areas, thus spearheading forest policies that include community forestry and agroforestry programs (Hall, 2003).

2.4.3 NURSERY MANAGEMENT

Successful nursery management is dependent on many aspects, requiring good planning, organization, and control in order to avoid oversupplies or deficits in planting material and planting stock production. Factors that must be considered in managing the nursery include the selection and development of a suitable site; efficient supervision and administration; adequate planning, forecasting, and control procedures; orderly timing of operations; use of appropriate cultural methods; and protection from pests, diseases, and other types of damage (Evans, 1982). The efficiency of nursery management can be assessed through productivity (the required number of healthy and quality seedlings are produced at the right time) and reasonable costs of production (in terms of labor and material) (Hall, 2003).

The main operation of a forest nursery may include:

(i) Planning, controlling, and record keeping from receipt of seeds to tree planting

Planning the nursery operation includes producing schedules and data collection with the help of tools such as nursery calendars, plant development registers, nursery inventories (or production records), and records of nursery experiments. These records will provide information for production management and research (Jaenicke, 1999), and will be useful for task allocation, daily monitoring, and, in the long term, reviewing the success and failures of the nursery method employed (Hall, 2003). The nursery calendar is used to plan necessary actions (e.g. purchase of seed, supplies, and equipment) and nursery activities (e.g. bed construction, preparation of potting soil, start of compost production, sowing, weeding, repairs to nursery, and timing of annual staff leave); the plant development register is used to record information about seed treatment, germination requirements and duration, plant development, potting substrate, watering, shading, or disease control, and to provide labels of the exact identities and locations of seed lots and seedlings; the nursery inventory is used to keep track of the species and number of seedlings in different stages of development,

FIGURE 2.8 Nursery staff lifting, discarding, and rearranging tubes of *Eucalyptus deglupta* germinant (left) under shaded nursery (left), and the growing-on or hardening area of *E. deglupta* seedlings (right).

their survival, and their distribution to detect under- or over-supply; and the experiment record is for documenting the methods and results of nursery experiments (Jaenicke, 1999; Hall, 2003).

(ii) Capacity building and staff training

Efficient nursery operation relies on having permanent staff who are professional, careful, and honest in their work; and on the availability of hard-working technical workers. Permanent staff should be well-trained in all aspects of the nursery's activities through scheduled courses or regular staff meetings that cover aspects that are important (e.g. production methods, safety issues); this should then be reinforced through periodic explanation during the working period (Jaenicke, 1999). Permanent staff can then engage with the technical workers of the methods learnt.

(iii) Seed storage and pre-treatment (Section 2.3.5)

(iv) Soil preparation in the seed-bed or container (Section 2.5.3)

(v) Basal fertilizer application and top-dressing to control seedling nutrition

(vi) Sowing seed and/or rooting cuttings (section Figure 2.7)

(vii) Operations of pricking out, standing out, undercutting, lifting, transplanting, etc. (Figure 2.8)

(viii) Control of weeds

(ix) Protection against climatic damage, irrigation, shading, rain, pests, and diseases

(x) Packaging and dispatch of plant (Section 2.5.4)

2.5 PLANTING STOCK PRODUCTION

A steady and sufficient supply of high-quality planting stock is important for the success of any tree- planting program. Forest nurseries in the tropics tend to source their own planting stock, especially when using native tree species. Direct seeding in tropical conditions is rarely possible for establishing stands. Wildings obtained from the forest can suffer in open conditions, and bare-rooted seedlings have shown rather poor survival (Appanah and Weinland, 1993). Initially, tropical forestry imitated the practice of temperate countries by raising seedlings in open beds and planting out bare-rooted seedlings. However, the survival rate was often poor because of the high temperatures, dry conditions, and unpredictable start to the wet season (Smith et al., 1997).

Field trials to determine the suitability of tree species using a direct seeding approach must be conducted to improve seedling establishment. For example, only a few species (*Artocarpus dadah, Callerya atropurpurea, Vitex pinnata, Peltophorum obovatum, Diospyros oblonga, Microcos paniculata, Cinnamomum iners*, and *Garcinia hombroniana*) with seeds that were large or intermediate-sized and round or oval-shaped, with low or medium moisture content and seed-coat thickness of more than 0.4 mm, showed early seedling survival in a direct-seeding trial in Thailand. Seeds of late-successional forest tree species tend to be associated with these seed characteristics (Tunjai and Elliott, 2012).

Most local forest nurseries raise plants through conventional planting stock production methods via wilding collection, seed, and rooted cuttings propagation. For many tropical tree species, propagation by seed is still the most effective, cost-effective, and commonly used in forestry, but it requires sufficient knowledge to manage dormancy or recalcitrant seeds, germination conditions, facilities, and the growth requirements of specific tree species. The germination and survival rates of seedlings depend on adequate water and aeration, a suitable sowing medium, temperature, humidity, shade conditions, regular maintenance, and pest and disease control to ensure optimal growing conditions (Krishnapillay, 2002). Species not easily propagated through conventional methods are purchased from reliable commercial producers set up by governmental organizations and private commercial laboratories that mass-produce the plants by tissue culture. Tissue-culture facilities normally produce planting stock according to the orders received from the planters.

2.5.1 Wilding Collection

Wildings are germinants from the forest floor that are used for forest planting when seedling stock is scarce (Ford-Robertson, 1983) or in the case of problems with viable seed supply, unacceptable planting stock quality, and inadequate nursery facilities (Krakowski, n.d.). In the tropics, wildings are a reliable source of planting stock that are found in large numbers on the forest floor near their parent trees (Appanah and Weinland, 1993), but they only constitute a small portion used of the planting stock produced (Krakowski, n.d.). In the past, forest departments in the tropics regularly collected wildings to establish seedling reservoirs. Before an anticipated heavy seeding period, the areas beneath fruiting trees were cleared to form tree nurseries using natural regeneration. The wilding yield was usually very good, and they were used for enrichment planting in forests lacking natural regeneration or in secondary vegetation (Appanah and Turnbull, 1998).

The efficacy of transplanting wildings can be limited because the plants can be injured during extraction, handling, transporting, re-potting, and re-planting. At the time of transplanting, leafy tropical wildings are actively growing and sometimes do not respond well to handling, and they must be planted promptly. However, the rapid change in the growing environment can kill the plants (Krakowski, n.d.). Limitations related to the use of wildings include unsuccessful transplanting of larger wildings, high mortality with the use of small-sized wildings, and cases in which the limited number of wildings on the forest floor cannot supply the demands of large-scale plantations (Wyatt-Smith, 1963).

Therefore, methods were developed so that wildings are collected and then cared for in the nursery to allow the plants root systems to recover. The wildings can be gradually hardened to adapt to the new growing conditions (Krakowski, n.d.). During moist and cool conditions after rain, wildings can be lifted; the roots should be wrapped with wet paper, the leaves trimmed to reduce moisture loss, and the plants kept moist in closed polythene bags (moist chamber) after collection and during transportation (Mohd. Affendi et al., 1996; Krishnapillay, 2002). At the nursery, the wildings are potted, watered, and placed in covered polyethylene chambers to maintain high humidity and partial shade (45%–55% incident light) until new roots are formed (Moura-Costa, 1993; Krishnapillay, 2002). If potting cannot be done immediately, the roots of the wildings can be immersed in water in plastic bottles for up to 5 days (Mohd. Affendi et al., 1996). High survival rates of the wildings (80-95%) were obtained when young or small-sized wildings of about 3–5cm tall were collected during the wet season (Mohd. Affendi et al., 1996; Krishnapillay, 2002). This was demonstrated in *Endospermum malaccese, Parashorea malaanonan, Shorea parvifolia*, and *Dryobalanops lanceolate* wildings. The wildings can then be out-planted in the field after 2–3 months (Appanah and Weinland, 1993).

Recommendations from past research to improve the survival of wildings (Table 2.1) include using small-sized seedlings (<20cm) (Lantion, 1938; Mauricio, 1957; Rayos, 1940), allowing only a short storage time in the nursery (Rayos, 1940), maintaining wildings in moist chambers and watering them daily (Moura-Costa, 1993), the use of planting patch cultivation (Barnard and Setten, 1955), and shoot and root pruning of wildings after collection (Wardani, 1989). For species such as *Hopea odorata,* pruning the wildings or seedlings prior to transplanting and maintaining 25°C growing temperature in the nursery increased their survival rates to 50% (PROSEA, 1993).

2.5.2 RAISING CONTAINERIZED SEEDLINGS

This section will focus more on raising containerized or potted seedlings from seeds, since bare-rooted seedlings in the tropics have proven to be unsuitable for tropical forest conditions. Containerized seedlings are those that are transferable and planted in to the field with the soils intact on the roots (Smith et al., 1997). Containerized seedlings are a necessity under tropical and arid conditions due to high temperatures and light and sometimes drying conditions. The introduction of cheap polythene for tubes and bags (Figure 2.9) has led to the almost exclusive use of container-grown seedlings in the tropics (Appanah and Weinland, 1993; Smith et al., 1997). Container seedlings make efficient use of the growing space due to the high densities (e.g. 96-plugs per crate), involve less losses to culling, use less water for irrigation, and offer more protection to seedlings. Container seedlings are more tolerant of handling and suffer less transplant shock because of the root protection given by the growing

Sowing *Eucalyptus* seeds   *Eucalyptus* seedlings   Hardening *Eucalyptus* seedlings

Acacia mangium seedlings   *Neolamarckia*   *Moluccana falcataria* seedlings
                            cadamba seedlings

FIGURE 2.9   Tree seedlings raised in polythene bags and containers.

medium and container (Krakowski, n.d.). Container seedlings are costly to produce, but seeds and other resources can be used more carefully, are systematic, and can be semi-mechanized. The individual seedlings are easily cared for in the nursery (Smith et al., 1997), and operations can be monitored closely.

For ITPs, soilless media such as cocopeat or a mixture of topsoil and cocopeat are treated with fungicide and mixed with slow-release fertilizer and filled into plug tubes. Then, tree seeds are directly sown into the containers. There are different types of polyethylene tubes. The first are plastic containers for producing plugs, often in cylindrical cavities that are closely spaced, and the inset vertical is located in reusable molded blocks. The second type, called a *book planter*, is made of hinged sheets of thin molded plastic that form cavities when folded together, and the book planter is opened to remove the plugs. The third type is made up of individual tubes than can be set into and taken out of fixed racks. A tool for punching out planting holes of the same dimensions as the seedling plugs is also available (Smith et al., 1997). These sorts of containers are commonly used to raise small seeds of fast-growing tree species (e.g. *Acacia, Eucalyptus*) that can be out-planted to the field at 2–3 months after sowing. As seedlings become larger, the removable plugs are distanced in alternating rows at about 6–8 weeks for fast-growing species to provide adequate space for the tops to develop and to reduce the possibility of pest and disease spread from water splash during watering (Smith et al., 1997) as practiced in Sabah. Lifting and sorting the tube plants is labor-intensive, but most of the time it requires few technical employees (Krakowski, n.d.).

In the past, the smallest polythene bags or tubes were used to keep growing and transporting costs low. Plug trays have been standardized for high-density packaging. However, this can be a disadvantage as the elongated roots tend to deform into

TABLE 2.1
Production of Tropical Wilding Planting Stocks

Species	Location	Reference
Parashorea plicata, Shorea contorta	N.A.	Mauricio (1957); Cappellan (1961)
Dryobalanops oblongifolia, D. aromatic,	N.A.	Barnard (1954)
S. macrophylla, S. multiflora,		
Dipterocarpus baudii, Neobalanocarpus		
heimii, Agathis alba, Casuarina		
equisetifolia, Dacrydium elatum,		
Gonystylus bancanus, Kompassia		
malaccensis		
Anisoptera laevis, S. curtisii, S. leprosula,	West Malaysia	Gill (1970)
S. parvifolia, S. platyclados		
Dipterocarpus caudiferus, Dryobalanops	West Malaysia	Fox (1973); Chai (1975)
lanceolate, Parashorea tomentella		
Anisoptera, Dipterocarpus, Hopea,	West Sumatra	Jafarsidik and Sutomo (1988),
Parashorea, Shorea	West Kalimantan	Wardani and Jafarsidik (1988)
Dipterocarpus grandifloras, S. teysmanniana	N.A.	Lantion (1938)
Dryobalanops oblongifolia (76% nursery),	West Malaysia	Anon. (1951)
D. aromatic (90% field survival)		
Hopea pierrei	N.A.	Rayos (1940);
		Siagian et al. (1989a)
Parashorea malaanonan, Shorea parvifolia,	N.A.	Moura-Costa (1993)
Dryobalanops lanceolate (95%)		
Vatica sumatrana	N.A.	Masano and Omon (1985)
Dipterocarpus retusus	N.A.	Omon and Masano (1986)
Shorea platyclados	N.A.	Napitupulu and Supriana (1987)
S. selanica	N.A.	Siagian et al. (1989b)
Endospermum malaccese	West Malaysia	Mohd. Affendi et al. (1996)
Parashorea malaanonan	West Malaysia	Krishnapillay (2002)
S. parvifolia		
Dryobalanops lanceolate		

Note: N.A. (Not Available)

corkscrew roots (Appanah and Weinland, 1993) as they stick to the walls of the container, and they tend to stay deformed. Tree seedlings can become root bound as they quickly fill the plug with roots, which can create trees that are predisposed to slow growth after out-planting (Anon, 2000). There are methods that can offset the problem. The first method involves forming vertical grooves or indentations into the walls of the containers that can direct the roots downward. The second method is air-pruning, where containers with holes at the bottom are supported to provide air space which can retard the growth of roots beyond the level of the container. When the seedlings are out-planted, they still have the ability to resume normal growth when planted in contact with the soil. A similar effect to air pruning can be obtained by treating the container walls with toxic copper salts (Smith et al., 1997).

Long, narrow, and long-lasting tubes can be used to develop symmetrical roots, together with the right timing of out-planting to the field. Generally, small seedlings between 15 and 30 cm tall suffer less from transplanting shock, root more quickly, are easier to handle, and are less endangered by drying conditions (Appanah and Weinland, 1993). Additionally, proper selection of planting stock type can enhance the survival and growth of trees following out-planting (Pinto et al., 2011). For example, ponderosa pine (*Pinus ponderosa*) seedlings raised in different container types exhibited differences in growth parameters, with the best-performing seedlings raised in larger containers. At sites with sufficient moisture (mesic), the *P. ponderosa* seedlings outgrew their initial container-induced characteristics and were affected by environmental and genetic factors. Conversely, growth was reduced on a xeric (dry) site, and traits affected by the container type persisted longer. Thus, xeric sites may benefit from deep-rooted seedlings grown in large containers (Pinto et al., 2011).

Large seeds (e.g. *Intsia palembanica*, *Durio zibethinus*, Dipterocarps) are preferably sown in large pots or polyethylene bags filled with compost, decayed sawdust, or rich forest topsoil (Krishnapillay, 2002). The best dipterocarps seeds should be large, and the wings should be removed before sowing. The large end can be inserted into the soil or sown in a horizontal position so that the radicle grows downward into the soil. The depth of sowing should be about the length of the seed. Dipterocarps require about 50% shading initially which can be gradually reduced to 25% to simulate the conditions in the forest. The hardening period of dipterocarp species is about 2–6 months (Appanah and Weinland, 1993; DENR, 1996; Krishnapillay, 2002).

2.5.3 NURSERY BED

The tree nursery sowing bed is usually about 1 m wide and is raised to 15 cm (Figure 2.10), with the length varying based on the space and topography of the site. The beds are separated by narrow walkways, and wood planks are used to maintain the shape of the beds (Krishnapillay, 2002). The soil in the sowing bed is ploughed so that it is smooth and fine textured (Lee et al., 2008). The sowing media varies according to availability of material and its suitability for the tree species. Very small-scale seed sowing may utilize germination trays. The media is solar sterilized by drying under the hot sun covered with plastic sheet for a week and is changed frequently to reduce the spread of soil-borne fungal diseases (e.g. *Fusarium* spp., *Rhizoctonia solani*). For a small community nursery, the media mixture can be heat sterilized on galvanized iron sheets over a fire for 60 minutes (Bertomeu and Sungkit, without dates). Soil and sand of 1:1 to 3:1 ratio or mixed with compost or organic material such as rice hulls, peat, sawdust, or cocopeat are commonly used as the sowing media (Krishnapillay, 2002). The Sabah Forest Research Centre uses a mixture of forest topsoil, sand, and composted sawdust (7:3:2) (Lee et al., 2008). The addition of organic matter to the sowing media improves soil aeration, which is essential for fast and consistent seed germination.

The nursery beds should be designed to allow for fast drainage of excess water during watering or rainfall. However, during the dry season, the nursery beds must be able to retain sufficient moisture in the soil and the surroundings to maintain the germinants. Therefore, nursery beds are normally partly shaded about 40%–70%

FIGURE 2.10 Sowing bed (left) or the use of sowing trays (right).

with sarlon nets or are established under the shade of trees and covered with transparent polythene sheets to maintain moisture for most tropical species. However, some pioneer tree species require full sunlight (e.g. *Tectona grandis*, *Gmelina arborea*) to germinate well (Krishnapillay, 2002). Industrial tree nurseries usually build permanent nursery beds raised above the ground in a net house fully equipped with irrigation, adjustable shading, polythene covers, and drainage. The sowing media and seeds can be treated with fungicide to reduce the incidence of disease, with fresh media used after each sowing.

Very fine seeds (e.g. *Octomeles sumatrana*, *Neolamarckia cadamba*, *Eucalyptus* spp.) are mixed with sand in a 1:10 ratio to ensure seed separation and avoid overcrowding of germinants. The fine seeds can be broadcast on the sowing bed and the germinants can be transferred into containers (Mbora et al., 2008). Seeds sown too closely together produce weaker and non-uniform seedlings with small root systems that may not grow well after transplanting (Krishnapillay, 2002). Medium and large-sized seeds (e.g. *Tectona grandis*, *Shorea* spp., *Swietenia macrophylla*, *Hopea odorata*) can be sown in lines that are about 5 cm wide, spaced about 10 cm apart, and with a sowing depth of 1–2 times the length of the seed. After sowing, a thin layer of sand (Appanah and Weinland, 1993; Krishnapillay, 2002) or other organic matter such as sawdust (PROSEA, 1993) or rice hull (Bortomeu and Sungkit, without dates) is spread on top. Sowing too deep could delay seed germination (Mbora et al., 2008).

Weeding is best carried out after the germinants have established their root systems to prevent uprooting the germinants. Mature leaves of older germinants can be removed to allow sunlight to reach smaller germinants and thus produce equal sized seedlings. If the nursery bed is overcrowded, some germinants can be pricked out to provide 3–4 cm growing space (Lee et al., 2008). When the germinants reach 4–5 cm tall, they are ready for pricking and potting (Appanah and Weinland, 1993) in forest soil or a mixture of forest soil and sand. Forest soils are commonly used for native species such as the dipterocarps to provide mycorrhizal inoculation, which can improve the health of the planting stock (Mohd. Affendi and Ang, 1994; Supriyanto et al., 1994). Transplanted seedlings are kept under partial shade and watered adequately. Recalcitrant seeds will start to germinate a week after sowing, and this is completed by the third week with quite a high germination percentage. Growing tropical *Eucalyptus deglupta* for community forestry involves sowing in a soil–sand mixture (3:1). The *E. deglupta* germinants which produced the first pair of leaves (30–50 days after sowing) were transplanted into polyethylene bags (3″ × 5″) containing a mixture of topsoil, rice hulls, and river sand (6:3:1). The *E. deglupta* transplants were placed in the shade for the first 5 days and then grown in full sunlight thereafter (Bertomeu and Sungkit, without dates).

2.5.4   MAINTENANCE AND HANDLING OF PLANTING STOCK

After sowing and the potting of the tree seedling, great care must be given during the tending phase. Watering must be controlled according to the temperature and humidity levels in the nursery. Too much moisture renders the plants prone to fungal disease, while insufficient moisture leads to drying out of the media. The seedlings must be protected against insects, mammals, fungi, and weed through good cultural practices and with the help of suitable chemical application (Appanah and Weinland, 1993; Krishnapillay, 2002). If producing different types of species, the shading requirement of the species during different development stages must be understood. Shade beds are maintained at 70% shade for very light-sensitive species and seedlings at the earliest stage of development. Transplanting beds are maintained at 50% shade, and the hardening areas can range from 30% shade to full sunlight exposure (Lee et al., 2008). As the seedlings are gradually transferred from the shade bed, then to the transplanting bed, and finally to the hardening area, the watering frequency and amount is also reduced. For example, the seedlings of *Hopea odorata* are kept under 50% shade in the nursery for 6–9 months, and the shade is gradually reduced until full exposure to sunlight for the last 2 months prior to field planting (Krishnapillay, 2002).

The shaded area must also have good ventilation (Bertomeu and Sungkit, without dates) to prevent the incidence of fungal diseases. Not all tree species show reduced transplantation shock despite the hardening process. In dry Mediterranean climates, the use of shelter tubes for the purpose of safeguarding seedlings against browsing had also improved seedling survival. The growth of shade-tolerant *Quercus ilex* L. and shade-intolerant *Pinus halepensis* Mill. in shelter tubes with varying light transmissivity demonstrated that the latter species had reduced root growth during the wet season. *Quercus ilex* grown in different shelter conditions showed that sheltered seedlings survived better in the field than unsheltered seedlings. This suggests that late successional shade-tolerant tree species planted in shelter tubes will benefit from reduced light stress without impairing root growth in the wet season (Puértolas et al., 2010).

Fertilizer application experiments on tree seedlings in the nursery have proven to be beneficial for planting stock even after transplanting to the field and affect plants produced through vegetative propagation. Fertilizer application on stock plants, for example 50g NPK (20:10:10) applied on *Allanblackia floribunda*, significantly increased the number of stem cuttings produced. Foliar NPK applied on leafy stem cuttings of *A. floribunda* did not significantly affect the number of roots produced by the cuttings but was best applied using foliar NPK 20:20:20 at the callus stage

(Tsobeng et al., 2016). Seedlings exhibit obvious macronutrient (N, P, K, Mg, Ca) deficiency symptoms above and below ground. Seedling growth and nutrient concentration in the tissue is reduced when grown in nutrient deficient soil and without fertilizer application (Jeyanny et al., 2009).

The growth performance of Aleppo pines (*Pinus halepensis* Mill.) and holm oak (*Quercus ilex* L.) nursery tree seedlings on dry sites is dependent on the tree species, seedling size, and N concentration. *Pinus halepensis* showed greater differentiated response to weed competition than *Q. ilex* due to functional differences. Large *P. halepensis* seedlings had better survival and growth rates than small seedlings in the field because of greater gas exchange, root growth, and N cycling (Cuesta et al., 2010). In the Mediterranean tree species *Quercus suber*, fertilization of seedlings in the nursery affected seedling traits but not field performance. Unfertilized seedlings had decreased total biomass accumulation and slenderness ratio (ratio of stem height to root collar diameter). However, seedling survival in the field was not correlated to nutritional regimes, seedling size, or root growth potential. The height of unfertilized seedlings was affected because of shoot dieback and slow growth (Trubat et al., 2010).

Tree species (*Hyeronima alchomeoides*, *Cedrela odorata*, *Cordia alliodora*) also demonstrated varying nutrient use efficiency (based on nutrient components, productivity, and the mean residence time of nutrients) and their ability to deal with N limitation. High nutrient productivity and longer nutrient retention in *Hyeronima* indicate the ability to thrive on infertile sites. *Cedrela* showed high nutrient productivity and brief nutrient retention, implying potential high productivity only on fertile soil (Hiremath et al., 2002). Pioneer (*Nauclea diderrichii*) and climax (*Entandrophragma angolense*) tree species have shown different photosynthetic characteristics when grown under high and low-proton flux densities and different levels of nutrient supply. The pioneer *Nauclea* showed greater assimilation and photosynthetic rate increase and anatomical response to both light and nutrient supply than *Entandrophragma* (Riddoch et al., 1991). The response of different tree species under varying conditions can differ and should be tested to determine the best and most cost-effective nutrient application.

Other experiments that tested different combinations of fertilizer application and the size of the containers the seedlings were grown in revealed that using small-size containers or polybags has negative effects on the growth of seedlings despite supplying fertilizer or using rich media substrate (Akpo et al., 2014). Container capacity affects seedling morphology, production costs, and post-planting growth of the different stock-types. For example, survival of *P. halepensis* and *P. pinea* seedlings after transplantation was lower for seedlings raised in the smallest containers. For minimal production cost per living seedling and per stem volume after 3 years in the field, the optimum stock-type was propagated in medium size containers (300 cm³) with additional late-season fertilization to improve post-planting performance (Puértolas et al., 2012).

Effective microorganisms (EM), especially mycorrhizal fungi, have been used in the production of good quality, healthy tree-planting stocks. Mycorrhizal inoculation can boost the growth performance of tree seedlings in the nursery and help the seedlings to survive transplantation shock (Urgiles et al., 2009). Mycorrhizae occur naturally on the forest floor and can be collected as fungal fruiting bodies (Tata et al., 2008) or from mycorrhizae roots of forest seedlings (Urgiles et al., 2009), prepared as slices of 100g of fungi in 1 L water and inoculated on the forest planting stocks (Tata et al., 2008). Commercial mycorrhizae in the forms of spore tablets are also available for purchase. Mycorrhizal fungi (*Boletus* spp., *Russula* spp., *Scleroderma* spp.) improved growth (diameter, dry weight) and increased the accumulation of micro-nutrients (Fe, Mn, Cu, Zn, and Al) in all portions of *Hopea odorata*, *Shorea compresa*, *S. pinanga*, *S. stenoptera*, and *Vatica sumatrana* seedlings (Santoso et al., 1989). Similarly, arbuscular mycorrhizal (AM) *Glomus clarum* and *Gigaspora decipiens* increased *Dyera polyphylla* and *Aquilaria filarial* seedling growth, and increased shoot N and P concentrations by 70%–153% and 135%–360% respectively. Survival rates were higher in the AM-colonized seedlings at 180 days after transplantation than in non-inoculated seedlings, indicating that AM fungi can accelerate the establishment of the planting stocks in the field (Turjaman et al., 2006).

Two-month-old rooted cuttings of *Shorea contorta* (Vid.) inoculated with vegetative mycelia of three ectomycorrhizae strains of *Pisolithus* collected under eucalypts and a strain of *Scleroderma* from dipterocarps had 11%–38% root colonization prior to out-planting. Two years after out-planting in Surigao, the Philippines, the growth (height and diameter) and survival of inoculated seedlings was higher by 7%–17% from the control seedlings. Field growth assessment at 2 and 8 years after out-planting indicated that the performance of seedlings and beneficial effects of ectomycorrhizae strains did not persist consistently (Aggangan et al., 2013). Some experiments that combined fertilizer application and mycorrhizal infection gave inconsistent results. Turner et al. (1993) reported that *Shorea macroptera* seedlings treated with NPK fertilizer (10g m^{-2}) and ectomycorrhizae showed infection to a significant extent. But unfertilized seedlings showed a better correlation between the extent of ectomycorrhizae infection and growth of seedlings, which suggests that the species is responsive to fertilizer addition when grown at low nutrient availabilities. Conversely, moderate fertilization did not suppress mycorrhizae development in *Cedrela montana* and *Heliocarpus americanus*. Growth performance of non-inoculated and non-fertilized plants was lower compared to inoculated and/or fertilized plants. However, mycorrhizae application improved the root collar diameter of the seedlings (Urgiles et al., 2009).

Rhizobacteria are another type of EM, but its use is still limited in forest nurseries. Some research has demonstrated the potential application of rhizobacteria in promoting seedling growth. For example, *Erwinia*, *Rhizobium*, *Enterobacter*, *Duganella*, *Alcaligeneceae bacterium*, *Oxalobacteraceae bacterium*, and yeasts isolated from *Shorea leprosula* were inoculated on *Shorea selanica* seedlings. The results were promising as the functional bacteria could potentially improve stand density in the plantation (Enebak et al., 1998), and assist in fungal mycorrhizal formation (mycorrhizal helper bacteria) (Sitepu et al., 2007). Screening for rhizobacteria has identified *Bacillus*, *Pseudomonas*, *Chruseobacterium*, *Mucilaginibacter*, and *Rhodococcos* sp. as reliably improving adventitious rooting of *E. nitens* × *E. globulus* mini-cuttings in Chile (González et al., 2018).

An experiment combining controlled-release fertilizer with *Pseudomonas fluorescens* rhizobacteria improved *P. halepensis* seedling growth and nutrient uptake compared to independent fertilizer or rhizobacteria application. The combined treatment could reduce pollutants and improve microbiota community in poor soils (Dominguez-Nuñez et al., 2015). Nadeau et al. (2018) demonstrated that selected mycorrhizal fungi and rhizobacteria can be applied simultaneously to improve the health and growth of white spruce seedlings to be grown on waste rocks or fine tailings in Abitibi, Canada. The mycorrhizae (*Cadophora finlandia*) promoted plant health and belowground development, while the rhizobacteria (*Pseudomonas putida*) enhanced aboveground plant biomass. In all of the mycorrhizae and rhizobacteria inoculation experiments, seedlings that were not inoculated were not colonized.

Seedlings must be graded based on size, health, and vigor for each species prior to field planting because of the different growth response of species in the nursery and in the field. Dipterocarp seedlings are usually maintained in the nursery between 3 and 9 months, and the height is about 25–50 cm tall, with younger planting stock more likely to survive transplanting in the field than older seedlings (Appanah and Weinland, 1993; Barnard, 1954; Mohd. Affendi, 1994). Furthermore, small seedlings from 15 to 25 cm tall are easier to transport and handle (Lamprecht, 1989). A manual on grading Dipterocarp nursery seedlings was produced based on research (Sheikh Ibrahim, 2006b). Fast-growing tree species such as *A. mangium* and *E. deglupta* seedlings are ready for field planting at 2–4 months after sowing depending on the size of the containers used. The choice of size of the planting stock is also subject to the condition of the planting site. Large planting stock is preferred on sites with intensive weed growth. Vigorous seedlings with low collar-to-shoot proportions tend to form roots faster and can tolerate drought stress (Moura-Costa, 1993). Large seedlings are not necessarily desirable in tropical plantations because they transpire more water, have over-grown roots in potted seedlings, are costly to transport, and can delay planting activities (Smith et al., 1997). Generally, only healthy, vigorous stock should be used for planting while weak, smaller than normal, badly shaped, or diseased stock should be culled periodically (Appanah and Weinland, 1993).

Tropical FRMs easily desiccate during transportation from nursery to field due to their delicate nature and the high temperatures and treacherous terrain of most forest and plantation areas in the tropics (Ng, 1996). Containerized seedlings are packed into crates or on racks designed to prevent damage to the seedlings. Containerized planting stock can be transferred to the field whenever it is suitable for out-planting. Prior to field planting, the seedlings should be watered, then shaded during transportation to prevent dehydration and wind and sun scorch damage (Krishnapillay, 2002). If large numbers of seedlings are transferred to the field in a big lorry, and the number of workers are limited, a temporary shaded shelter close to a water source should be set up to store seedlings for large-scale replanting. Transplanting is normally done early in the morning or in the late evening during the wet season to reduce seedling desiccation. The root of the seedling is inserted into the planting hole up to the root-collar region, and the soil around the hole is then compressed to ensure good contact between the soil and the root collar (Lee et al., 2008).

2.5.5 PRODUCING VEGETATIVE PLANTING STOCKS

Rooting juvenile stem and shoot cuttings is the most commonly used technique because it is a relatively easy and convenient compared to grafting, budding, and layering, which require skill and are expensive, labor intensive, and unsuitable for mass propagation (Krishnapillay, 2002). Therefore, this section will focus on reviewing literature related to the techniques and the subsequent improvements that have successfully increased the production of FRMs.

Vegetative propagation methods for producing forest planting stock are necessary for species that are not amenable to seed production, and when there is a need to multiply individuals with highly desirable genetic qualities. Standardized planting stocks in large quantities can be produced much quicker to meet the demand of forest plantation (Appanah and Turnbull, 1998). The process of tree selection based on phenotype or genotype and natural breeding takes a considerably long time. Vegetative propagation is not able to improve the planting stock produced (Smith et al., 1997), but it is useful for multiplying improved propagules from tree improvement programs.

The selection of superior clones of tree species for producing better planting stock enables rapid genetic gain (Naidu and Jones, 2009) compared to raising planting stocks from seeds of improved or unimproved trees. The clones selected are those that have rapid growth, are high yielding, and are pest and disease resistant. The genetic uniformity of trees grown under similar environmental conditions is beneficial for commercial tree plantations because the stock is less costly to maintain and harvest (Appanah and Turnbull, 1998). Planting stock produced through vegetative propagation can reach maturity much quicker and produce seeds earlier, which is extremely useful to speed up the breeding cycle in tree improvement programs. For example, Bergmann (1998) reported that *Paulownia elongata, P. fortunei,* and the hybrid *Paulownia* 'Henan 1' produced from shoot cuttings and micro-propagated shoots had better survival and growth than those produced from seeds a year after field planting in the southeastern United States. Significant variation in survival and growth among clones within all species allows for the selection of superior clones for vegetative propagation.

Despite the potential advantages of vegetative propagation, some tropical tree species are difficult to propagate vegetatively (Aminah et al., 1995; Aminah and Lokmal, 2002; Sheikh Ibrahim, 2006a) due to low rooting ability (Brondani et al., 2012; Gimenes et al., 2015). Problems related to the low rooting success of dipterocarps include that old clonal multiplication gardens must be replaced to provide juvenile materials required for the production of stem cuttings. Nursery facilities need to be upgraded and sufficient facilities installed for commercial-scale production. Trained personnel are required to handle the large-scale production of rooted stem cuttings of the Dipterocarp species (Sheikh Ibrahim, 2006a). The procedure for selecting superior stock plants and proper techniques must be established prior to large-scale application (Krishnapillay, 2002), as those developed for dipterocarp species in Indonesia (Tata et al., 2008).

New plants can be regenerated from stock plants, with some investment in a mist propagation system (Krishnapillay, 2002). Stock plants that will be used to harvest vegetative shoots can be raised in the ground or in containers. Shade-tolerant species (e.g. dipterocarps) require partial shading (30%), while shade-intolerant species (e.g. *A. mangium*) can be raised in full sunlight. Stock plants raised in the ground are also known as *clonal hedge gardens* or *clonal multiplication gardens*. Raising stock plants generally involves the same care and maintenance as planting stock in the nursery, such as watering, weeding, fertilizer, and pesticide and fungicide application to ensure healthy growth. Periodic pruning of the stock plants is required to produce new coppice shoots and juvenile growth for cuttings (Hartmann et al., 1990). Stock plants of dipterocarps or other tropical hardwoods are usually young nursery seedlings (< 1 year) or from seedling orchards (Tata et al., 2008).

For the propagation of macro-cutting, juvenile material is taken from the shoots of seedlings, rooted cuttings, or orthotropic shoots of pruned stock plants (Aminah, 1991). Woody, lignified stems and those from mature trees are not suitable for vegetative propagation as they are less likely to produce root or take too long to root. The appropriate length of macro-cuttings differs according to species. For many species, cuttings are cut at a right angle to the stem containing a single nodal, and they are usually about 5 cm in length but can be to up to 30 cm in length for other species. The leaves of the cuttings are usually removed or trimmed to reduce water loss while still allowing photosynthesis. The base of the cutting is treated with exogenous auxins to promote rooting, and, based on the literature, usually with indolebutyric acid (IBA) (Noraini and Liew, 1994; Brodie, 2003; Castellanos-Castro and Bonfil, 2013).

Many research studies have tested different types of rooting media (e.g. rice hull, commercial rooting media, topsoil mixed with organic matter) for different tropical tree species. Some experiments, including recent efforts, reported cleaned river sand as the best medium (Assis, 2011; Ajik and Kimjus, 2013). Generally, light intensity of 15%–30% and humidity above 80% within the propagator is desirable for good rooting. Regular inspection of the cuttings is necessary to remove dead and deteriorated cuttings and to pot rooted cuttings (Krishnapillay, 2002). The simplest propagators are non-misted and built as an enclosure with plastic sheet with the base filled with separate layers of sand (2 cm), stones (10 cm), gravel (3 cm), and rooting medium (10 cm). Water is poured into the base until just below the rooting medium which provides moisture to the cuttings through capillary action (Krishnapillay, 2002). The mist propagator is the most common propagation method, maintaining humidity by producing mist at specified time periods. The rooting media should be changed with every new batch of cuttings. Species such as the dwarf rootstock of persimmon (*Diospyros kaki* Thunb.) show very low rooting percentages under mist irrigation. When placed under a fog system, the cuttings rooted better (80%) as the relative humidity was maintained at 100%, whereas with the mist system, humidity drops to less than 50% at midday (Tetsumura et al., 2017).

Factors that can influence the rooting ability of stem and shoot macro-cuttings include the age of stock plant, season of harvesting the cuttings, type and height of cutting, presence or absence of vegetative buds, number of leaves on the cutting (Hartman et al., 1990), rooting facilities, rooting media, source of cutting material, treatment of cuttings (Krishnapillay, 2002), rooting environment conditions, and genetic nature of the species (Naidu and Jones, 2015). Most studies reported that rooting success was dependent on the propagation medium and exogenous application of auxins (Aminah et al., 1995; Brodie, 2003). The combination of medium-auxin, and fungicide (Benlate)-medium showed significant effects on the rooting percentage and mortality of *Shorea splendida* and *Dipterocarpus oblongifolia*. *Dipterocarpus oblongifolia* had better rooting in perlite/vermiculite with 500 ppm

IBA, while *S. splendida* performed better in locally available material combination (sand–cocopeat–gravel) with 1500 ppm IBA, and sand–cocopeat without IBA. The two species responded differently to these factors because of different metabolic rates (Brodie, 2003).

Fertilizing *Dyera costulata* stock plants using NPK with trace elements ($12N:12P_2O_5:17K_2O:2MgO$) can improve its growth compared to unfertilized plants. However, it does not have an observable positive impact on the rooting of stem cuttings in *D. costulata* (Aminah and Lokmal, 2002). In three *Bursera* species, rooted cuttings and the number of roots per cutting varied between species, IBA concentration, and the age of the stock plants. The rooting percentage was higher when cuttings were taken from young stock plants, but were less responsive to IBA treatment at 11 weeks after sowing (Castellanos-Castro and Bonfil, 2013). The length of *E. grandis* × *E. urophylla* cuttings affected rooting percentage and growth, with the optimal cutting length being from 8 to 10 cm (Naidu and Jones, 2009).

Vegetative propagation has become an essential tool for exploiting hybrid heterosis by cloning hybrids with spontaneous resistance to diseases, thereby improving the productivity of forest plantation in Brazil (Assis, 2011). The micro-cutting technique, which is the rooting of apical shoots from micro-propagated plants, was developed after the macro-cutting method. Subsequently, the mini-cuttings technique, which is the rooting of axillary sprouts of rooted stem-cuttings, was invented (Assis, 2011). The benefits of mini-cuttings include the higher production of juvenile and herbaceous cuttings, better control of the hedge environment, increased productivity per square meter, intensive management within a small area, superior rapid rooting and quality root systems, nursery efficiency, enhanced nursery capacity, high-quality plants, and good initial field performance (Naidu and Jones, 2015).

Micro-cuttings and mini-cuttings are the propagation technology used for large-scale cloning of *Eucalyptus* species. Both techniques start with the establishment of mini-clonal hedges that are raised in hydroponic systems consisting of river sand beds with drip irrigation or periodic flooding. The mini-hedges are supplied with specific nutrient requirements to produce sprouts at the commercial level (Assis, 2011). Mini-hedges of *Eucalytpus grandis* × *E. urophylla* can be planted at 10 cm × 15 cm, in comparison to a 60 cm × 80 cm space for a macro-hedge. For micro-hedges, 66 hedges were established, and 264 cuttings were produced per square meter. For macro-hedges, two hedges were established, and 24 cuttings were produced per square meter. Therefore, mini-hedges offer an 11-fold increase in the number of cuttings per unit area. Mini-cuttings of *E. grandis* x *E. urophylla* are smaller at about 4–7 cm. The mini-cuttings had better rooting percentage by 8.3%–24.9 %, better root quality (i.e. cumulative root length, root dry mass, shoot dry mass), and better shoot growth over macro-cuttings (Naidu and Jones, 2015). Low temperatures can promote adventitious root establishment with dispersed vascular connections in *Eucalyptus benthamii* mini-cuttings (Brondani et al., 2012). Therefore, the production cost of mini-cuttings is lower, but they have high rooting ability with superior root systems. Effective rooting of mini-cuttings can reduce the turnaround time of the plantation program, minimize contact with pathogenic fungi, and use less fungicide (Assis et al., 2004).

The improved mini-cutting method has been adapted for other tropical trees such *Azadirachta indica* (Gehlot et al., 2014) and *Cabralea canjerana* (Gimenes et al., 2015) with good success compared to macro-cutting methods. Gehlot et al. (2014) reported that IBA resulted in the best rooting percentage, number of roots, root length, and number of leaves per rooted mini-cutting when grown in sand compared to vermiculite and soil. Mini-cutting allowed for the extension of propagation season in the months that were free of monsoon, gave higher yields of rooted cuttings per stock plant, and resulted in a high rooting percentage through IBA application (Gehlot et al., 2014). The combination of commercial substrate, coarse sand, and carbonized rice husks maximized mini-cuttings rooting of *C. canjerana* due to air-filled pore space that was closer to the optimal value of 30%. The application of 3000 mg/L of IBA improved rooting differentiation and growth of the mini-cuttings. Nodal mini-cuttings had higher rooting capability than apical mini-cuttings, and the clones vary in rooting ability and survival rates. The production of mini-cuttings increased from the first shoot harvest to the third harvest, indicating that harvesting period of the mini-cuttings can enhance the productivity of the mini-stumps. The maintenance of mini-stumps in greenhouse conditions over time may increase the survival and rooting of mini-cuttings (Gimenes et al., 2015).

ACKNOWLEDGEMENT

We gratefully acknowledge Acacia Forest Industries Sdn. Bhd. (AFISB), Sabah Forest Industries Sdn. Bhd. (SFI), Sabah Forestry Department Authority (SAFODA), Universiti Kebangsaan Malaysia, and Universiti Malaysia Sabah for permitting field work and providing technical support.

REFERENCES

Adjers, G., Hadengganan, S., Kuusipalo, J., Otsamo, A. and Vesa, L. 1998. Production of planting stock from wildings of four *Shorea* species. *New Forests*, 16(3): 185–197.
Aggangan, N.S., Pollisco, M.A.T., Bruzon, J.B. and Gilbero, J.S. 2013. Growth of *Shorea contorta* Vid. inoculated with Eucalypt ectomycorrhizal fungi in the nursery and in a logged-over dipterocarp forest in Surigao, Philippines. *American Journal of Plant Sciences*, 4: 896–904.
Ajik, M. and Kimjus, K. 2013. An attempt to produce rooted cuttings from some selected dipterocarp species. *Sepilok Bulletin*, 17 and 18: 57–68.
Akpo, E., Stomph, T.J., Kossou, D.K., Omore, A.O. and Struik, P.C. 2014. Effects of nursery management practices on morphological quality attributes of tree seedlings at planting: The case of oil palm (*Elaeis guineensis* Jacq.). *Forest Ecology and Management*, 324: 28–36.
Aminah, H. 1991. Rooting ability of stem cuttings of eleven dipterocarp species. *Journal of Malaysian Applied Biology*, 20: 155–159.
Aminah, H. and Lokmal, N. 2002. Effects of fertilizer treatments on growth of *Dyera costulata* stock plants and rooting ability of their stem cuttings. *Journal of Tropical Forest Science*, 14(3): 414–420.
Aminah, H., Dick, J.McP., Leakey, R.R.B., Grace, J and Smith, R.I. 1995. Effect of indole butyric acid (IBA) on stem cuttings of *Shorea leprosula*. *Forest Ecology and Management*, 72(2–3): 199–206.

Anon. 1951. Artificial regeneration. Report of the Forestry Administration 1950/1951. Kuala Lumpur, pp. 9–10.

Anon. 1998. *Malaysia-ITTO Project, Sustainable Forest Management and Development in Peninsular Malaysia PD 185/91 Rev 2(F)—Phase I, Guidelines to Propagate Dipterocarp Species by Stem Cuttings*. Forest Department Peninsular Malaysia, Kuala Lumpur.

Appanah, S. and Turnbull, J.M. 1998. *A Review of Dipterocarps: Taxonomy, Ecology and Silviculture*. Center for International Forestry Research, Bogor Indonesia, pp. 57–99, 151–186.

Appanah, S. and Weiland, S. 1993. Planting quality timber trees in Peninsular Malaysia—A review. Malayan Forest Records No. 38. Forest Research Institute Malaysia, Kepong.

Ashton, P.S., Givnish, T.J. and Appanah, S. 1988. Staggered flowering in the Dipterocarpaceae: New insights into floral induction and evolution of mast fruiting in the aseasonal tropics. *The American Naturalist*, 132(1): 44–66.

Assis, T.F. 2011. Hybrids and mini-cutting: A powerful combination that has revolutionized the *Eucalyptus* clonal forestry. *BMC Proceedings*, 5(Suppl 7): 118.

Assis, T.F., Fett-Neto, A.G. and Alfens, A.C. 2004. Current techniques and prospects for the clonal propagation of hardwoods with emphasis on *Eucalyptus*. *In*: Walter, C. and Carson, M. (eds) *Plantation Forest Biotechnology for the 21st Century*. Research Signpost, Trivadrum, India, pp. 303–333.

Barnard, R.C. 1954. A manual of Malayan silviculture for inland lowland Forestry. Res. Pam. No. 14. Forest Research Institute Malaysia, Kepong, Malaysia.

Barnard, R.C. and Setten, G.G.K. 1955. 'Notch' and 'patch' planted wildings. *Malayan Forester*, 18: 85–88.

Barner, H. 1975. Classification of sources for procurement of forest reproductive material. *In*: *Report on the FAO/DANIDA Training Course on Forest Seed Collection and Handling. Chiang Mai, Thailand, Vol. 2*. FAO, Rome.

Barrett, W.H.G. 1985. Seed orchards. *In*: *FAO/DANIDA Training Course on Forest Tree Improvement*. FAO, Rome: 1980.

Bergmann, B.A. 1998. Propagation method influences first year field survival and growth of *Paulownia*. *New Forests*, 16(3): 251–164.

Bertomeu, M.G. and Sungkit, R.L. n.d. *Propagating Eucalyptus Species—Recommendations for Smallholders in the Philippines*. ICRAF, Philippines.

Bhat, D.M., Murali, K.S. and Ravindranath, N.H. 2001. Assessment of propagation techniques for forest species of the Western Ghat region of India. *Forests, Trees and Livelihoods*, 11(3): 233–250.

Bonfil, C. and Trejo, I. 2010. Plant propagation and the ecological restoration of Mexican tropical deciduous forests. *Ecological Restoration*, 28(3): 369–376.

Boulay, A., Tacconi, L. and Kanowski, P. 2013. Financial performance of contract tree farming for smallholders: The case of contract eucalypt tree farming in Thailand. *Small-Scale Forestry*, 12(2): 165–180.

Brand, D. 2015. Balancing the need for timber plantations and forest conservation. *In*: International Conference on Enhancing Biodiversity Towards No Net Loss and Beyond within the Heart of Borneo Landscape, 10th November 2014, Kota Kinabalu, Malaysia. Sabah Forestry Department, Sandakan, Malaysia, pp. 73–75.

Broadhurst, L., Hopley, T., Li, L. and Begley, J. 2017. A genetic assessment of seed production areas (SPAs) for restoration. *Conservation Genetics*, 18(6): 1257–1266.

Brodie, J.F. 2003. Factors affecting the rooting ability of *Dryobalanops oblongifolia* and *Shorea splendida* (Dipterocarpaceae) stem cuttings. *Journal of Tropical Forest Science*, 15(1): 109–116.

Brondani, G.E., Baccarin, F.J.B., de Wit Ondas, H.W., Stape, J.L., Gonçalves, A.N. and de Almeida, M. 2012. Low temperature, IBA concentrations and optimal time for adventitious rooting of *Eucalyptus benthamii* mini-cuttings. *Journal of Forestry Research*, 23(4): 583–592.

Brotto, L., Pettenella, D., Cerutti, P. and Pirard, R. 2016. Planted forests in emerging economies: Best practices for sustainable and responsible investment. Occasional paper 151. CIFOR, Bogor, Indonesia.

Camirand, R. 2002. Guidelines for forest plantation establishment and management in Jamaica. Jamaica trees for tomorrow project, Phase II. CIDA Contract No. 504/15808. TECSULT International, Québec, Canada.

Cappellan, N.M. 1961. Possibilities of bagtikan (*Parashorea plicata* Brandis), white lauan (*Pentacme contorta* Merr. and Rolfe), amugi (*Koordersiodendron pinnatum* (Blanco) Merr.), rain tree (*Samanea saman*) (Jacq.) Merr.) and Spanish cedar (*Cedrela odorata* L.) wildings as a nursery planting stock. *Philippine Journal of Forestry*, 17: 101–112.

Castellanos-Castro, C. and Bonfil, C. 2013. Propagation of three *Bursera* species from cuttings. *Botanical Sciences*, 91(2): 217–224.

Chai, D.N.P. 1975. Enrichment planting in Sabah. *Malaysian Forester*, 38: 271–277.

Chechina, M. and Hamann, A. 2015. Choosing species for reforestation in diverse forest communities: Social preference versus ecological suitability. *Ecosphere*, 6(11): 240.

Chin, H.F. and Roberts, E.H. 1980. *Recalcitrant Crop Seeds*. Tropical Press Sdn. Bhd., Kuala Lumpur, Malaysia.

Chong, K.Y., Chong, R., Tan, L.W.A., Yee, A.T.K., Chua, M.A.H. and Wong, K.M. 2016. Seed production and survival of four dipterocarp species in degraded forest in Singapore. *Plant Ecology and Diversity*, 9(5–6): 486–490.

Copeland, L.O. and McDonald, M.B. 1995. *Seed Science and Technology*. Chapman and Hall, New York, USA.

Cuesta, B., Villar-Salvador, P., Puertolas, J., Jacobs, D.F. and Benayas, J.M.R. 2010. Why do large, nitrogen rich seedlings better resist stressful transplanting conditions? A physiological analysis in two functionally contrasting Mediterranean forest species. *Forest Ecology and Management*, 260(1): 71–78.

Daisuke, H., Tanaka, K., Jawa, K.J., Ikuo, N. and Katsutoshi, S. 2013. Rehabilitation of degraded tropical rainforest using dipterocarp trees in Sarawak, Malaysia. *International Journal of Forestry Research*, 2013(683017). doi://10.1155/2013/683.

Darus, H.A., Thompson, S. and Pirrie, A. 1989. Vegetative propagation of *Acacia mangium* by stem cuttings: The effect of seedling age and phyllode number on rooting. *Journal of Tropical Forest Science*, 2(4): 274–279.

DENR 1996. Propagation protocols for indigenous/endangered and exotic forest tree species and some fruit trees. Infor. Bull. 1 (June 2011). DENR, Ecosystems Research and Development Bureau, Laguna, Philippines.

Dominguez-Nuñez, J.A., Delgado-Alvez, D., Berrocal-Lobo, M., Anriquez, A. and Albanesi, A. 2015. Controlled-release fertilizers combined with *Pseudomonas fluorescens rhizobacteria* inoculum improve growth in *Pinus halepensis* seedlings. *iForest— Biogeosciences and Forestry*, 8(1): 12–18 [online 2014-05-12].

Elliott, S., Navakitbumrung, P., Kuarak, C., Zangkum, S., Anusarnsunthorn, V. and Blakesley, D. 2003. Selecting framework tree species for restoring seasonally dry tropical forests in northern Thailand based on field performance. *Forest Ecology and Management*, 184(1–3): 177–191.

Elliott, S.D., Blakesley, D. and Hardwick, K. 2013. *Restoring Tropical Forests: A Practical Guide*. Royal Botanical Gardens, Kew, p. 344.

Ellis, R.H., Hong, T.D. and Roberts, E.H. 1990. An intermediate category of seed storage behavior? *Journal of Experimental Botany*, 41(9): 1167–1174.

Enebak, S.A., Wei, G. and Kloepper, J.W. 1998. Effect of plant growth-promoting rhizobacteria on loblolly and slash pine seedlings. *Forest Science*, 44(1): 139–144.

Evans, J. 1982. *Plantation Forestry in the Tropics*. Oxford University Press, New York, USA,
pp. 121–164.

Farrant, J.M., Pammenter, N.W. and Berjak, P. 1986. Recalcitrance—A current assessment.
Seed Science and Technology, 16(1): 155–166.

Finkeldey, R. 2005. *An Introduction to Tropical Forest Genetics*. Institute of Forest Genetics
and Forest Tree Breeding, Georg-August University Göttingen, Germany, p. 219.

Fisher, R.F. 1995. Amelioration of degraded rain forest soils by plantations of native trees.
Soil Science Society of American Journal, 59(2): 544–549.

Ford, G.A., McKeand, S.E., Jett, J.B. and Isik, F. 2015. Effects of inbreeding on growth and
quality traits in loblolly pine. *Forest Science*, 61(3): 579–585.

Ford-Robertson 1983. *Terminology of Forest Science, Technology, Practice and Products,
No. 1*. Society of American Foresters, p. 370.

Fox, J.E.D. 1973. Dipterocarp seedling behavior in Sabah. *Malaysian Forester*, 36: 205–214.

Francis, J.K. 1998. Tree species for planting in forest, rural and urban areas of Puerto Rico.
General technical report IITF-3, USDA For. Serv., International Institute of Tropical
Forestry, Rio Piedras.

Gehlot, A., Gupta, R.K., Arya, I.D., Arya, S. and Tripathi, A. 2014. *De novo* adventitious
root formations in mini-cuttings of *Azadirachta indica* in response to different rooting
media and auxin treatments. *iForest*, 8: 558–564 [online 2014-12-09].

Gill, L.S. 1970. Enrichment planting and its future in hill forest silviculture. *Malaysian
Forester*, 33: 135–143.

Gimenes, E.S., Kielse, P., Haygert, K.L., Fleig, F.D., Keathley, D.E. and Bisognin, D.A.
2015. Propagation of *Cabralea canjerana* by mini-cutting. *Journal of Horticulture and
Forestry*, 7(1): 8–15.

Goh, D., Galiana, A. and Monteuuis, O. 2014. Options for mass producing *Acacia mangium*
and *A. Mangium x A. auriculiformis* superior planting materials. *In*: International
Conference W.P. 2.08.07: Genet. and silviculture of Acacia, sustaining the future of
Acacia Plantation Forestry.

Goh, D. and Monteuuis, O. 2016. Teak. *In*: Yill-Sung, P., Jan, B. and Hyeung-Kyu, M. (eds.)
Vegetative Propagation of Forest Trees (Online edn.). National Institute of Forest
Science, Korea.

González, P., Sossa, K., Rodríguez, F. and Sanfuentes, E. 2018. Rhizobacteria strains as
promoters of rooting in hybrids of *Eucalyptus nitens x Eucalyptus globulus*. *Chilean
Journal of Agricultural Research*, 78(1): 3–12.

Granhof, J. 1991. *Mass Production Improved Material (2) Seed Orchards: Concepts, Design
and Role in Tree Improvement*. Danida Forest Seed Centre. Lecture Note D-8, Uni. of
Copenhagen, Denmark.

Griffin, A.R. 2014. Clones or improved seedlings of Eucalyptus? Not a simple choice.
International Forestry Review, 16(2): 216–224.

Grippin, A., Nor Aini, A.S., Nor Akhirrudin, M., Hazandy, A.H., Sures Kumar, M. and
Ismail, P. 2018. The prospect of micropropagating *Gonystylus bancanus* (Miq.) Kurz,
a tropical peat swamp forest timber species through tissue culture technique-Review.
Journal of Forest Science, 64(1): 1–8.

Guarino, L., Ramantha, R. and Reid, R. 1995. *Collecting Plant Genetic Diversity*. CABI
Publishing, Wallingford, UK.

Hall, K.C. 2003. *Manual on Nursery Practices*. Forestry Department, Jamaica, p. 77.

Harrington, J.F. 1972. Seed storage and longevity. *In*: Kozlowski, T.T. (ed.) *Seed Biology*, vol.
3. Academic Press, New York, USA, pp. 145–245.

Hartmann, H.T., Kester, D.E. and Davis, Jr., F.T. 1990. *Plant Propagation—Principles and
Practices*, 5th edition. Prentice-Hall Inter. Edn., Eaglewood Cliffs, NJ, USA, p. 647.

Harwood, C.E., Thinh, H.H., Quang, T.H., Butcher, P.A. and Williams, P.A. 2004. The effect of inbreeding on early growth of *Acacia mangium* in Vietnam. *Silvae Genetica*, 53(2): 65–69.

Hiremath, A.J., Ewel, J.J. and Cole, T.G. 2002. Nutrient use efficiency in three fast-growing tropical trees. *Forest Science*, 48(4): 662–672.

Indufor. 2012. *Forest Stewardship Council (FSC) Strategic Review on the Future of Forest Plantations*. Indufor, Helsinki, Finland.

Intongkaew, W. and Liu, J. 2017. Development of economic forest plantation in Thailand. *International Journal of Sciences*, 6(10): 52–62.

Itoh, A., Yamakura, T., Kanzaki, M., Ohkubo, T., Palmiotto, P.A., LaFrankie, J.V., Kendawang, J.J. and Lee, H.S. 2002. Rooting ability of cuttings relates to phylogeny, habitat preference and growth characteristics of tropical rainforest trees. *Forest Ecology and Management*, 168(1–3): 275–287.

Jaenicke, H. 1999. *Good Tree Nursery Practices, Practical Guidelines for Research Nurseries*. International Centre for Research in Agroforestry, Nairobi, Kenya, p. 94.

Jafarsidik, Y. and Sutomo, S. 1988. Description and illustrations of a number of dipterocarp wildings from Sijunjung production forest, west Sumatra. *Buletin Penelitian Hutan*, 501: 13–58.

Jansson, G., Hansen, J.K., Haapanen, M., Kvaalen, H. and Steffenrem, A. 2017. The genetic and economic gains from forest tree breeding programmes in Scandinavia and Finland. *Scandinavian Journal of Forest Research*, 32(4). doi:10.1080.02827581.2016.1242770

Jeyanny, V., Rasip, A.A., Wan Rasidah, K. and Ahmad Zuhaidi, Y. 2009. Effects of macronutrient deficiencies on the growth and vigour of *Khaya ivorensis* seedlings. *Journal of Tropical Forest Science*, 21(2): 73–80.

Krakowski, M.J. n.d. Producing planting stock in forest nurseries. *Forests and Forest Plants— Vol. III*. Encyclopedia of Life Support Systems (EOLSS). http://www.eolss.net/Eolss

Krishnapillay, B. 2002. A manual for forest plantation establishment in Malaysia. Malayan For. Rec. No. 45. Forest Research Institute Malaysia, Kuala Lumpur, Malaysia.

LaFrankie, Jr, J.V. and Chan, H.T. 1991. Confirmation of sequential flowering in *Shorea* (Dipterocarpaceae). *Biotropica*, 23(2): 200–203.

Lampela, M., Jauhiainen, J., Sarkkola, S. and Vasender, H. 2017. Promising native tree species for reforestation of degraded tropical peatlands. *Forest Ecology and Management*, 394: 52–63.

Lampela, M., Jauhiainen, J., Sarkkola, S. and Vasender, H. 2018. To treat or not to treat? The seedling performance of native tree species for reforestation on degraded tropical peatlands of SE Asia. *Forest Ecology and Management*, 429: 217–225.

Lamprecht, H. 1989. *Silviculture in the Tropics. Tropical Forest Ecosystems and their Tree Species. Possibilities and Methods for their Long-term Utilization*. Deutsche Gesellschaft für Technische Zusammenarbeit (GTZ), p. 296.

Lantion, D.C. 1938. Wild forest seedlings as planting stock. *Philippine Journal of Forestry*, 1: 199–210.

Lee, Y.F., Anuar, M. and Chung, A.Y.C. 2008. *A Guide to Plantation Forestry in Sabah*. Sabah For. Rec. No. 16. Forestry Department, Sandakan, Malaysia.

Lu, Y., Ranjitkar, S., Jian-Chu, X., Xiao-Kun, O., Ying-Zai, Z., Jian-Fang, Y., Xun-Feng, W., Weyerhaeuser, H. and He, J. 2016. Propagation of native tree species to restore subtropical evergreen broad-leaved forests in SW China. *Forests*, 7(1): 12.

Maimunah, Siti, Syed Ajijur, R., Yusuf, B.S., Yustina, A., Trifosa, I.S., Sarah, A., Soo, M.L., Himlal, B. 2018. Assessment of suitability of tree species for bioenergy production on burned and degraded peatlands in Central Kalimantan, Indonesia. *Land*, 7(115). doi://10.3390/land7040115

Marzalina, M. 1995. Penyimpanan biji benih mahogany (*Swietenia macrophylla* King). Ph.D. Thesis, Universiti Kebangsaan Malaysia, Bangi.

Marzalina, M., Abd. Khalim, A.S., Siti Hasanah, M.S., Abd. Rahman, A.J. and Kassim, A. 1998. Mobile seed-seedling chamber: mechanism to maintain germplasm viability over long journey. *In*: Marzalina, M., Khoo, K.C., Jayanthi, N., Tsan, F.Y. and Krishnapillay, B. (eds.) *Proceedings of the of International Union of For. Res. Org. Seed Symposium*, Kuala Lumpur, Malaysia.

Masano and Omon, R.M. 1985. The effect of NAA hormone on growth of *Vatica sumatrana* stumps and wildings in Darmaga, Bogor. *Buletin Penelitian Hutan*, 468: 1–7.

Mason, J. 2004. *Nursery Management*. 2nd edition. Landlinks Press, Collingwood, Australia.

Mauricio, F.P. 1957. A preliminary study on the behavior of wild dipterocarp seedlings when transplanted in the forest. *Philippine Journal of Forestry*, 13: 147–159.

Mbora, A., Lillesø, J.-P.B. and Jamnadass, R. 2008. *Good Nursery Practices: A Simple Guide*. World Agroforestry Centre, Nairobi, Kenya.

Mohd. Affendi, H. and Ang, L.H. 1994. Nursery technique for dipterocarps—Part I: Planting stock production of *Shorea assamica* (Meranti pipit), *Shorea parvifolia* (Meranti sarang punai), Shorea leprosula (Meranti tembaga), *Dryobalanops aromatic* (Kapur) and *Hopea odorata* (Merawan siput jantan). FRIM Tech. Info. No. 49. FRIM Kepong, Malaysia, p. 4.

Mohd. Affendi, H., Ab. Rasip, A.G., Mohd. Noor, M. and Mohd. Jaffar, S. 1996. Teknik pengeluaran anan benih sesenduk melalui kutipan anak liar dan keratan batang. FRIM Tech. Info. No. 54. FRIM Kepong, Malaysia, p. 6.

Moura-costa, P. 1993. Large scale enrichment planting with dipterocarps, methods and preliminary results. Proceedings of the of Inter. Workshop BIO-REFOR, Yogyakarta, pp. 72–77.

Munjuga, M.R., Gachuiri, A.N., Ofori, D.A., Mpanda, M.M., Muriuki, J.K., Jamnadass, R.H. and Mowo, J.G. 2013. *Nursery Management, Tree Propagation and Marketing: A Training Manual for Smallholder Farmers and Nursery Operators*. World Agroforestry Centre, Nairobi Kenya, p. 60.

Nadeau, M.B., Laur, J. and Khasa, D.P. 2018. Mycorrhizae and rhizobacteria on Precambrian Rocky Gold Mine Tailings: I. Mine-adapted symbionts promote white spruce health and growth. *Frontiers in Plant Science*, 9: 1267.

Naeim, M., Widiyatno, Al-Fauzi, M.Z. 2014. Progeny test of *Shorea leprosula* as key point to increase productivity of secondary forest in Pt. Balik Papan Forest Industries, east Kalimantan Indonesia. *In: 4th Inter. Conference on Sus. Future for Human Security, Sustain 2013. Procedia Environmental Sciences*, 20: 816–822.

Naidu, D. and Jones, N. 2015. Evaluation of mini-cuttings as a propagation system for *Eucalyptus* hybrids. IPPS 18h Annual Conference, Saint Ives. March 2015.

Naidu, R.D. and Jones, N.B. 2009. The effect of cutting length on the rooting and growth of subtropical *Eucalyptus* hybrid clones in South Africa. *Southern Forests: A Journal of Forest Science*, 71(4): 297–301.

Napitupulu, B. and Supriana, N. 1987. A preliminary experiment on the use of growth hormone on meranti (*Shorea platyclados*) seedlings at Purba Tongah, north Sumatra. *Buletin Penelitian Kehutanan*, 3: 45–52.

Ng, F.S.P. 1996. *High Quality Planting Stock—Has Research Made a Difference?* CIFOR Occasional Paper No. 8. (Nov 1996). CIFOR, Bogor, Indonesia, pp. 1–13.

Nickles, G. and Spidy, T. 1984. Seedling seed sources of *Pinus caribaea* var. hondurensis in Queensland. *In*: Barnes, R.D. and Gibson, G.L. (ed.) *Prof. Conference IUFRO on Provenance and Genetic Improvement Strategies in Tropical Forest Trees, Mutare, Zimbabwe 9–14 April1984*. P.C. and Newton, R.S., pp. 579–581.

Noraini, A.B. and Liew, T.S. 1994. Effects of plant materials cutting positions, rooting media and IBA on rooting of *Shorea leprosula* (Dipterocarpaceae) cuttings. *Pertanika Journal of Tropical Agricultural Science*, 17(1): 49–53.

Norisada, M., Hitsuma, G., Kuroda, K., Yamanoshita, T., Masumori, M., Tamge, T., Yagi, H., Nuyim, T., Sasaki, S., Kojima, K. 2005. *Acacia mangium*, a nurse tree candidate for reforestation on degraded sandy soils in the Malay Peninsula. *Forest Science*, 51(5): 498–510.

Omon, R.M. and Masano 1986. The effect of NAA hormone on wilding and stump growth of *Dipterocarpus retutus* in Darmaga, Bogor. *Bulein Penelitian Hutan*, 479: 28–35.

Parsada, N. 2013. Multi storied forest management system as an enrichment planting arrangement on poorly stocked inland forest: An experience in Peninsular Malaysia. *International Journal of Sciences*, 2(11): 28–42.

Perumal, M., Mohd. Effendi, W., Soo-Ying, H., Lat, J. and Hamsawi, S. 2017. Survivorship and growth performance of *Shorea macrophylla* (de Vriese) after enrichment planting for reforestation purpose at Sarawak, Malaysia. *OnLine Journal of Biological Sciences*, 17(1): 7–17.

Pinto, J.R., Marshall, J.D., Dumroese, R.K., Davis, A.S. and Cobos, D.R. 2011. Establishment and growth of container seedlings for reforestation: A function of stocktype and edaphic conditions. *Forest Ecology and Management*, 261(11): 1876–1884.

PROSEA. 1993. *In:* Soerianegara, I. and Lemmens, R.H.M.J. (eds.) *Plant Resources of Southeast-Asia. Timber Trees: Major Commercial Timbers*, Pudoc Scientific Publishers, Wageningen, The Netherlands, p. 610.

Puertolás, J., Jacobs, D.F., Benito, L.F. and Peñuelas, J.L. 2012. Cost-benefit analysis of different container capacities and fertilization regimes in *Pinus* stock-type production for forest restoration in dry Mediterranean areas. *Ecological Engineering*, 44: 210–215.

Puertolás, J., Oliet, J.A., Jacobs, D.F., Benito, L.F. and Peñuelas, J.L. 2010. Is light the key factor for success of tube shelters in forest restoration plantings under Mediterranean climates? *Forest Ecology and Management*, 260(5): 610–617.

Rayos, J.A. 1940. Preliminary study on the survival of wild bare-rooted Dalingdan seedlings stores in sawdust. *Philippine Journal of Forestry*, 3: 417–423.

Riddoch, I., Lehto, T. and Grace, J. 1991. Photosynthesis of tropical tree seedlings in relation to light and nutrient supply. *New Phytologist*, 119(1): 137–147.

Roberts, E.H. 1973. Predicting the storage life of seeds. *Seed Science and Technology*, 1: 499–514.

Sandewall, M., Ohlsson, B., Sandewall, R.K. and Viet, L.S. 2010. The expansion of farm-based plantation forestry in Vietnam. *Ambio – A Journal of the Human Environment*, 39(8): 567–579.

Santoso, E., Hadi, S., Soeseno, R. and Koswara, O. 1989. Akumulasi unsur mikro oleh lima jenis Dipterocarpaceae yang diinokulasi dengan beberapa fungi mikoriza. *Buletin Penelitian Hutan*, 514: 11–17.

Schmidt, L. 2000. *Guide to Handling of Tropical and Subtropical Forest Seed*. DANIDA Forest Seed Centre, Humlebaek, Denmark.

Selvaraj, P. and Muhammad, A.B. 1980. *A Checklist of Plantation Trials in Peninsular Malaysia*. FRIM Res. Pamphlet 79. Kepong, Malaysia.

Sheikh Ibrahim, S.I. 2006a. Project completion report Malaysia—International Tropical Timber Organisation Joint Project: PD 185/91 Rev. 2 (F) – Phase II. For. Dept. Peninsular Malaysia, Kuala Lumpur/ITTO.

Sheikh Ibrahim, S.I. 2006b. A manual of grading of nursery seedlings. Malaysia—International Tropical Timber Organisation Joint Project: PD 185/91 Rev. 2 (F) – Phase II. For. Dept. Peninsular Malaysia, Kuala Lumpur / ITTO, p. 22.

Siagian, Y.T., Masano, Harahap, R.M.S., Alrasyid, H. 1989a. The effect of rootone F on stem cuttings and stumps of commercial forest trees. *Buletin Penelitian Hutan*, 505: 41–52.

Siagian, Y.T., Masano, Harahap, R.M.S., Alrasyid, H., Soemarna, K. 1989b. Effect of application of Rootone F and length of storage on survival rate of potted *Shorea selanica* and *Diospyros celebica*. *Buletin Penelitian Hutan*, 508: 27–36.

Sitepu, I.R., Aryanto, Ogita, N., Osaki, M., Santoso, E., Tahara, S., Hashidoko, Y. 2007. Screening of rhizobacteria from dipterocarp seedlings and saplings for the promotion of early growth of *Shorea selanica* seedlings. *Tropics*, 16(3): 245–252.

Smith, D.M., Larson, B.C., Kelty, M.J. and Ashton, P.M.S. 1997. *The Practice of Silviculture: Applied Forest Ecology*. John Wiley and Sons Inc., Toronto, Canada, pp. 265–273.

Stanton, T., Echavarria, M. and Hamilton, K. 2010. *State of Watershed Payments: An Emerging Marketplace*. Forest Trends. http://www.forest-trends.org/documents.files/doc_2438.pdf. Accessed on 3 March 2018.

Subiakto, A., Rachmat, H.H. and Sakai, C. 2016. Choosing native tree species for establishing man-made forest: A new perspective for sustainable forest management in changing world. *Biodiversitas, Journal of Biological Diversity*, 17(2): 620–625.

Supriyanto, Setiawan, I., Mylayana, O., Santosa, E. 1994. Effectiveness of *Schleroderma dictyosporum* obtained by protoplast culture in accelerating the growth of *Shorea selanica* and *Shorea leprosula* cuttings. *In*: Wickneswari, R., Ahmad Zuhaidi, U., Amir Husni, M.S., Darus, A., Khoo, K.C., Kazuo, Suzuki, K., Sakurai, S., Ishii, K. (eds.) *Proceedings of the Inter. Bio-Refor Workshop, 28 Nov.–1 Dec. 1994*. Kangar, Malaysia, pp. 170–176.

Tata, H.L., Wibawa, G. and Joshi, L. 2008. *Enrichment Planting with Dipterocarp Species in Rubber Agroforests: Manual*. World Agroforestry Centre (ICRAF) Southeast Asia Regional Program; Indonesian Res. Ins. For Estate Crops, Bogor, Indonesia, p. 33.

Tchoundjeu, Z., Avana, M.L., Leakey, R.R.B., Simons, A.J., Asaah, E., Duguma, B. and Bell, J.M. 2002. Vegetative propagation of *Prunus africana*: Effects of rooting medium, auxin concentrations and leaf area. *Agroforestry Systems*, 54(3): 183–192.

Tetsumura, T., Ishimura, S., Honsho, C. and Chijiwa, H. 2017. Improved rooting of softwood cuttings of dwarfing rootstock for persimmon under fog irrigation. *Scientia Horticulturae*, 224: 150–155.

Toda, R. 1974. Notes on Japanese stage government forest tree breeding project. *In*: *Forest Tree Breeding in the World*. Yamatoya Ltd, Tokyo, pp. 161–169.

Trubat, R., Cortina, J. and Vilagrosa, A. 2010. Nursery fertilization affects seedling traits but not field performance in *Quercus suber* L. *Journal of Arid Environments*, 74(4): 491–497.

Tsobeng, A., Ofori, D., Tchoundjeu, Z., Asaah, E. and Van Damme, P. 2016. Improving growth of stockplants and rooting ability of leafy stem cuttings of *Allanblackia floribunda* Oliver (Clusiaceae) using different NPK fertilizers and periods of application. *New Forests*, 47(2): 179–194.

Tunjai, P. and Elliott, S. 2012. Effects of seed traits on the success of direct seeding for restoring southern Thailand's lowland evergreen forest ecosystem. *New Forests*, 43(3): 319–333.

Turjaman, M., Tamai, Y., Santoso, E., Osaki, M. and Tawaraya, K. 2006. Arbuscular, mycorrhizal fungi increased early growth of two nontimber forest product species *Dyera polyphylla* and *Aquilaria filaria* under greenhouse conditions. *Mycorrhiza*, 16(7): 459–464.

Turner, I.M., Brown, N.D. and Newton, A.C. 1993. The effect of fertilizer application on dipterocarp seedling growth and mycorrhizal infection. *Forest Ecology and Management*, 57(1–4): 329–337.

Uhl, C., Buschbacher, R. and Serrao, E.A.S. 1988. Abandoned pastures in eastern Amazonia, I. Patterns of plant succession. *The Journal of Ecology*, 76(3): 663–681.

Urgiles, N., Loján, P., Aguirre, N., Blaschke, H., Günter, S., Stimm, B. and Kottke, I. 2009. Application of mycorrhizal roots improves growth of tropical tree seedlings in the nursery: A step towards reforestation with native species in the Andes of Ecuador. *New Forests*, 38(3): 229–239.

Wang, B.S.P. 1975. Tree and shrub trees. *In*: *Advances in Research and Technology of Seeds, Part 1*. PUDOC, Wageningen, pp. 34–43.

Wardani, M. 1989. The effect of root wrenching and leaf stripping on the growth of *Dipterocarpus hasseltii* wildings. *Buletin Penelitian Hutan*, 514: 19–23.

Wardani, M. and Jafarsidik, Y. 1988. Description and illustrations of a number of dipterocarp wildings from a woodlot in Sanggau, West Kalimantan. *Buletin Penelitian Hutan*, 502: 27–48.

Willan, R.L. 1985. *A Guide to Forest Seed Handling with Special Reference to the Tropics*. DANIDA Forest Seed Centre, Denmark, p. 394.

WWF. 2018. https://www.worldwildlife.org/industries/timber. Accessed on 16 October 2018.

Wyatt-Smith, J. 1963. *Manual of Malayan Silviculture for Inland Forests*. Malayan Forest Record No. 23. Forest Research Institute Malaysia, Kepong.

Zobel, B.J., Barber, J., Brown, C.L. and Perry, T.O. 1958. Seed orchards-their concept and management. *Journal of Forestry*, 56(11): 815.

3 Nursery Practices on *Neolamarckia cadamba* Seedlings' Growth Performance

Julius Kodoh and Mandy Maid

CONTENTS

3.1   INTRODUCTION

Neolamarckia cadamba (ROXB.) BOSSER is in the *Rubiaceae* family, of the subfamily *Cinchonoideae*. Commonly known as Laran (Malaysia), it is a deciduous tropical tree species that originated in and is widely distributed from South Asia and Southeast Asia (Orwa et al., 2009) to Australia (Australian Tropical Rainforest [Plants, 2010]). It was also successfully introduced into tropical and subtropical areas outside of its native range. It is a fast-growing species, has good silvicultural characteristics (straight cylindrical bole), is able to grow in diverse soil conditions and is free from serious pests and diseases. The species is increasingly important for various wood-based industries as smallholders and plantation concessionaires expand their plantings in the region (Mojiol et al., 2018). The species is distributed naturally from Sri Lanka, India, Nepal, Bhutan, Bangladesh, southern China, Indo-China, but throughout the Malesian region, Papua New Guinea, Australia, African and Central American countries as ornamental and industrial plantation trees. The tree commonly grows in broad-leaved primary and secondary forests at altitudes of between 100 and 1000 m above sea level on a variety of soil conditions; it can grow on river banks and in the transitional zone between swampy, permanently flooded areas and drier loams, and in areas that are periodically flooded (Krisnawati et al., 2011; Soerianegara and Lemmens, 1993; Tuan et al., 2011), and is very light demanding and intolerant to frost (Joker, 2000a,b). The tree has a broad umbrella-shaped crown; a straight cylindrical bole (Lee et al., 2008), the height reaching up to 45 m with stem diameter at breast height between 100 and 160 cm; and horizontally spreading branches that are characteristically arranged in tiers, flattening and becoming subterete and glabrescent (Tuan et al., 2011).

    The genetic resources of *N. cadamba* are secure as they are widespread and abundant (Soerianegara and Lemmens, 1993) in the natural forest. Their economic importance saw the establishment of seed stands in natural settings and in trial and commercial plantations. Several seed orchards and seedling seed orchards have been established in Sabah, Malaysia, since 2001 to serve as seed sources to meet local seed demands (Sabah Forestry Department, 2008). The *N. cadamba* tree has good silviculture characteristics that make it very suitable for the purposes of forest plantation and replanting (Matra et al., 2011). In Malaysia, commercial and trial planting has been established at Kanowit Sarawak, Sandakan and Tawau Sabah and Setul Forest Reserve (F.R.), Batu Lenggong F.R., and within the Forest Research Institute of Malaysia (FRIM). In Sabah, agroforestry trials of inter-cropping oil palm and Laran have been highly recommended (Lee et al., 2005).

The purpose of this chapter is to show the results of studies of some nursery practices on the growth performance of *N. cadamba* seedlings. The study was conducted at Forestry Complex Nursery, Faculty of Science and Natural Resources, Universiti Malaysia Sabah, to identify the effects of different types of growing media, intensity of shading, frequency of watering, and fertilization rate at nursery stage for a period of 10–12 weeks.

3.2 BACKGROUND TO EXPERIMENTAL WORK

3.2.1 EFFECT OF GROWING MEDIA ON *NEOLAMARCKIA CADAMBA* SEEDLINGS' GROWTH PERFORMANCE

Good growing media will give healthy roots, which will in turn result in a strong plant, good survival, and good growth in the nursery and in the field. The important properties of a growing media that influence plant growth and development can be divided into three categories: chemical properties (fertility, acidity, cation exchange capacity, and buffer capacity); physical properties (water-holding capacity, porosity, plasticity, and bulk density); and biological properties (presence of beneficial, harmful, or other micro-organisms) (Jaenicke, 1999).

A study on the growth performance of *N. cadamba* seedlings using different growing media was carried out to identify a suitable growing media for *N. cadamba* seedlings to grow at nursery stage for a period of 10 weeks. Six treatments using different types of growing media were used, namely, topsoil (100%) as a control (T1), cocopeat (70% fine: 30% coarse) (T2), coco peat (30% fine: 70% coarse) (T3), coco peat (50% fine : 50% coarse) (T4), Jiffy Pellet (T5), and sawdust of *N. cadamba* mixed with vermicompost (T6). Data on the seedlings' height, collar diameter, number of leaves, and leaf area were taken every 7 days. The shoot:root ratio was measured at the end of week 10. Polybags measuring 6×9 cm were used as planting pots. The seedlings were watered manually twice a day in the morning and afternoon, and no fertilizer was applied, while pests and weeds were monitored every day.

3.2.2 EFFECT OF INTENSITY OF SHADING ON *NEOLAMARCKIA CADAMBA* SEEDLINGS' GROWTH PERFORMANCE

Krisnawati et al. (2011) stated that light is an important component in the growth of seedlings. However, the tolerance level of *N. cadamba* seedlings of the rate of light intensity has not yet been stated in any current published journal. Light intensity is able to change a plant's phenotype, flower development, and size and color of the leaves in both herbaceous plants (Jeong et al., 2009; Vendrame et al., 2004) and woody species (Hampson et al., 1996). Abdul Khaliq (2001) explains that shading will affect the morphological and physiological performance of the plant.

A study on the growth performance of *N. cadamba* seedlings in different intensities of shading was carried out to identify the suitable intensity of shading for *N. cadamba* seedlings to grow at nursery stage for a period of 10 weeks. Polybags measuring 6×9 cm were used as planting pots. Four treatments of different intensities of shading were used, namely, 0% intensity of shading as a control (T1), 50% intensity of shading (T2), 70% intensity of shading (T3) and 90% intensity of shading (T4). Data on the seedlings' height, collar diameter, number of leaves, and leaf area were taken every 7 days interval. The shoot:root ratio was measured at the end of week 10. The seedlings were watered manually twice a day in the morning and afternoon, and no fertilizer was applied, while pests and weeds were monitored every day. The data collected were analyzed using one-way ANOVA and correlation analysis.

3.2.3 Effect of Watering on *Neolamarckia cadamba* Seedlings' Growth Performance

Water helps plants with essential nutrients movement into the plant. If there is not enough water in the plant cells, the plant will wither away. The water helps the plant to remain strong. The water solution in plants contains sugar and other nutrients. Without the right balance of water, plants will not only suffer malnutrition but will also be physically weakened and unable to accommodate their own weight (Agus and Imam, 2006).

A study on the growth performance of *N. cadamba* seedlings using different watering frequencies was carried out to identify the suitable watering frequency for *N. cadamba* seedlings to grow at nursery stage for a period of 12 weeks. Polybags measuring 6×9 cm were used as planting pots. Six treatment of different watering frequencies were used, namely, no watering applied as a control (T1), watering every day (T2), watering every 2 days (T3), watering every 4 days (T4), watering every 6 days (T5), and watering using a water reservoir (T6) (Elhadi et al., 2013). Data on the seedlings' height, collar diameter, number of leaves, and leaf area were taken every 7 days. The shoot:root ratio was measured at the end of week 12, while pests and weeds were monitored every day. The data collected were analyzed using one-way ANOVA and correlation analysis.

3.2.4 Effect of Fertilizing on *Neolamarckia cadamba* Seedlings' Growth Performance

A plant's or seed's fertility is dependent on the source of light, water, and nutrients. The type of fertilizer and the quantity of fertilizer are important to determine the growth performance of seedlings from the early stages. Fertilization aims to restore the nutrient content in the soil that has been absorbed by plants through the process of physiology. A lack of nutrients can cause the growth of seedlings in the nursery to be stunted, and plant leaves to become brittle and yellow (McCauley et al., 2011). Thus, fertilizer can be defined as a source of plant nutrition that can be added to the growing media to supply nutrients that are needed by seedlings (Cole and

Gessel, 1992). Agroblen is a slow release fertilizer type which has been used widely for the early stages of seedling growth to provide nutrients to the seedlings (Termizi, 2015). By using slow-release fertilizers such as Agroblen, nutrients can be provided to the *N. cadamba* seedlings and the loss of nutrients caused by leaching and runoff minimized. Agroblen also can supply nutrients to the seedlings for a long period of time and release nutrients slowly to the seedlings along the enlargement of the seed (Shaviv, 2005).

A study on the growth performance of *N. cadamba* seedlings using different quantities of Agroblen fertilizer was carried out to identify the suitable quantity of Agroblen fertilizer for *N. cadamba* seedlings to grow at nursery stage for a period time of 10 weeks. Polybags size measuring 6×9 cm were used as planting pots. Four treatments of different quantities of Agroblen fertilizer were used, namely, 0 kg/m^3 (T1), 8 kg/m^3 (T2), 12 kg/m^3 (T3), and 16 kg/m^3 (T4). Data on the seedlings' height, collar diameter, number of leaves, and leaf area were taken every 7 days. The shoot:root ratio was measured at the end of week 10. The seedlings were watered manually twice a day in the morning and afternoon, and no fertilizer was applied, while pests and weeds were monitored every day. The data collected were analyzed using one-way ANOVA and correlation analysis.

3.3 SURVIVAL RATE GROWTH

3.3.1 DIFFERENT GROWING MEDIA

Table 3.1 shows the survival percentage of *N. cadamba* seedlings for a period of 10 weeks. The average of survival percentage was 94.5%. Some seedlings grown in media T2, T5, and T6 died. The deaths of seedlings in media T2 were due to the high absorption water capacity, and in media T5, this might have been due to the ability of Jiffy Pellet (peat pellet) to store more capacity of water that could cause the roots of seedlings to rot, especially in areas that receive high water content (Mohd Ashraf, 2008). While for media T6, the death of seedlings might be caused by the high temperature of the vermicompost (Pttnaik and Reddy, 2010).

TABLE 3.1

Survival Percentage of *N. cadamba* Seedlings in Different Growing Media

Treatment	Growing Media	Survival (%)
T1	Topsoil (100%)	100
T2	Coco peat (70% fine : 30% coarse)	89
T3	Coco peat (30% fine : 70% coarse)	100
T4	Coco peat (50% fine : 50% coarse)	100
T5	Jiffy Pellet	89
T6	Sawdust of *N. cadamba* mix with vermicompost	89
	Average	94.5

3.3.2 Different Intensities of Shading

After 10 weeks, the survival rate of *N. cadamba* seedlings under all treatments was 100%, except for T1 (0% light intensity), with an 89% survival rate due to defoliator insects. The average survival rate was 97.25%. Based on observation, grasshoppers were found to be among the insects that contribute to the damage to leaves. Grasshoppers are a general feeder and tend to feed on young leaves (Flint, 2013). Defoliation of leaves can cause plant mortality due to the interference that occurs in the food-making process that mainly occurs in leaves (Wargo, 1979).

3.3.3 Different Watering Treatments

Table 3.2 shows the survival percentage of *N. cadamba* seedlings for a period of 62 days. The survival percentage for all treatments was 100%, except for control treatment T1, where the survival percentage was 0% because no watering was applied to the seedlings. If plant cells do not contain enough water, the plant will die (Ismail and Badri, 1995). Water is among the most important aspects in maintaining the growth of seedings so that they grow well and healthy (Kramer, 1980). However, in this study, it was found that *N. cadamba* seedlings can survive for 35 days without watering, although the growth rate was low. Plants that lacked water closed their stomata, and photosynthetic activity slowed (Rick, 2000). The ability of *N. cadamba* seedlings to survive for 35 days was possible because the growing media used was a mixture of 70% coco peat + 30% topsoil. According to Yahya et al. (2009), coco peat has water-holding capacity and contains the elements calcium (Ca), magnesium (Mg), potassium (K), sodium (Na), and phosphorus (P) (Charles, 2008). The selection of growing media for *N. cadamba* seedlings is very important because the growing media allows the seedlings to get enough nutrients in addition to promoting the growth of seedlings (Mhango et al., 2008). The type of soil or growing media can affect the ability of plants to absorb water from the soil (Rick, 2003). If water is not absorbed, the activity of cells and tissues that perform the functions of a plant's normal physiology is reduced or stopped altogether, and as a result, the plants die or their growth is impaired (Yunasfi, 2002).

TABLE 3.2
Survival Percentage of *N. cadamba* Seedlings Treated with Different Watering Frequencies

Treatment	Watering Frequency	Survival (%)
T1	No watering applied as a control	0
T2	Watering every day	100
T3	Watering every 2 days	100
T4	Watering every 4 days	100
T5	Watering every 6 days	100
T6	Watering using a water reservoir	100

3.3.4 DIFFERENT FERTILIZING TREATMENTS

Table 3.3 shows the survival percentage of *N. cadamba* seedlings for a period of 10 weeks. Treatment T1 (0 kg/m^3 Agroblen), treatment T3 (12 kg/m^3 Agroblen), and treatment T4 (16 kg/m^3 Agroblen) recorded a 100% survival rate, but treatment T2 (8 kg/m^3 Agroblen) resulted in a 92% survival rate. The mortality of the seedlings was caused by disease that attacks the seedlings, in which types of white fungus with a texture like fine threads (mycelium) grow on the surface of the media in a polybag and spread up to the trunk of *N. cadamba* seedlings. A study conducted by Ai Rosah (2014) explains that the morphology of the white fungus that has a texture like fine threads (mycelium) that attack *N. cadamba* seedlings in nurseries is from *Botryodiplodia* spp., *Fusarium* spp., and *Colletotrichum* spp.

3.4 HEIGHT GROWTH

3.4.1 GROWING MEDIA TREATMENT

The average height of *N. cadamba* seedlings was significantly different (F = 7.637; p ≤ 0.05) depending on the different media used in this study. Figure 3.1 shows that the highest average height of *N. cadamba* seedlings was achieved using media T1 and T6, at 6.3 cm; this was followed by media using T2, T3, and T4, with an average height of 5.2 cm; and the lowest average height was achieved using media T5, at 3.8 cm. Figure 3.2 shows the growth performance of *N. cadamba* seedlings treated with different growing media. The topsoil used as a growing media in T1 featured an upper layer of soil profile known as Horizon O, which is darker in color (Normah and Mojiol, 2001), rich in nutrients, an a more productive soil than the other layer (Normah and Mojiol, 2001). The sawdust has a high wet weight that can help in absorbing water; in addition it contains a high level of nutrients, and the addition of vermicompost helps to increase the soil's ability to absorb and store water to enhance absorption of nutrients (Sallaku et al., 2009), highly nutritive 'organic fertilizer', and more powerful 'growth promoter'.

TABLE 3.3
Survival Percentage of *N. cadamba* Seedlings Treated with Different Quantities of Agroblen Fertilizer

Treatment	Quantity of Agroblen Fertilizer	Survival (%)
T1	0 kg/m^3	100
T2	8 kg/m^3	92
T3	12 kg/m^3	100
T4	16 kg/m^3	100

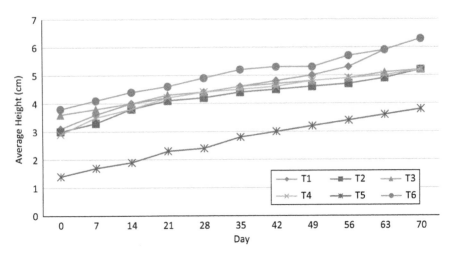

FIGURE 3.1 Average height of *N. cadamba* seedlings. [Note: (T1) – topsoil (100%), (T2) – coco peat (70% fine : 30% coarse), (T3) – coco peat (30%fine : 70% coarse), (T4) – coco peat (50% fine : 50% coarse), (T5) – Jiffy Pellet, (T6) – sawdust of *N. cadamba* mix with vermicompost.]

FIGURE 3.2 Growth performance of *N. cadamba* seedlings treated with different growing media. [Note: (T1) – topsoil (100%), (T2) – coco peat (70% fine : 30% coarse), (T3) – coco peat (30% fine : 70% coarse), (T4) – coco peat (50% fine : 50% coarse), (T5) – Jiffy Pellet, (T6) – sawdust of *N. cadamba* mixed with vermicompost.]

3.4.2 INTENSITY OF SHADING TREATMENT

Figure 3.3 shows the increment average height of *N. cadamba* seedlings in each treatment for 10 weeks (63 days). T3, with 90% intensity of shade, showed the highest average of height at 13.1 cm, followed by T2 (50% intensity of shading) and T4 (90% of intensity of shading) with average heights of 12.9 cm and 12.5 cm, respectively. The lowest average height was 1.0 cm on T1 with 0% of intensity of shading. There was a significant difference ($F = 3.94$; $p \leq 0.05$) in the average height of *N. cadamba* seedlings treated with different intensities of shading. In the nursery stage, seedlings require shading (Prastowo et al., 2006); seedlings should be raised under the shade to provide a suitable environment during the day to protect them from high temperatures and conserve soil moisture. According to Guedes and Pivetta (2014), the suitable intensity of shading for fast-growing species to grow better was under 70% of intensity of shading followed by 50% of intensity of shading.

3.4.3 WATERING TREATMENT

Figure 3.4 showed the average height growth of *N. cadamba* seedlings within 63 days treated with different watering frequencies. The seedlings that were treated using a water reservoir (T6) had the highest average height growth at 9.31 cm, followed by T5 (watered every 6 days), T3 (watered every 2 days), and T2 (watered every day) at 7.01 cm, 6.83 cm, and 6.78 cm, respectively. The lowest was T4 (watered every 4 days) with 6.41 cm. Meanwhile, at day 36, all seedlings treated with no watering died. There was a significant difference ($F = 2.969$; $p \leq 0.05$) in the average height of *N. cadamba* seedlings treated with different watering frequencies. In this study, the reservoir provided a constant source of water to the seedlings. Water was provided in the container and the seedlings were placed in the reservoir. According to Santoso (2010), when a plant suffers from a lack of water resources, its growth will be stunted so it becomes a dwarf. But if the plant has adequate water resources, its growth will increase.

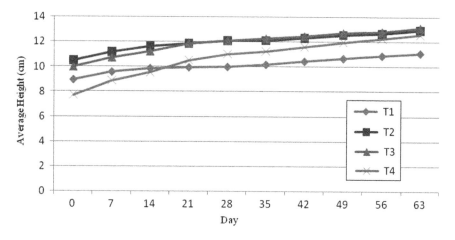

FIGURE 3.3 Average height of *N. cadamba* seedlings. [Note: T1 (0% intensity of shading as a control), T2 (50% intensity of shading), T3 (70% intensity of shading) and T4 (90% intensity of shading).]

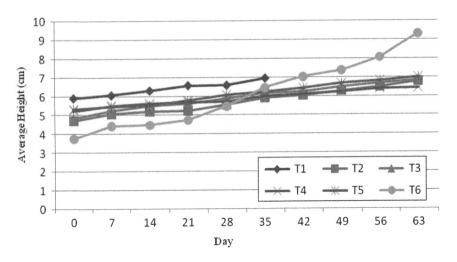

FIGURE 3.4   Average height of *N. cadamba* seedlings. [Note: T1 (no watering applied as a control (T1), watering every day (T2), watering every 2 days (T3), watering every 4 days (T4), watering every 6 days (T5), and using a water reservoir (T6).]

3.4.4   FERTILIZING TREATMENT

Based on Figure 3.5, T4 (16 kg/m^3 Agroblen) has the highest average height of *N. cadamba* seedlings, which is 48.45 cm, followed by T3 (12 kg/m^3 Agroblen) andT2 (8 kg/m^3 Agroblen), with an average height of 47.63 cm and 44.26 cm, respectively. Treatment T1 (0 kg/m^3 Agroblen) showed a very slow growth rate and a height of 7.64 cm. Figure 3.6 shows the differences in growth height of *N. cadamba* seedlings after being treated with different quantities of Agroblen fertilizer for 10

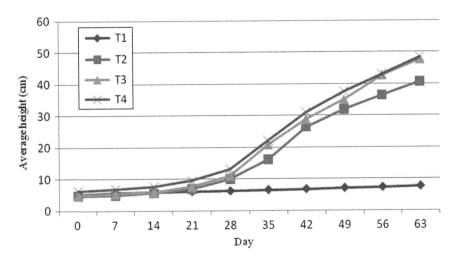

FIGURE 3.5   Average height of *N. cadamba* seedlings. [Note: T1 (0 kg/m^3), T2 (8 kg/m^3), T3 (12 kg/m^3) and T4 (16 kg/m^3).]

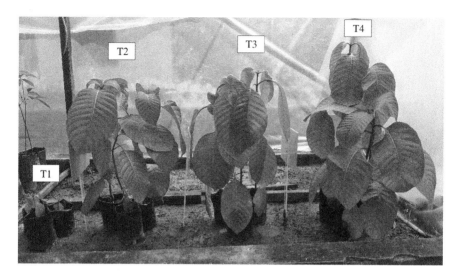

FIGURE 3.6 Height growth of *N. cadamba* seedlings after treatment with different quantities of Agroblen fertilizer for 10 weeks. [Note: T1 (0 kg/m³), T2 (kg/m³), T3 (12 kg/m³) and T4 (16 kg/m³)]

weeks. There was a significant difference (F = 48.141; p ≤ 0.05) in the average height of *N. cadamba* seedlings treated with different quantities of Agroblen fertilizer. According to Lobell (2007), the growth performance of seedlings will improve with each addition of NPK value, but at a certain stage the seedlings will reach a constant rate of growth at which the addition of NPK will not affect the growth process because it has reached its optimum.

3.5 COLLAR DIAMETER GROWTH

3.5.1 GROWING MEDIA APPLICATION

There was a significant difference (F = 3.33; p ≤ 0.05) in the average collar diameter of *N. cadamba* seedlings treated with different growing media. The highest average collar diameter of *N. cadamba* seedlings was using growing media T1, with a collar diameter of 2.4 mm, followed by growing media T6, T5, T4, and T3, with 2.15 mm, 2.12 mm, 2.07 mm, and 1.86 mm, respectively. The lowest average collar diameter was achieved using growing media T2 with 1.82 mm (Figure 3.7). As a growing media, topsoil can provide organic matter (Darmody et al., 2009) and nutrients in order to support the growth of seedling (Koenig and Isman, 2002; Darmody et al., 2009).

3.5.2 INTENSITY OF SHADING APPLICATION

Based on Figure 3.8, the highest average collar diameter of *N. cadamba* seedlings was achieved under 70% intensity of shading (T3), with an average collar diameter of 3.05 mm, followed by 50% intensity of shading (T2) and 0% intensity of shading

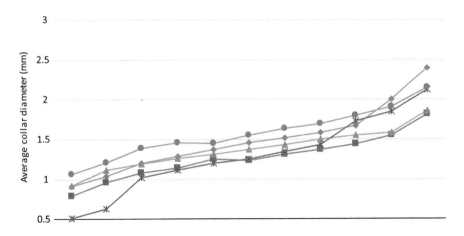

FIGURE 3.7 Average collar diameter of *N. cadamba* seedlings. [Note: (T1) – topsoil (100%), (T2) – coco peat (70% fine : 30% coarse), (T3) – coco peat (30% fine : 70% coarse), (T4) – coco peat (50% fine : 50% coarse), (T5) – Jiffy Pellet, (T6) – sawdust of *N. cadamba* mixed with vermicompost.]

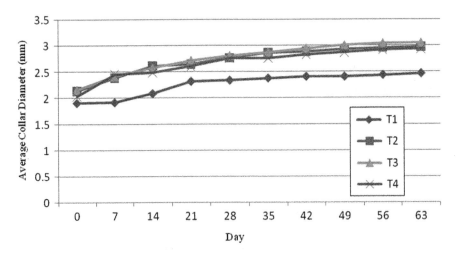

FIGURE 3.8 Average collar diameter of *N. cadamba* seedlings. [Note: T1 (0% intensity of shading as a control), T2 (50% intensity of shading), T3 (70% intensity of shading), and T4 (90% intensity of shading).]

(T1) with 2.97 mm and 2.46 mm, respectively. The lowest average collar diameter was achieved under 90% intensity of shading (T4), with 1.9 mm. There was a significant difference ($F = 5.01$; $p \leq 0.05$) in the average collar diameter of *N. cadamba* seedlings under different intensities of shading. The collar diameter growth for *N. cadamba* seedlings increased especially under 70% intensity of shading. This could be related to the increasing rate of cambium activity in the stem under suitable shade (Guedes and Pivetta, 2014), which is higher than in plants grown without shading (Lopes et al., 2015).

3.5.3 WATERING APPLICATION

Figure 3.9 shows the average collar diameter of *N. cadamba* seedlings treated with different watering frequencies for 63 days. The seedlings treated with a water reservoir (T6) had the highest average collar diameter growth with 3.48 mm, followed by T4 (watered every 4 days), T3 (watered every 2 days), and T5 (watered every 6 days) with 2.88 mm, 2.50 mm, and 2.46 mm, respectively. The lowest average collar diameter growth was T2 (watered every day) with 2.33 mm. There was a significant difference (F=4.846; $p \leq 0.05$) in the average height of *N. cadamba* seedlings treated with different watering frequencies.

3.5.4 FERTILIZING APPLICATION

Based on Figure 3.10, T4 (16 kg/m^3 Agroblen) had the highest average collar diameter of *N. cadamba* seedlings, which was 8.00 mm, followed by T3 (12 kg/m^3 Agroblen) and T2 (8 kg/m^3 Agroblen) with an average collar diameter of 7.84 mm and 7.30 mm, respectively. The lowest average collar diameter was 3.07 mm for treatment T1 (0 kg/m^3 Agroblen). There was a significant difference (F=33.736; $p \leq 0.05$) in average of collar diameter growth of *N. cadamba* seedlings treated with different quantities of Agroblen fertilizer. Hannah (1999) stated that low availability of NPK can cause seedling growth to become slow. Nitrogen deficiency will cause the growth of seedlings to be stunted and seeds to become yellowish.

3.6 NUMBER OF LEAVES

3.6.1 GROWING MEDIA

Based on Figure 3.11, the highest average number of leaves was achieved using growing media T1 with 12 leaves, followed by T5 and T2 with 9 leaves and T4 and T6 with 8 leaves. The lowest average of number of leaves was T3 with 7 leaves. There

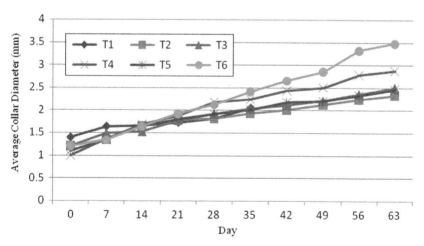

FIGURE 3.9 Average collar diameter growth of *N. cadamba* seedlings. [Note: T1 (no watering applied as a control (T1), watering every day (T2), watering every 2 days (T3), watering every 4 days (T4), watering every 6 days (T5), and using a water reservoir (T6).]

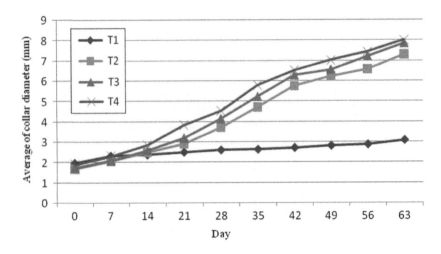

FIGURE 3.10 Average collar diameter growth of *N. cadamba* seedlings. [Note: T1 (0 kg/m³), T2 (8 kg/m³), T3 (12 kg/m³), and T4 (16 kg/m³).]

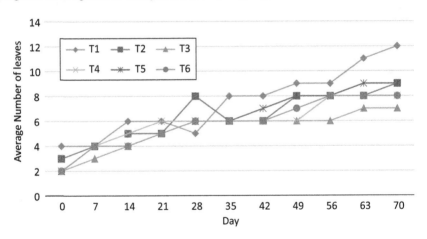

FIGURE 3.11 Average number of leaves of *N. cadamba* seedlings. [Note: (T1) – topsoil (100%), (T2) – coco peat (70% fine : 30% coarse), (T3) – coco peat (30% fine : 70% coarse), (T4) – coco peat (50% fine : 50% coarse), (T5) – Jiffy Pellet, (T6) – sawdust of *N. cadamba* mixed with vermicompost.]

was a significant difference ($F = 3.52$; $p \leq 0.05$) in the average number of leaves on *N. cadamba* seedlings treated with different growing media. One of the factors that affect the growth of seedlings is the number of leaves. The more leaves, the faster the plant will grow, because the leaves are the source of the nutrients the plant needs to grow well (Goldworthy and Fisher, 1996).

3.6.2 INTENSITY OF SHADING

Based on Figure 3.12, the average number of leaves on each *N. cadamba* seedling was 5. The highest average number of leaves was achieved under 70% intensity of

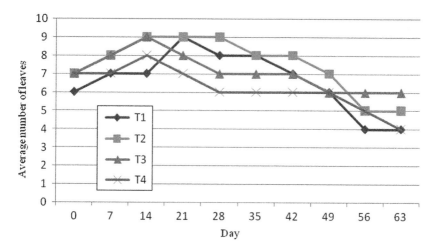

FIGURE 3.12 Average number of leaves of *N. cadamba* seedlings. [Note: T1 (0% intensity of shading as a control), T2 (50% intensity of shading), T3 (70% intensity of shading), and T4 (90% intensity of shading).]

shading (T3) with 6 leaves, followed by T2 (50% intensity of shading) with 5 leaves, and T1 (0% of intensity of shading) and T4 (90% intensity of shading). However, there was no significant difference ($F = 2.65$; $p \geq 0.05$) in the average number of leaves on *N. cadamba* seedlings under different of intensities of shading. This showed that the intensity of shading did not affect leaf growth.

3.6.3 WATERING

Figure 3.13 shows the average number of leaves of *N. cadamba* seedlings treated with different watering frequencies for 62 days. The highest average number of

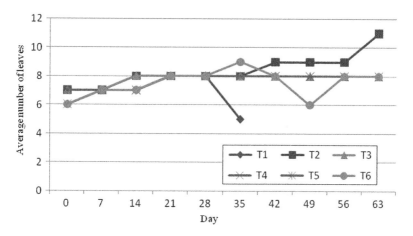

FIGURE 3.13 Average number of leaves of *N. cadamba* seedlings. [Note: T1 (no watering applied as a control (T1), watering every day (T2), watering every 2 days (T3), watering every 4 days (T4), watering every 6 days (T5), and using a water reservoir (T6).]

leaves was achieved with watering every day, with 11 leaves, while other treatments achieved an average number of 8 leaves. There was a significant difference ($F = 12.699$; $p \leq 0.05$) in the average height of *N. cadamba* seedlings treated with different watering frequencies.

3.6.4 FERTILIZING

Based on Figure 3.14, T4 (16 kg/m^3 Agroblen) achieved the highest average number of leaves on *N. cadamba* seedlings, which was 12 leaves, followed by T3 (12 kg/m^3 Agroblen) and T2 (8 kg/m^3 Agroblen) with an average of 11 and 10 leaves, respectively. Treatment T1 (0 kg/m^3 Agroblen) showed the lowest average number of leaves, which was 6. There was a significant difference ($F = 23.160$; $p \leq 0.05$) in the average height of *N. cadamba* seedlings treated with different quantities of Agroblen fertilizer.

3.7 LEAF AREA

3.7.1 GROWING MEDIA

Figure 3.15 showed the different average leaf areas of *N. cadamba* seedlings between the different treatments of growing media. The highest average leaf area was achieved by T1 with 2270.3 mm^2, followed by T5, T4, T6, and T2 with 929.2 mm^2, 806 mm^2, 717.1 mm^2, and 438.1 mm^2, respectively. The lowest average of leaf area was achieved by T3, with 420.3 mm^2. There was a significant difference ($F = 7.66$; $p \leq 0.05$) in the average leaf area of *N. cadamba* seedlings treated with different growing media. According to Bischoff et al. (2013), the larger the leaf area, the more surface for photosynthesis. Thus, growing media T1 (100% topsoil) showed the best growth performance compared to other growing media.

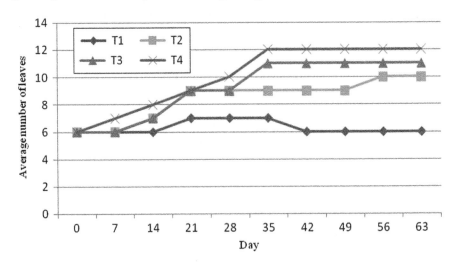

FIGURE 3.14 Average number of leaves of *N. cadamba* seedlings. [Note: T1 (0 kg/m^3), T2 (8 kg/m^3), T3 (12 kg/m^3) and T4 (16 kg/m^3).]

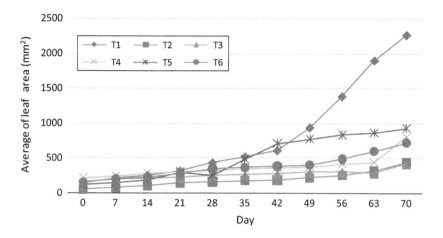

FIGURE 3.15 Average of leaf area of *N. cadamba* seedlings. [Note: (T1) – topsoil (100%), (T2) – coco peat (70% fine : 30% coarse), (T3) – coco peat (30% fine : 70% coarse), (T4) – coco peat (50% fine : 50% coarse), (T5) – Jiffy Pellet, (T6) – sawdust of *N. cadamba* mixed with vermicompost.]

3.7.2 INTENSITY OF SHADING

Figure 3.16 shows a significant difference (F = 5.04; p ≤ 0.05) in the average leaf area of *N. cadamba* seedlings under different intensities of shading. The highest average leaf area was achieved under 70% of intensity of shading (T3) with 2439.89 mm² followed by 50% of intensity of shading (T2) and 90% of intensity of shading (T3) with an average leaf area of 1583.5 mm² and 1565.3 mm², respectively (see Figure 3.17). The lowest average leaf area was achieved without shading (T1), with 1098.89 mm². Leaves that are tolerant to light usually tend to have a bigger area than leaves tolerant to shading (Groninger et al., 1996). Light intensity does influence the plant form, flower, leaf size, and color on both herbaceous plants (Jeong et al., 2009; Vendrame et al., 2004) and woody species (Hampson et al., 1996). Plants may be of the same age and species, but each plant has a different rate of growth (Evans, 1972; Hunt, 1982; Metcalf et al., 2003).

3.7.3 WATERING

Based on Figure 3.18, the highest average leaf area of *N. cadamba* seedlings over 62 days of the experiment was achieved by treatment using a water reservoir (T6) with 4871.00 mm², followed by T2 (watered every day), T3 (watered every 2 days), and T4 (watered every 4 days), with 2877.11 mm², 2360.75 mm², and 2348.75 mm², respectively. The lowest average leaf area was achieved under T5 (watered every 6 days) with 2139.10 mm². There was a significant difference (F = 19.145; p ≤ 0.05) in the average height of *N. cadamba* seedlings treated with different intensities of shading.

3.7.4 FERTILIZING

Based on Figure 3.19, T4 (16 kg/m³ Agroblen) achieved the highest average leaf area in *N. cadamba* seedlings, which was 43095.00 mm², followed by T3 (12 kg/m³

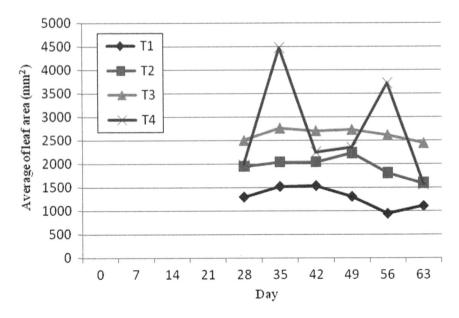

FIGURE 3.16 The average of leaf area of *N. cadamba* seedlings. [Note: T1 (0% intensity of shading as a control), T2 (50% intensity of shading), T3 (70% intensity of shading) and T4 (90% intensity of shading).]

FIGURE 3.17 Growth performance of *N. cadamba* seedlings treated with different intensities of shading for 10 weeks. [Note: T1 (0% intensity of shading as a control), T2 (50% intensity of shading), T3 (70% intensity of shading), and T4 (90% intensity of shading).]

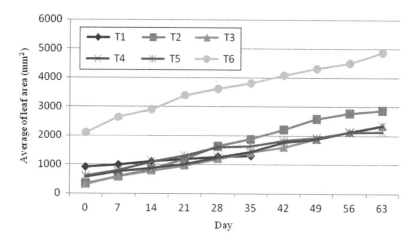

FIGURE 3.18 Average leaf area of *N. cadamba* seedlings. [Note: T1 (no watering applied as a control (T1), watering every day (T2), watering every 2 days (T3), watering every 4 days (T4), watering every 6 days (T5), and using a water reservoir (T6).]

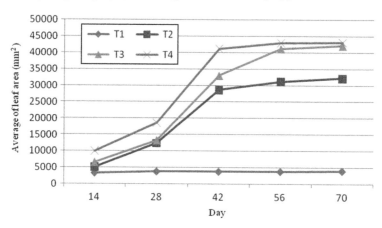

FIGURE 3.19 Average leaf area of *N. cadamba* seedlings. [Note: T1 (0 kg/m³), T2 (8 kg/m³), T3 (12 kg/m³), and T4 (16 kg/m³).]

Agroblen) and T2 (8 kg/m³ Agroblen), with an average leaf area of 42022.00 mm² and 32150.00 mm², respectively. Treatment T1 (0 kg/m³ Agroblen) showed the lowest average leaf area, which was 3880 mm². There was a significant difference ($F=41.768$; $p \leq 0.05$) in the average height of *N. cadamba* seedlings treated with different quantities of Agroblen fertilizer.

3.8 SHOOT:ROOT RATIO

3.8.1 GROWING MEDIA

Based on Table 3.4, the highest shoot:root ratio was achieved using growing media T1 with a ratio of 1:3, followed by T6 with a ratio of 1:2. The lowest shoot:root

TABLE 3.4
Growing Media and Shoot:Root Ratio for _N. cadamba_ Seedlings

Treatment	Growing Media	Shoot:Root Ratio
T1	Topsoil (100%),	1 : 3
T2	Coco peat (70% fine : 30% coarse)	1 : 1
T3	Coco peat (30% fine : 70% coarse)	1 : 1
T4	Coco peat (50% fine : 50% coarse)	1 : 1
T5	Jiffy Pellet	1 : 1
T6	Sawdust of _N. cadamba_ mix with vermicompost	1 : 2

ratio was achieved under T2, T3, T4, and T5 with a ratio of 1:1. Figure 3.20 showed the shoot:root ratio of _N. cadamba_ seedlings after being treated with different growing media for 10 weeks. Based on correlation analysis, there was a significant positive correlation between the dry weight of shoot with dry weight of shoot of _N. cadamba_ seedlings ($r = 0.65$; $p \leq 0.01$). According to Putri and Nurhasbi (2010), the highest dry weight of shoot and root, the better growth rate of the plant to grow well.

3.8.2   Intensity of Shading

Based on Table 3.5, the highest ratio of shoot to root was achieved under T4 (90% of intensity of shading) with a ratio of 1:2.5, followed by T3 (70% of intensity of shading) with ratio 1:1.9. The lowest shoot:root ratios were achieved under T1 and

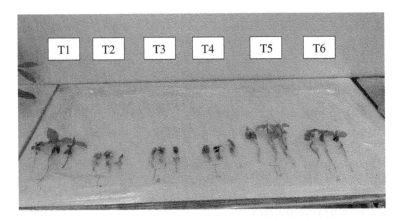

FIGURE 3.20 Shoot:root ratio of _N. cadamba_ seedlings after treatment with different growing media for 10 weeks. [Note: (T1) – topsoil (100%), (T2) – coco peat (70% fine : 30% coarse), (T3) – coco peat (30% fine : 70% coarse), (T4) – coco peat (50% fine : 50% coarse), (T5) – Jiffy Pellet, (T6) – sawdust of _N. cadamba_ mix with vermicompost.]

TABLE 3.5
Shading Intensity and Shoot:Root Ratio for *N. Cadamba* Seedlings

Treatment	Shading Intensity	Shoot:Root Ratio
T1	0% intensity of shading as a control	1 : 1.4
T2	50% intensity of shading	1 : 1.4
T3	70% intensity of shading)	1 : 1.9
T4	90% intensity of shading	1 : 2.5

T2 with a ratio of 1:1.4. The shoot:root ratio of this study was found to be similar to those achieved in previous studies by Kennedy et al. (2007), where the biomass of the plant reduced under high shade compared to plants that were not shaded. The increase in the ratio of root mass to shoot can be seen in plants that are receiving more light due to plant modification to increase the transpiration rate (Sultan, 2003). Based on correlation analysis, there was a significant positive correlation between dry weight of the shoot and dry weight of shoot of *N. cadamba* seedlings ($r = 0.747$; $p \leq 0.01$). The increments in shoot weight were parallel with the root weight increments. The higher the dry weight of shoot and root, the better the growth rate of the plant (Putri and Nurhasbi, 2010) (Figure 3.21).

FIGURE 3.21 Shoot:root ratio of *N. cadamba* seedlings after treatment with different watering frequencies for 62 days. [Note: watering every day (T2), watering every 2 days (T3), watering every 4 days (T4), watering every 6 days (T5), and using a water reservoir (T6).]

3.8.3 WATERING

Based on Table 3.6, the highest ratio of shoot to root was attained when using a water reservoir (T6) with a ratio of 1:0.69, followed by T3, T5, and T4 with ratios of 1:0.52, 1:0.49, and 1:0.40 respectively. The lowest ratio of shoot to root was T2 with a ratio of 1:0.37. Based on correlation analysis, there was a significant positive correlation of dry weight of shoot with dry weight of shoot of *N. cadamba* seedlings ($r = 0.979$; $p \leq 0.01$). According to Putri and Nurhasbi (2010), the higher the dry weight of shoot and root, the better the plant's growth rate.

3.8.4 FERTILIZING

Based on Table 3.7, the highest ratio of shoot to root was attained under T3 and T4 with a ratio of 1:3.1, followed by T2 and T1 with ratios of 1:2.9 and 1:2.3, respectively. Based on correlation analysis, there was significant ($p \leq 0.01$) and a very strong positive correlation, $r = 0.992$ (Fowler et al., 1998) of dry weight of shoot with dry weight of shoot of *N. cadamba* seedlings. According to Fita et al. (2011), phosphorus is an element that influences the development of a plant's root system. Based on the results obtained, the presence and different quantities of Agroblen fertilizer are a major factor contributing to the growth and formation of roots. It can be seen clearly that where Agroblen fertilizer is not used in the T1 treatment, there is a lack of formation and development in the root system of the seedlings. The growth of and ration between

TABLE 3.6
Watering and Shoot:Root Ratio for *N. cadamba* Seedlings

Treatment	Watering	Shoot:Root Ratio
T1	No watering applied as a control	–
T2	Watering every day	1 : 0.37
T3	Watering every 2 days	1 : 0.52
T4	Watering every 4 days	1 : 0.40
T5	Watering every 6 days	1 : 0.49
T6	Using a water reservoir	1 : 0.69

TABLE 3.7
Fertilizer and Shoot:Root Ratio
for *N. cadamba* Seedlings

Treatment	Fertilizer	Shoot:Root Ratio
T1	0 kg/m³	1 : 2.3
T2	8 kg/m³	1 : 2.9
T3	12 kg/m³	1 : 3.1
T4	16 kg/m³	1 : 3.1

shoots and roots is influenced by nitrogen and phosphorus elements in which the elements of macronutrients such as nitrogen, phosphorus, and potassium and micronutrient elements such as calcium and manganese are important for the growth of roots and shoots of seedlings in the early stages (Wang et al., 2016; Harris, 1992).

3.8.5 COST OF TREATMENT PREPARATION

The cost of treatment preparation was associated with the effect on *N. cadamba* seedlings' growth rate for each treatment. The objective of this cost preparation calculation is to determine the economical cost accordance with the good growth rates *N. cadamba* seedling in the nursery. The costs of treatment preparation were different for each treatment where it was based on the different quantities of Agroblen fertilizer used. The most expensive treatment preparation in this study was T4 (16 kg/m^3 Agroblen), which was RM 30.02/treatment, followed by T3 (12 kg/m^3 Agroblen) and T2 (8 kg/m^3 Agroblen) with RM 22.81/treatment and RM 15.61/treatment, respectively. T1 (0 kg m^3 Agroblen) incurred no cost because it did not involve the use of any amount of fertilizer. According to Mattson et al. (2009), it is a cost waste to use large quantities of fertilizer to try to make a significant difference in the growth performance of seedlings at the nursery level, since the plants reach the optimum requirement of nutrient absorption just by using a small quantity.

3.9 CONCLUSIONS

The usage of different growing media, intensities of shading, watering frequencies, and quantities of Agroblen fertilizer affected the growth performance of *N. cadamba* seedlings. This can be seen in improved growth increments in height, collar diameter, leaf area, number of leaves, and the better ratio of shoot to root. The most suitable growing media for raising *N. cadamba* seedlings in the nursery was topsoil; the most suitable intensity of shading for *N. cadamba* seedlings' growth in the nursery was 70%; using a reservoir was the most suitable watering method for the *N. cadamba* seedlings growth in the nursery; and the most cost-effective growth performance of *N. cadamba* seedlings in the nursery was achieved with application of 12 kg/m^3 of Agroblen fertilizer.

ACKNOWLEDGMENTS

The authors would like to express their sincere gratitude and thanks to Mr. Alexsius Carlos Loibin, Miss Evianna Alai, Miss Jackquelyn Anak Jalai, and Miss Sharifah Nur Ezzati Sh Bokrata for their invaluable assistance.

REFERENCES

Abdul, K.C. 2001. Effect of shade on growth performance of four tree species: Nursery stage. *Pakistan Journal of Agricultural Science f1j1*, 38(1–2):69–72.

Agus, A. and Imam, I. 2006. *Fisiologi Lingkungan Tanaman*. Gadjah Mada University Press, Indonesia.

Ai Rosah, A. 2014. *Identifikasi dan Patogenisitas Cendawan Penyebab Primer Penyakit Mati Pucuk pada Bibit Jabon (Anthocephalus cadamba (Roxb.).* Institut Pertanian Bogor, p. 71.

Australian Tropical Rainforest Plants. 2010. *Neolamarckia cadamba.* Centre for Australian National Biodiversity Research, Canberra. http://keys.trin.org.au/key-server/data/0e0f05040-0103-430d-800460d07080d04/media/Html/taxon/Neolamarckia_cadamba.htm

Bischoff, A., Streinger, T., and Bramasto, Y. 2013. Manual Budidaya Jabon Putih. Badan Penelitian dan Pengembangan Kehutanan, Pusat Penelitian dan Pengembangan Kehutanan dan Direktorat Jenderal Bina Pengelolaan Daerah Aliran Sungai dan Perhutanan Sosial, Direktorat Bina Perbenihan Tanaman Hutan.

Charles, L.B. 2008. *Nutritional Properties of Agrocoir Horticulture Soils and Nutrition.* Consulting, Michigan.

Cole, D.W., and Gessel, S.P. 1992. Fundamentals of tree nutrition. *College of Forest Resources,* 2:7–16.

Darmody, R.G., Daniels, W.L., Marlin, J.C., and Cremeens, D.L. 2009. Topsoil: What is it and who cares. In *Revitalizing the Environment: Proven Solutions and Innovative Approaches.* National Meeting of The American Society of Mining and Reclamation, Billings, MT.

Elhadi, A.M., Ibrahim, A.K., and Magid, A.T.D. 2013. Effect of different watering regimes on growth performance of five tropical trees in the nursery. *Jonares,* 1:14–18.

Evans, G.C. 1972. *The Quantitative Analysis of Plant Growth.* Blackwell Scientific, Oxford.

Flint, M.L. 2013. *Pest Notes: Grasshoppers.* 3332:276.

Fowler, J., Cohen, L., and Jarvis, P. 1998. *Practical Statistic for Field Biology* (2nd Ed). John Wiley and Sons Ltd.

Goldsworthy, P.R., and Fisher, N.M. 1996. *Fisiologi Tanaman Budidaya Tropik.* Gajah Mada University Press, Yogyakarta, Indonesia.

Groninger, J.W., Seiler, J.R, Peterson, J.A., and Kreh, R.E. 1996. Growth and photosynthetic responses of four Virginia Piedmont tree species to shade. *Tree Physiology,* 16(9):773–778.

Guedes, R.B.M., and Pivetta, K.F.L. 2014. Initial growth of Bauhinia variegata trees under different colored shade nets and light conditions. *Revista Arvore Journal,* 38:1133–1145.

Hampson, C.R., Azarenko, A.N., and Potter, J.R. 1996. Photosynthetic rate, flowering and yield component alteration in hazelnut in response todifferent to different light environments. *Journal of the American Society for Horticultural Science,* 121(6):1103–1111.

Hannah, J. 1999. *Good Tree Nursery Practices. Practical Guidelines for Research Nurseries.* International Centre for Research in Agroforestry, Nairobi, Kenya, pp. 7–49.

Harris, R.W. 1992. Root-shoot ratios. *Journal of Arboriculture,* 18(1): 39–42.

Hunt, R. 1982. *Plant Growth Curves: A Functional Approach to Plant Growth Analysis.* Edward Arnold, London.

Jaenicke, H. 1999. Good Tree nursery practices: Practical guidelines for research nurseries. http://www.cgiar.org/icraf. Accessed on 29 May 2018.

Jeong, K.Y., Pasian, C.C., McMahon, M., and Tay, D. 2009. Growth of six Begonia species under shading. *Open Horticultural Journal,* 2(1):22–28.

Joker, D. 2000a. Seed leaflet: Neolamarckia Cadamba (ROXB.) BOSSER. https://curis.ku.dk/ws/files/20648324/neolamarckia_cadamba_int.pdf. Accessed on 10 October 2018.

Joker, D. 2000b. Neolamarckia cadamba (Roxb.) Bosser (Anthocephalus chinensis (Lam.) A.Rich. Ex Walp.). Seed Leaflet, No. 17, Danida Forest Seed Center, Denmark.

Kennedy, S., Kevin, B., and O'Reilley, C. 2007. The effect of shade on morphology, growth, and biomass allocation in Picea sitchensis, Larix X eurolepis and Thuja plicata. *New Forests,* 33:139–153.

Koenig, R., and Isaman, V. 2002. Topsoil quality guidelines for landscaping. All archived publication. Paper 42.

Kramer, P.J. 1980. *Plant and Soil Water Relationship: A Modern Synthesis.* McGraw-Hill Publication Co., New Delhi.

Krisnawati, H., Kallio, M., and Kanninen, M. 2011. *Anthocephalus Cadamba MIQ: Ecology, Silviculture and Productivity.* CIFOR, Bogor.

Lee, Y.F., Anuar, M., and Chung, A.Y.C. 2008. *A Guide to Plantation Forestry in Sabah.* Sabah Forest Record. 16. Sabah Forestry Department, Sandakan.

Lee, Y.F., Chia, F.R., Anuar, M., Ong, R.C., and Ajik, M. 2005. The use of Laran and Binuang for forest plantations and intercropping with oil palm in Sabah. *Sepilok Bulletin,* 3:1–13.

Lobell, D.B. 2007. The cost of uncertainty for nitrogen fertilizer management: A sensitivity analysis. *Field Crops Research,* 100(2–3):210–217.

Lopes, M.J.D.S, Dias-Filho, M.B., Neto, M.A.M., and Cruz, E. 2015. Morphological behaviour, and cambial activity in seedlings of two Amazonian tree species under shade. *Journal of Botany,* 2015:1–10.

Matra, N., Bulan, P., and Seng, H.W. 2011. Seed germination and DNA Genotyping of Neolamarckia Cadamba (Roxb). Progenies (Half-Sib Family). 9th Malaysia Genetics Congress Proceeding, 28–30 September 2011, Kuching, Sarawak.

Mattson, N.S., Leatherwood, W.R., and Peters, C. 2009. *Save on Fertilizer Costs.* Department of Horticulture, GrowerTalks, Pennsylvania, pp. 52–56.

McCauley, A., Jones, C., and Jacobsen, J. 2011. Plant nutrition functions and deficiency and toxicity symptons. *Nutrient Management Module.*

Metcalf, J.C., Rose, K.E., and Rees, M. 2003. Evolutionary demography of monocarpic perennials. *Trends in Ecology and Evolution,* 18(9):471–480.

Mhango, J., Akinnifesi, F.K., Mng'omba, S.A., and Sileshi, G. 2008. Effect of growing medium on early growth and survival of Uapaca kirkiana Muell Arg. Seedlings in Malawi. *African Journal of Biotechnology,* 7(13): 2197–2202.

Mohd Ashraf, Hj.S. 2008. Panduan asas tanaman cili secara fertigasi. http://www.mohdashraf.com.

Mojiol, A.R., Lintangah, W., Maid, M., and Kodoh, J. 2018. *Neolamarckia Cadamba III-4.* Enzyklopadie der Holzgewachse-70.Erg.Lfg.01/18. John Wiley and Sons Ltd.

Normah, A.B., and Mojiol, A.R. 2001. *Hubungan Tanah dan Pokok-Pokok Bandar.* Universiti Malaysia Sabah, Kota Kinabalu.

Orwa, C., Mutua, A., Kindt, R., Jamnadass, R., and Anthony, S. 2009. Agroforestry tree database: A tree reference and selection guide version 4.0. http://www.worldagroforestry.org/treedb2/AFTPDFS/ Anthocephalus cadamba.

Pattnaik, S., and Reddy, M.V. 2010. Nutrient status of vermicompost of urban green waste processed by three earthworm species *Eisenia foetida, Eudrilus eugeniae,* and *Perionyx excavates. Applied and Environmental Soil Science,* 2010.

Prastowo, N.H., Roshetko, J.M., Maurung, G.E.S., Nugraha, E., Tukan, J.M., and Harum, F. 2006. *Teknik pembibitan dan Perbanyakan Vegetatif Tanaman Buah.* World Agroforestry Centre (IGRAF) and Winrock International, Bogor, Indonesia, p. 100.

Putri, K.P., and Nurhasybi, N. 2010. Pengaruh Jenis Media Organik Terhadap Kualitas Bibit Takir (Duabanga moluccana). *Jurnal Penelitian Hutan Tanaman,* 7(3):141–146.

Rick, P. 2003. *Introduction to Plant Science.* Delmar Publishers, Albany, NY.

Sabah Forestry Department. 2008. *Annual Report.* Sabah Forestry Department, Sandakan, pp. 277–298.

Sallaku, G., Babaj, I., Kaciu, S., and Balliu, A. 2009. The influence of vermicompost on plant growth characteristics of cucumber (*Cucumis sativus* L) seedlings under saline conditions. *Journal of Food Agriculture and Environment,* 7(3 and 4):869–872.

Santoso, B. 2010. *Faktor-faktor pertumbuhan dan penggolongan tanaman hias.* Fakultas Pertanian, Universitas Gajah Mada, Yogyakarta.

Shaviv, A. 2005. Controlled release fertilizer. IFA International Workshop on Enhanced-Efficiency Fertilizer, Frankfurt. International Fertilizer Industry Association, Paris, France.

Soerianegara, I., and Lemmens, R.H.M.J. 1993. *Plant Resources of South-East Asia 5 (1): Timber Trees: Major Commercial Timbers.* Pudoc Science Publication, Wageningen.

Sultan, S.E. 2003. Phenotypic plasticity in plants: A case study in ecological development. *Evolution and Development,* 3(2):138–144.

Tarmizi, S. 2015. Lawatan ke tapak semaian getah dan kelapa sawit. http://tarmizisayuti.com/blog/lawatan-ke-tapak-semaian-getah-dan-kelapasawit/. Accessed on 13 April 2018.

Tuan, H.S., Chen, T., and Taylor, C.M. 2011. Neolamarckia Bosser, Bull. Mus. Natl., B., Adansonia 6:247, 1985. In: Wu, Z.Y., Raven, P.H. and Hong, D.Y. (eds.) *Flora of China. Vol. 19, Cucurbitaceae through Valerianaceae, with Annonaceae and Berberidaceae.* Science Press, Beijing and Missouri Botanical Garden Press, St. Louis.

Vendrame, W., Moore, K.K., and Broschat, T.K. 2004. Interaction of light intensity and controlled-release fertilization rate on growth and flowering of two New Guinea impatiens cultivars. *Horticultural Technology,* 14:491–495.

Wang, Y., Kristensen, K.T., Jensen, L.S., and Magid, J. 2016. Vigorous root growth is a better indicator of early nutrient uptake than root hair traits in spring wheat grown under low fertility. *Frontiers in Plant Science,* 7:865. doi:10.3389/fpls.2016.00865

Wargo, P.M. 1979. Insects have defoliated my tree- now what's going to happen. *Ornamental Northwest Archives,* 3:13–18.

Yahya, A., Shaharom, A.S., Mohamad, R.B., and Selamat, A. 2009. Chemical and physical characteristic of cocopeat-based media mixtures and their effects on the growth and development of *Celosia cristata. America Journal of Agricultural and Biological Science,* 4(1):63–71.

Yunasfi. 2002. Faktor-faktor Mempengaruhi Perkembangan Penyakit dan Penyakit yang Disebabkan oleh Jamur. Fakultas Pertanian Jurusan Ilmu Kehutanan, Universitas Sumatera Utara. https://en.m.wikipedia.org/wiki/Neolamarckia_cadamba. Accessed on 18 May 2018.

4 Tree Planting Techniques of Batai (*Paraserianthes falcataria*) and Its Soil Nutrients

Affendy Hassan and Nurlaila Fauriza Abdul Rahman

CONTENTS

4.1 · INTRODUCTION

Paraserianthes falcataria (L.) Nielsen, also known as Batai, is one of the fastest growing and most important pioneer multipurpose tree species in some Southeast Asian countries. It is one of the tree species preferred for industrial forest plantations because of its ability to grow on a variety of soils, its favourable silvicultural characteristics and its acceptable quality of wood for the panel and plywood industries. *P. falcataria* also plays an important role in both commercial and traditional farming systems.

This species, like other fast-growing tree species (Walters, 1971), is expected to become increasingly important for wood industries as supplies for plywood from natural forests decrease. The numbers of large-scale and smallholder *P. falcataria* plantations have increased steadily in recent years. The main *P. falcataria* cultivation areas are in Sumatra, Java, Bali, Flores and Maluku (Charomaini and Suhaendi, 1997), as well as some forest management unit (FMU) and forest plantation companies in Sabah, Malaysia.

4.2 DESCRIPTION OF THE SPECIES

4.2.1 TAXONOMY

Botanical name: *Paraserianthes falcataria* (L.) Nielsen
Family: Fabaceae
Subfamily: Mimosoideae
Synonyms: *Adenanthera falcata* Linn., *Adenanthera falcataria* Linn., *Albizia falcata* (L.) Backer, *Albizia falcata sensu* Backer, *Albizia falcataria* (L.) Fosberg, *Albizia moluccana* Miq., and *Falcataria moluccana* (Miq.) Barneby and J.W. Grimes (Soerianegara and Lemmens, 1993).
Vernacular and common names:
Common names in other countries: puah (Brunei); albizia, sengon laut (Java), batai, Indonesian albizia, moluca, paraserianthes, peacock plume, white albizia (England); kayu machis (Malaysia); white albizia (Papua New Guinea); falcata, moluccan sau (Philippines) (Soerianegara and Lemmens, 1993).

4.2.2 BOTANY

P. falcataria is a large tree that can grow up to 40 m tall with the first branch at a height of up to 20 m (Figure 4.1). The tree can grow to 100 cm or sometimes more in diameter, with a spreading flat crown (Figure 4.2). When grown in the open, trees form a large umbrella-shaped canopy. The buttress is small or absent (Figure 4.3). The leaves are alternate, bipinnately compound and 23–30 cm in length (Figure 4.4). Leaflets are opposite and many, with 15–20 pairs on each axis; they are stalkless, small and oblong (6–12 mm long, 3–5 mm wide) and short-pointed at the tip. The topside of the leaf is a dull-green colour and hairless, and the underside is pale with fine hair (Soerianegara and Lemmens, 1993, Arche et al., 1998). The bark surface is white, grey or greenish; smooth or slightly warty; and sometimes shallowly fissured with longitudinal rows of lenticels (Figure 4.5).

FIGURE 4.1   Straight bole of *P. falcataria* at 20 years old.

FIGURE 4.2   Canopy of *P. falcataria* at 5 years old.

Inflorescence is axillary, consisting of pedunculate spikes or racemes; the spikes are sometimes arranged in panicles. The flowers are bisexual, 12 mm long, regular pentamerous, subtended to bracts, and funnel or bell shaped; their colour is cream to yellowish. The fruit is flat, straight pod, 10–13 cm long and 2 cm wide. It is not segmented and is dehiscented along both sutures and winged along the ventral suture with many seeds (15–20). Seeds are subcircular to oblong, 6 mm long, flat to convex, without aril; their colour is dull to dark brown and they are not winged (Soerianegara and Lemmens, 1993).

FIGURE 4.3   Trunk and buttress of *P. falcataria* at 5 years old.

FIGURE 4.4   Flowering and seed of *P. falcataria*.

FIGURE 4.5   Morphology trunk of *P. falcataria*.

4.2.3 DISTRIBUTION

P. falcataria is native to Indonesia, Papua New Guinea, the Solomon Islands, Australia (Soerianegara and Lemmens, 1993, Orwa et al., 2009) and Haiti (Orwa et al., 2009). Natural stands of *P. falcataria* are scattered around the eastern part of Indonesia (i.e. South Sulawesi, Maluku and Papua) (Martawijaya et al., 1989). In Maluku, *P. falcataria* is found on the islands of Taliabu, Mangolle, Sasan, Obi, Bacan, Halmahera, Seram and Buru. In Papua, it is found in Sorong, Manokwari, Kebar, Biak, Serui, Nabire and Wamena. In addition, it is planted in Java (Martawijaya et al., 1989).

The species is also widely planted throughout the tropics including in Brunei, Cambodia, Cameroon, Cook Islands, Fiji, French Polynesia, Japan, Kiribati, Laos, Malaysia, Marshall Islands, Myanmar, New Caledonia, Norfolk Island, the Philippines, Samoa, Thailand, Tonga, the United States of America, Vanuatu and Vietnam (Orwa et al., 2009).

4.2.4 ECOLOGICAL RANGE

P. falcataria can grow on a wide range of soils. It does not require fertile soil; it can grow well on dry soils, on damp soils and even on salty to acid soils as long as drainage is sufficient (Soerianegara and Lemmens, 1993). In plantations in Java, it has been reported to grow on various soil types with the exception of grumusols (Charomaini and Suhaendi, 1997). On latosols, andosols, luvial and red-yellow podzolic soils, its growth is very robust. On marginal sites, fertiliser may be needed to accelerate initial growth. However, growth will be faster thereafter as the ability to fix nitrogen increases.

P. falcataria is categorised as a pioneer species that occurs in primary forest but more characteristically in secondary lowland rainforest and in light montane forest, grassy plains and along roadsides near the coast.

In the species' natural habitat, annual rainfall ranges from 2000 to 2700 mm and can be up to 4000 mm with a dry season of more than 4 months (Soerianegara and Lemmens, 1993). The species is highly evapotranspiring, which requires a wet climate; therefore, an annual rainfall of 2000–3500 mm is considered optimal (Parrotta, 1990). Rainfall of less than 2000 mm/year will result in dry conditions, and rainfall of more than 3500 mm/year will create very high humidity, together with very low light intensity, which will possibly create fungal problems. The optimal temperature range is between 22°C and 29°C with a maximum of 30–34°C and a minimum of 20–24°C (Soerianegara and Lemmens, 1993). Ideally, during the dry months, there will be rain for at least 15 days. On very dry sites, growth can be drastically reduced and the risk of stem borer attack may increase.

The altitudinal range of the species' natural habitat is up to 1600 m above sea level but may extend up to 3300 m above sea level (Soerianegara and Lemmens, 1993). In Papua, the species is found at an elevation of 55 m above sea level at the lowest site in Manokwari (Charomaini and Suhaendi, 1997).

4.2.5 Wood Characteristics

P. falcataria wood is generally lightweight and soft to moderately soft. The colour of the heartwood ranges from whitish to a pale pinkish-brown or a light yellowish- to reddish-brown; the heartwood of the younger trees is not clearly demarcated from the sapwood (pale coloured), but it is more distinct in older trees (Soerianegara and Lemmens, 1993). The wood density is between 230 and 500 kg/m^3 at 12–15% moisture content. The grain of the wood is straight or interlocked, and the texture is moderately coarse but even. The wood is not durable when used outside; it is often highly vulnerable to various kinds of insects and fungal attacks. Graveyard tests in Indonesia showed an average service life in contact with the ground of 0.5–2.1 years. However, the wood treated with preservatives can have an average life in contact with the ground of 15 years in tropical conditions (Soerianegara and Lemmens, 1993).

4.2.6 Prospects and Utilization

P. falcataria is one of the important commercial timber species used for both the pulp and paper industries and for furniture. The wood is also suitable for general purposes such as light construction (e.g. rafters, panelling, interior trim, furniture and cabinetwork), lightweight packing materials (e.g. packages, boxes, cigar and cigarette boxes, crates, tea chests and pallets), matches, wooden shoes, musical instruments, toys, novelties and general turnery (Soerianegara and Lemmens, 1993). Tomimura et al. (1988) found that *P. falcataria* has good promise as a suitable raw material for the manufacture of fibreboard. The wood is an important source of lightweight veneer and plywood and is very suitable for the manufacture of light- and medium-density particleboard, wood–wool board and hardboard as well as block-board (Parrotta, 1990). The wood is extensively used for the manufacture of rayon and for supplying pulp for the manufacture of paper (Soerianegara and Lemmens, 1993). *P. falcataria* is also used as a shade tree for tea, coffee and other crops in Java (Parrotta, 1990).

As a nitrogen-fixing species, *P. falcataria* is also commonly planted for reforestation and afforestation to improve soil fertility (Heyne, 1987). The natural drop of leaves and small branches contributes nitrogen, organic matter and minerals to the upper layers of soil (Orwa et al., 2009). The trees are sometimes interplanted with agricultural crops such as corn and cassava and fruit trees (Charomaini and Suhaendi, 1997). They are often planted in home gardens for fuelwood (charcoal), and the leaves can be used as fodder for chickens and goats. The trees are also planted as windbreaks and firebreaks. They have been used as ornamental and shade trees planted along the sides of highways.

4.3   PROPAGATION AND NURSERY

4.3.1   SOWING

In the nursery, *P. falcataria* seed can be sown by broadcasting the seeds onto the seedbed or sow directly into mini pot of tray. Liew et al. (2016) discovered the potential of using bioplastic pots for *P. falcataria* seedlings. Normally, coco peat is used for sowing (Figure 4.6) and is mixed with slow-release fertiliser using a machine (Figure 4.7). After that, the mixture put into trays manually (Figure 4.8) and is then ready for sowing (Figure 4.9). Rollen et al. (2017) found that the use of arbuscular mycorrhizal fungi (AMF) and carbonized rice hull (CRH) had the potential to promote plant health even when grown in Cu-contaminated soil. Before sowing, the soil or medium germination should be sterilised to avoid damping-off. The seeds are

FIGURE 4.6   Coco peat used for sowing in the nursery.

FIGURE 4.7   Mixing coco peat with fertiliser in a machine.

FIGURE 4.8   Preparation of coco peat to be mixed with slow-release fertiliser into medium tray.

pressed gently into the soil and then covered with a layer of fine sand up to 1.5 cm thick (Figure 4.10). The soil in the seedbed must be loose and well drained. Application of a surface layer of mulch is recommended, and excessive shade should be avoided (Figure 4.11). Germination usually takes places 5–10 days after sowing. Untreated seeds germinate irregularly; germination may start after 5–10 days but can be delayed for up to 4 weeks. Watering at this stage is also crucial for the seedlings, as is dry weather. A moist watering system and automatic control is recommended for the nursery (Figure 4.12).

The seedlings can be exposed to sunlight gradually (Figure 4.13) at 3–4 weeks after germination (Figure 4.14). At week 4–6, the seedlings then can be spaced to

FIGURE 4.9   Coco peat with slow-release fertiliser ready for sowing.

FIGURE 4.10   Sowing of seed of *P. falcataria* by hand in a medium tray.

allow the seedlings room to grow (Figure 4.15). At the same time, watering can be reduced to just once per day for the hardening process (Figure 4.16).

4.3.2   PREPARATION AND PLANTING OUT

P. falcataria can be raised by planting out nursery-raised seedlings, container seedlings or stump cuttings. Wildings of *P. falcataria* are sometimes collected and potted for planting, but they are delicate and must be handled carefully. In the nursery,

FIGURE 4.11   *P. falcataria* seedlings after germination at the nursery.

FIGURE 4.12   Moist watering system.

seedlings are usually retained until they are about 8–10 weeks old before planting out. The seedlings can be transplanted when they have reached a height of 20–25 cm with a woody stem and a good fibrous root system (Figure 4.17).

Before the planting, legume cover crops should be firstly established in the field (Figure 4.18). A frequently used cover crop is *Mucuna bracteata* because its broad leaves and high deep rooting system enhance the moisture content of the soil. It also grows very fast and prevents weeds (Figure 4.19).

FIGURE 4.13   *P. falcataria* seedlings at the nursery.

FIGURE 4.14   Transfer of *P. falcataria* to direct sunlight at 3–4 weeks.

FIGURE 4.15　Spacing of *P. falcataria* at 4–6 weeks old.

FIGURE 4.16　Hardening process of *P. falcataria* by reducing watering (once per day).

FIGURE 4.17   Seedlings ready for planting.

FIGURE 4.18   Establishing of legume cover crops.

Prior to planting, land clearance or land preparation needs to be carried out. All the residual plants from the previous rotation should be collected or removed. The wild grass in the planting area should be removed using chemical spraying (Figure 4.20). After that, lines should be laid out in the planting area using rope (Figure 4.21). The planting design should either be a triangle or square pattern. Normally, a square planting design is commonly used for *P. falcataria*. Once the lines are finished the planting holes are marked using a pole stick (Figure 4.22). The workers will make holes for the seedlings according to the lining hole (Figure 4.23).

4.4   FIELD PLANTING

P. falcataria seedlings should be planted at the beginning of the rainy season. Before planting, all weeds that could hinder seedling growth and survival should be

FIGURE 4.19 Legume cover crops after establishment in the field.

FIGURE 4.20 Weed control during land preparation before planting.

removed. The seedlings are usually planted in fields at a spacing of 2×2 m–6×6 m. The spacing of planting is depend on the management objectives. A common spacing for pulpwood production is 3×3 m. For saw-log production, the trees are spaced at 6×6 m on fertile sites. For premium log production, trees may be planted in rows 10 m apart, with 1 m spacing between trees. For farm woodlots, trees may be planted in blocks at a spacing of 2×2 m. Sometimes, trees are planted along fence lines or boundaries to grow for timber.

FIGURE 4.21   Lining and marking for planting.

It is very important to select the condition of seedlings. The uniformity of *P. falcataria* seedlings include height, health and nutrient symptoms (Figure 4.24). After the seedlings have been selected at the nursery, they have to be watered to avoid plant stress during transplanting (Figure 4.25). The transplanting should be done on the same day and not in the dry season. Before planting, an amount of rock phosphate (RP) fertilizer should be put into the hole before the seedling is put in (Figure 4.26). The RP is very important to support the rooting system of the seedling at initial stage in the field.

4.5   HARVESTING

In forest plantations, harvesting to extract the timber from the forest is a very important task. *P. falcataria* trees will be felled in a particular area depending on the purpose of production (Figure 4.27). After the stands are felled, the timber will be

FIGURE 4.22 Land preparation for planting of *P. falcataria*.

FIGURE 4.23 Holing and planting of *P. falcataria* in the field.

FIGURE 4.24   Seedling of *P. falcataria* suitable for planting.

FIGURE 4.25   Transplanting of *P. falcataria* to the field.

FIGURE 4.26   Holing, fertilising and planting seedlings of *P. falcataria*.

FIGURE 4.27   Tree felling before log extraction.

moved to a nearby road. At this stage, a cable yarding system is normally used to reduce the impact of logging in the forest (Figure 4.28). The purpose of using a cable yarding system is to reduce the impact of logging and to reduce soil disturbance and residual trees in any particular area. Logs will be transported to the mill for processing either for sawn timber or chip wood (Figure 4.29). For harvesting, Listyanto (2018) recommends harvesting *P. falcataria* in a minimum rotation of 7 years, which is suitable for light construction.

FIGURE 4.28   Log harvesting using cable yarding system.

4.6   TREE MAINTENANCE

4.6.1   Weeding

P. falcataria plantations should be kept weed-free during the first 2 years (Figure 4.30). Weeding should be conducted after the first 2 months and then at 3-month intervals. Weeding at the initial stage of planting is important to support the growth of *P. falcataria*. When the weeding is done, the growth of *P. falcataria* trees is better and it is easy to monitor their performance (Figure 4.31).

4.6.2   Pruning

As *P. falcataria* trees have a tendency to fork, pruning is recommended at an early stage of stand development. Pruning is usually carried out starting at the age of 6 months, and then at 6-month intervals until 2 years of age.

4.6.3   Thinning

The principal objective of thinning is to improve the growth of remaining trees to obtain an acceptable form for the final crop. Trees selected for thinning should consist of diseased or pest-infested trees, deformed or poorly shaped trees and suppressed

FIGURE 4.29   Logs of *P. falcataria* at 10 years old, ready for the mill.

FIGURE 4.30   *P. falcataria* stands before weeding.

trees. The thinning activities should start 2 years after planting and then every year up to 10 years. When the stand for timber production is 4–5 years old, the stand can be thinned to a density of 250 trees per ha and then reduced progressively to 150 trees per ha after 10 years.

4.6.4   FERTILIZING

To improve the growth of *P. falcataria*, 100 g of NPK fertiliser should be applied to each seedling, either during or immediately after planting. Fertiliser may be placed

FIGURE 4.31   *P. falcataria* stands after weeding control.

in the planting hole or applied in a ring around the seedling. Depending on soil fertility, it is often recommended to fertilise plantations again 5 years after planting. However, the standard procedure for a fertilisation programme is also dependent on the plantation.

4.6.5 Pest and Disease Control

Several findings on pests and disease have been recorded. Stem borer (*Xystrocera festiva*), small bagworm (*Pteroma plagiophleps*) and yellow butterfly (*Eurema* spp.) are the main threats (Nair and Sumardi, 2000). A common method for controlling *X. festiva* is to cut or remove infested trees to prevent the spread of the borer population. In state-owned plantations, this operation is usually carried out together with regular thinning operations by removing infested trees instead of systematic thinning. This method reduces the infestation rate to 4–10% of trees, although this may not be sufficient (Kasno and Husaeni, 1998),

Small bagworm infestation causes defoliation. Small bagworm is a sporadic pest, but some companies in Sumatra have reported severe infestations (Nair and Sumardi, 2000). Another pest, yellow butterfly larvae, occasionally defoliate nursery seedlings in Sumatra and Java (Irianto et al., 1997), but the infestation is not economically important. The pest can be controlled by hand picking them and destroying them.

In the nursery, seedlings are occasionally damaged by damping-off fungi of *Pythium*, *Phytophthora* and *Rhizoctonia* (Nair and Sumardi, 2000). Cultural methods such as providing seedbeds with a roof and reducing watering frequency and intensity may prevent infection (Anino, 1997). Root rot caused by *Botryodiplodia* sp. occurs in young plantations in South Kalimantan and Jambi (Aggraeni and Suharti,

1997). Widyastuti (1999) reported incidence of root rot (infected by Ganoderma fungi) in older *P. falcataria* trees. In general, *P. falcataria* trees do not suffer from these diseases, except for trees older than 10 years (Nair and Sumardi, 2000).

The disease most recently reported to affect *P. falcataria* trees is gall rust (*Uromycladium tepperianum*), which has damaged some *P. falcataria* plantations in some parts of Java (Rahayu, 2008). Anino (1997) indicated that the disease was successfully managed by not planting trees at elevations above 250 m, where the disease has been very severe. Pruning and burning of infected parts, conversion of cleared areas to other suitable tree species and use of bio-control agents such as *Penicillium italicum, Acremonium recifei* and *Tuberculina* spp. may be effective in preventing the disease.

4.7   GROWTH AND YIELD

P. falcataria can grow rapidly (Parrotta, 1990), particularly young stands (Figure 4.32). Under favourable conditions, the trees can reach a height of 7 m in 1 year, 16 m in 3 years and 33 m in 9 years (Bhat et al., 1998). This can be seen in pictures of trees at 18 months and 5 years old (Figure 4.33 and Figure 4.34). Another study also reported that *P. falcataria* trees in 3–5-year-old stands growing in state-owned plantations in Kediri (East Java) have a mean diameter of 11.3–18.7 cm (maximum diameter 25.8 cm) with a mean height of 11.7–20.5 m (maximum height 23.5 m) (Kurinobu et al., 2007).

Krisnamwati et al. (2011) recorded that the mean diameter of *P. falcataria* trees ranges from 3.4 to 16.7 cm with a maximum diameter of 36.0 cm for trees younger than 4 years old. The mean height of the corresponding stands ranges from 3.9 to

FIGURE 4.32   *P. falcataria* plantation at 3 months old.

FIGURE 4.33 *P. falcataria* plantation at 18 months old.

FIGURE 4.34 Size of *P. falcataria* tree at 5 years old.

19.6 m with a maximum value of 27.0 m. The mean diameter for trees older than 5 years (but less than 10 years) growing on the same sites is between 8.7 and 40.1 cm, and the mean height is between 9.9 and 27.9 m. For older stands, the trees in 12-year stands are recorded to have a diameter of 24.6–74 cm and a height of 15.3–36.2 m. The wide variations in mean diameter and height as recorded in our study are probably due to differences in growing conditions, including site quality, altitude, slope and silvicultural management.

Sumarna (1961) noted that, up to the age of 5 years, the mean annual increment (MAI) in height of *P. falcataria* trees growing on sites of average quality is about 4 m; growth then slows as the trees age. At the age of 8–9 years, the height increment is about 1–1.5 m, and at 10 years of age it is only about 1 m. A similar trend was observed for the diameter increment; however, the MAI in diameter fluctuates around 4–5 cm until the age of 6 years. At the age of 8–9 years, the diameter increment is still high, about 3–4 cm; it then decreases slowly thereafter.

4.8   SOIL CHEMICAL PROPERTIES UNDER *P. FALCATARIA* STANDS

In our case of study, soil chemical properties of *P. falcataria* are measured at two different age stands. In stands of *P. falcataria* under 3 years old, the content of potassium (K), magnesium (Mg) and cation exchangeable capacity (CEC) are increased in subsoil whereas the other elements are almost the same especially for total nitrogen (N), available phosphorus (P) and organic carbon (OC) (Table 4.1).

For stands of 5 years old, the soil pH is increased for both depths. For the total N it seems that decreased, as the N is more mobile in the soil. However, it is clear that the OC compared to 3 years old stand is increased. It shows that the soil fertility in the *P. falcataria* stand (Table 4.2).

TABLE 4.1
Soil Chemical Properties under 3-year-old *P. falcataria*

Soil Layer	pH	N (%)	P (ppm)	K (ppm)	Mg (ppm)	Ca (ppm)	CEC (cmol$_+$/kg)	OC (%)
0–15 cm	4.2	0.17	6.63	9.97	9.65	15.93	4.88	25.3
15–30 cm	4.4	0.12	5.03	13.82	11.32	13.87	5.53	24.5

TABLE 4.2
Soil Chemical Properties under 5-Year-Old *P. falcataria*

Soil Layer	pH	N (%)	P (ppm)	K (ppm)	Mg (ppm)	Ca (ppm)	CEC (cmol$_+$/kg)	OC (%)
0–15 cm	4.8	0.11	6.36	7.64	9.23	14.7	2.47	30.9
15–30 cm	4.7	0.09	6.80	5.77	6.59	15.6	3.66	30.6

4.9 CONCLUSIONS

P. falcataria is one of the fast growing species and most important pioneer multi-purpose tree species in some Southeast Asian countries. It is one of the tree species preferred for industrial forest plantations because of its ability to grow on a variety of soils, and it is increasingly important for wood industries as supplies for plywood from natural forests. As the species is a legume tree, it is also valuable because the tree can extract nitrogen from the atmosphere through the nitrogen fixation as well as increasing the soil fertility of the particular area. Hence, it is highly recommended to expand forest plantation using *P. falcataria* because of its high economic value in the timber industries.

ACKNOWLEDGEMENTS

The authors wish to thank Sabah Softwood Berhad for providing facilities and pictures during the field visit to the *P. falcataria* plantation in Brumas, Tawau, Sabah Malaysia; and also the Forest Management Unit (FMU) under the company of Maxland Sdn. Bhd. (Sg. Pinangah Forest Reserve – Partly – FMU 17B) Sabah, Malaysia. They are also grateful to Sabah Forestry Department, especially in Kota Marudu district, Malaysia, for its cooperation during the field visit, particularly for the observation of *P. falcataria* trees.

REFERENCES

Aggraini, I. and Suharti, M. 1997. Identifikasi Beberapa Cendawan Penyebab Penyakit Busuk Akar pada Tanaman Hutan. *Buletin Penelitian Hutan* 610: 17–35.

Anino, E. 1997. Commercial plantation establishment, management, and wood utilization of *Paraserianthes falcataria* by PICOP Resources, Inc. pp. 131–139.

Arche, N., Anin-Kwapong, J.G. and Losefa, T. 1998. Botany and ecology. In: Roshetko, J.M. (ed.) *Albizia* and *Paraserianthes* production and use: a field Manual, 1–12. Winrock International, Morrilton, Arkansas, AS.

Bhat, K.M., Valdez, R.B. and Estoquia, D.A. 1998. Wood production and use. In: Roshetko, J.M. (ed.) *Albizia* and *Paraserianthes* production and use: a field manual. Winrock International, Morrilton, Arkansas, AS.

Charomaini, M. and Suhaendi, H. 1997. Genetic variation of *Paraserianthes falcataria* seed sources in Indonesia and its potential in tree breeding programs. In: Zabala, N. (ed.) Workshop international tentang spesies *Albizia* dan *Paraserianthes*, 151–156. Prosiding workshop, 13–19 November 1994, Bislig, Surigao del Sur, Filipina. Forest, Farm, and Community Tree Research Reports (tema khusus). Winrock International, Morrilton, Arkansas, AS.

Heyne, T. 1987. *Tumbuhan Berguna Indonesia*. Badan Penelitian dan Pengembangan Kehutanan, Jakarta, Indonesia.

Irianto, R.S.B., Matsumoto, K. and Mulyadi, K. 1997. The yellow butterfly species of the genus Eurema Hubner causing severe defoliation in the forestry plantations of *Albizia* and *Paraserianthes falcataria* (L.) Nieilsen, in the western part of Indonesia. *JIRCAS Journal* 4: 41–49.

Kasno, H.N. and Husaeni, E.A. 1998. An integrated control of sengon stem borer in Java. Paper to IUFRO Workshop on Pest Management in Tropical Forest Plantations. Chanthaburi, Thailand, 28–29 Mei, 1998.

Krisnawati, H., Varis, E., Kallio, M. and Kanninen, M. 2011. *Paraserianthes falcataria* (L.) Nielsen—Ecology, silviculture and productivity. Center of International Forestry Research, Bogor Barat, Indonesia.

Kurinobu, S., Daryono, P., Naiem, M. and Matsune, K. 2007. A provisional growth model with a size–density relationship for a plantation of *Paraserianthes falcataria* derived from measurements taken over 2 years in Pare, Indonesia. *Journal of Forestry Research* 12(3): 230–236.

Liew, K.C., Chang, C.Y., Kodoh, J., Russel, M.A. and Crispin, K. 2016. Growth performance of *Paraserianthes falcataria* (BATAI) planted in bioplastic pot. *International Journal of Agriculture, Forestry and Plantation* 2(February): 111–114.

Listyanto, T. 2018. Wood quality of *Paraserianthes falcataria* L. Nielsen Syn wood from three year rotation of harvesting for construction application. *Wood Research* 63(3): 497–504.

Martawijaya, A., Kartasujana, I., Mandang, Y.I., Prawira, S.A. and Kadir, K. 1989. *Atlas Kayu Indonesia Jilid II*. Pusat Penelitian dan Pengembangan Hasil Hutan, Bogor, Indonesia.

Nair, K.S.S. and Sumardi, S. 2000. Insect pests and diseases of major plantation species. Dalam: Nair, K.S.S. (ed.) *Insect Pests and Diseases in Indonesian Forests: An Assessment of the Major Treats, Research Efforts and Literature*. CIFOR, Bogor, Indonesia.

Orwa, C., Mutua, A., Kindt, R., Jamnadass, R. and Anthony, S. 2009. Agroforestree Database: a tree reference and selection guide version 4.0 (http://www.worldagroforestry.org/sites/treedbs/treedatabases.asp).

Parrotta, J.A. 1990. *Paraserianthes falcataria (L.)* Nielsen. *Batai. Moluccan sau. Leguminosae (Mimosoideae) Legume Family*. USDA Forest Service, Southern Forest Experiment Station, Institute of Tropical Forestry; 5 p. (SO-ITF-SM; 31).

Rahayu, S. 2008. Penyakit Karat Tumor pada Sengon. Makalah Workshop Penanggulangan Serangan Karat Puru pada Tanaman Sengon 19 November 2008. Balai Besar Penelitian Bioteknologi dan Pemuliaan Tanaman Hutan, Yogyakarta, Indonesia.

Rollon, R.J.C., Galleros, J.E.V., Galos, G.R., Villasica, L.J.D. and Garcia, C.M. 2017. Growth and nutrient uptake of *Paraserianthes falcataria* (L.) as affected by carbonized rice hull and arbuscular mycorrhizal fungi grown in an artificially copper contaminated soil. *AAB Bioflux* 9(2): 57–67.

Soerianegara, I. and Lemmens, R.H.M.J. 1993. Plant resources of South-East Asia 5(1): Timber trees: major commercial timbers. Pudoc Scientific Publishers, Wageningen, Belanda.

Sumarna, K. 1961. *Tabel Tegakan Normal Sementara untuk Albizia falcataria*. Pengumuman No. 77. Lembaga Penelitian Kehutanan, Bogor, Indonesia.

Tomimura, Y., Khoo, K.C., Ong, C.L. and Lee, T.W. 1988. Medium density fibreboard from *Albizia falcataria*. Forest Research Institute Malaysia, Kepong, Kuala Lumpur, Malaysia.

Walters, G.A. 1971. A species that grew too fast – *Albizia falcataria*. *Journal of Forestry* 69(3): 168.

Widyastuti, S.M., Sumardi, S. and Harjono, S. 1999. Potensi Antagonistik Tiga Trichoderma spp. terhadap Delapan Penyakit Busuk Akar Tanaman Kehutanan. *Buletin Kehutanan Universitas Gadjah Mada* 41: 2–10.

5 Tropical Tree Planting and Forest Communities

Prospects for Enhancing Livelihoods and Involvement in Sabah Forests, Malaysia

Hardawati Yahya

CONTENTS

5.1   INTRODUCTION

Community involvement in tree planting has been introduced in many developing countries, including Sabah, Malaysia. This initiative aims to solve various issues related to forest resources and the local community. In 2006, the Sabah Forestry Department identified that community forestry or social forestry* is the key to sustainable forest management. Social forestry aims to increase local community involvement in forest management, mitigate poverty, and reduce forest land encroachment (Sabah Forestry Department, 2006). One of the key activities in community forest management is tropical tree planting. In this activity, the local community is involved in planting tropical forest tree species such as rubber trees, dipterocarps, fruit trees, and other native tree species. Planting tropical trees is seen as providing an alternative income for local communities while they are waiting for their crop harvesting season to begin (Schroth and Ruf, 2014; Watkins et al., 2018). Furthermore, tree planting through community forestry management is seen as a significant way to mitigate poverty (Moktan et al., 2016). Several studies have demonstrated the benefits of tree planting, which include supporting the local community's social needs (Maryudi et al., 2012; Watkins et al., 2018), increasing income (Kulindwa and Ahlgren, 2018; Sunderlin, 2006), and reducing pressure on forest resources (Gebreegziabher and Kooten, 2013). In addition, the development of community forestry risks failure if the community deems the program as being non-profitable for them (Hajjar et al., 2016; Rasolofoson et al., 2015).

Moving on, approximately 3.5 million hectares (48%) of Sabah's forests are gazetted under forest reserve (Sabah Forestry Department, 2016), and approximately 25,000 people live in the forest area (Toh and Grace, 2006). These figures show that many local communities are dependent on forest land to accommodate their needs, settlement, and livelihood. In Sabah, the boundaries of forest reserves are clearly marked. Nonetheless, the boundaries between community settlements and the forest areas are still unclear due to issues of overlapping of native land. This has become an issue between the local community and the forestry departments, especially when the forest areas overlap with native land (Azima et al., 2015). A land can be claimed as a community's traditional land only if the legal perquisite can be fulfilled (Bernama, 2012). Conflicts arise when the end results are unsatisfactory for any of the parties involved (Emanuel and Ndimbwa, 2013; Lunkapis, 2015). In some cases, the local community is given a permit for settlement inside the forest reserve area by the Sabah Forestry Department (Sabah Forestry Department, 2013). Most communities that settle in the forest reserve area are underprivileged local farmers (Yahya et al., 2012).

This chapter analyses the local community's perspective on community involvement in tree planting. This includes the planning phase, site preparation, discussion, planting, and monitoring. Furthermore, the effect of local community involvement in tree planting was evaluated based on the Sustainable Livelihood Approach (SLA)

---

* "Social forestry is defined as the joint management and utilization of certain forest areas by indigenous people and local communities together with various stakeholders, in order to sustain forest, local economic resources, survivability and socio-cultural endurance" (Sabah Forestry Department, 2016, p. 257).

introduced by Scoones (1988). This approach was taken to explain the benefits of tree planting. Five indicators of sustainable livelihood were considered in this chapter: natural, social, human, physical, and economic capital. The results suggested that the local community's intentions to plant trees could be influenced by these five major indicators. It is important to highlight that other studies (Barnes et al., 2017; Chinangwa et al., 2016; Dehghani Pour and Milad, 2018; Dev et al., 2003; Feurer et al., 2018; Kaskoyo et al., 2017; Obodai et al., 2014) have investigated the indicators of the SLA framework with regards to the contribution of forest management and tree planting to community livelihood.

5.2 ANALYZING THE TROPICAL TREE PLANTING OUTCOMES: SUSTAINABLE LIVELIHOOD APPROACH (SLA)

The concept of SLA has been used by various scholars to evaluate the impact of community development programs, including forest management and community forestry. Dehghani Pour and Milad (2018), Dev et al. (2003), Feurer et al. (2018), Harbi et al. (2018), and Scoones (1988) explained that SLA can be used to assess the program outcome, livelihood resources, institutional process and livelihood strategies. The conceptual framework of SLA illustrates the five key indicators of assessment (Figure 5.1) (adapted and modified from (Knutsson, 2006; Scoones, 1988; Serrat, 2017).

Social capital can be defined as something that individuals require to enhance their livelihood. This includes social networking, relations, membership, and association (Lee et al., 2017; Waltkins et al., 2018). *Human capital* refers to the people skills, knowledge, labor, and health conditions that allow individuals to pursue their desired livelihoods (Colombo et al., 2018). *Physical capital* is the infrastructures, production equipment, technologies, and other physical goods that assist individuals

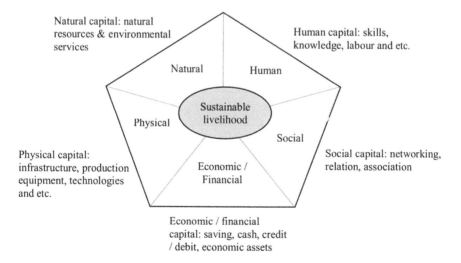

FIGURE 5.1   Sustainable Livelihood Approach (SLA) framework.

in their daily lives (Kibria et al., 2018). *Natural capital* is the natural resources and environmental services such as land, forest resources, and other environmental goods (Scoones, 1988). Finally, *financial* or *economic capital* refers to all financial resources such as cash, credit or debit, employment, and other types of economic development needed to fulfil the objectives of individuals (Colombo et al., 2018).

The involvement of local community in sharing their knowledge, perceptions, and interest using the Sustainable Livelihood Approach (SLA) can help to determine the factors and significant outcomes to improve the conservation development (Dehghani Pour and Milad, 2018). The concept of sustainable livelihood is a step in the right direction in addressing environmental degradation (Dehghani Pour and Milad, 2018; Colombo et al., 2018), community livelihood (Feurer et al., 2018; Kibria et al., 2018), environmental income (Angelsen et al., 2014), and environmental policy (Massoud et al., 2016; Serrat, 2008).

5.2.1　THE CONCEPT OF TREE PLANTING

The selection of tree species is important to ensure that the planting activity is effective, beneficial, and reflects the institutional aims (FAO, 2004). Figure 5.2 illustrates the aspects to be considered before selecting the trees (adapted and modified from (FAO, 2004)). Site selection is a crucial step in the tree-planting planning phase, which also includes seedling quality, species selection, species diversity, planting schedule, harvesting methods, and rotation (He et al., 2015; Rollan et al., 2018).

In addition, local community involvement combined with the power of reforestation* or tree-planting activity can contribute to a project's conservation strategy and outcomes (Baynes et al., 2015; Ostrom and Nagendra, 2006). Moreover, legal agreement can help increase the local community's trust and enhance their

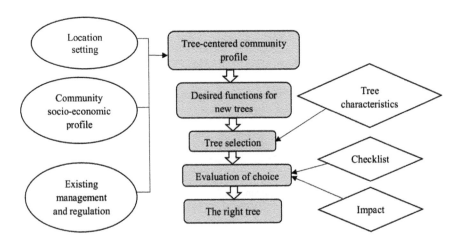

FIGURE 5.2　The tree planting framework.

---

* Reforestation is defined as "forests re-established artificially by reforestation or tree planting on unstocked land that is considered as forests or man-made forest" (FAO, 2012, p. 6).

participation in forest management (Ayine, 2008). A study done by Permadi et al. (2018) indicated that local communities prefers no-contract partnership in tree-planting projects. Following this, the community's socioeconomic status and people's perceptions must be considered before any tree-planting program begins (Malkamäki et al., 2018; Raintree, 1991; Rollan et al., 2018). Tree-planting policy and regulations may also provide advice and guidance for government agencies when planning tree-planting programs by addressing the reforestation goal (Kulindwa and Ahlgren, 2018; Valmassoi et al., 2017).

5.3 COMMUNITY FORESTRY MANAGEMENT IN SABAH

5.3.1 STUDY SITES

This study was conducted in four out of ten forest reserved areas that managed by community forestry management in Sabah. The forest reserves are Mangkuwagu Forest Reserve, Tamparuli Forest Reserve, Gana-Lingkabau Forest Reserve, and Deramakot Forest Reserve. The case study areas are presented in Figure 5.3. The selection was done based on the status of community forestry management (the case study must be under community forestry management). In addition, previous field research has been done on the area, adding relevant empirical findings from the previous study to the current study.

Mangkuwagu Forest Reserve is located in Tongod District, Sandakan. The Mangkuwagu Forest Reserve is gazetted under Class II (production forest), with

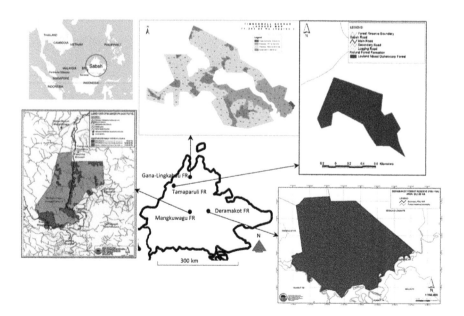

FIGURE 5.3   The case study areas. Map of forest reserve areas in Sabah, Malaysia. (Adapted and modified from Sabah Forestry Department, 2008, 2014, 2015, and Timberwell Berhad, 2015.)

20% of the 8,335 hectares forest areas being harvested under the Forest Management Unit 17C (FMU17C) (Sabah Forestry Department, 2008). It is important to note that this forest has been under threat of illegal encroachment and shifting cultivation practices for the past 20 years. Most of the area was encroached and burned to allow agricultural cultivation activities by the villagers. Agriculture is the primary source of income in the area, and most of the villagers cultivated hilly paddy and cash crops. The communities involved are from the Sungai and Dusin ethnic groups. There are four villages located inside and near the forest reserve, which are Mangkuwagu, Alitang, Saguon, and Tampasak villages.

Tamparuli Forest Reserve is located in Tamparuli District, which is 45 kilometers from Kota Kinabalu main town. The forest reserve is the smallest forest reserve in Sabah with 60 hectares of forest area (Sabah Forestry Department, 2008)). The forest is gazetted as Class I (protection forest). There is one village located within the forest reserve (Bukit Komunsi village) and one village located at the fringe of the forest (Botung village). Most of the villagers are of Kadazan Dusun ethnicity and work as farmers and rubber tappers. Unlike other forest reserves, Tamparuli Forest Reserve faces less threat of land encroachment and degradation (Sabah Forestry Department, 2010). The villagers are very protective of the forest. They assist the forestry department in reforestation efforts by planting tropical forest trees such as dipterocarp trees, fruit trees, rubber trees, and fast-growing trees such as *Peronema canescens* (Batai) and *Octomeles sumatrana* (Binuang) (Figure 5.4).

Gana-Lingkabau Forest Reserve is the combination of Gana Forest Reserve (Class I, protection forest) and Lingkabau Forest Reserve (Class II, production forest). These forests are located in Kota Marudu District and are 130 kilometers

FIGURE 5.4   Integrated tree planting in Tamparuli Forest Reserve.

from Kota Kinabalu. The total forest land is 72,184 hectares, with the Lingkabau Forest Reserve having the larger land area of 71,300 hectares (Sabah Forestry Department, 2006)). Nine villages have been relocated outside the forest reserve by the forestry department for Gana Resettlement and Integrated Development initiative: Kipapogong, Lingkahan, Sonsogon Paliu, Toguhu, Garung, Gouton, Nasapu, Minsusurud, and Makatol Darat villages. The majority of the villagers are of Dusun ethnicity. Agriculture is the major source of income for the community. The villagers plant fruit trees, rubber trees, and other cash crops to support their daily needs and income. Nonetheless, the soil fertility in the village home-garden is not optimal for their crops. As a result, land encroachment occurs as the villagers illegally plant crops inside the forest reserves for their own benefit.

Finally, the Deramakot Forest Reserve is located in Kinabatangan District and is a well-known model forest for sustainable forest management and Reduced Impact Logging*. The forest area is 55,507 hectares and is gazetted under Class II (production forest) (Sabah Forestry Department, 2010). According to the Sabah Forestry Department (2014), the forest reserve area is divided into three parts: 49,711 hectares for timber production, 5,778 hectares for conservation, and 18 hectares for community forestry purposes. There are four village settlements outside the forest reserve: Balat, Desa Permai, Kuamut, and Tulang-Tulang. It should be noted that more villagers in Balat (the nearest village) are actively involved in forest management activities compared to members of other villages. This is due to infrastructure facilities and distance. Most of the villagers from the four villages are of Sungai ethnicity and are dependent on river resources, fishing, and agriculture to earn their livelihood.

5.3.2 Framework on Sustainable Livelihood Approach (SLA)

This study was based on field research into four case studies of community forestry and literature data from 2010 to 2018, with a particular focus on tropical tree planting activity in community forestry management in Sabah. This study applied the SLA concept to elaborate on the impact of tropical tree planting on the local community's livelihood. This method was also taken to promote community involvement in forest management (Table 5.1). The indicators were based on Quandat et al. (2017), wherein the five key indicators of the SLA approach – human, social, natural, physical, and financial or economic capital – were used to evaluate the benefits of tree planting for the local community's livelihood.

For natural capital, indicators such as farm holding, the size of the farm, collection of forest products, involvement in agroforestry, and the distance to the forest area were determined. Farm land and forest products can provide direct benefits to the local community in terms of income and subsistence (Dhruba Bijaya et al., 2016) and can improve tenure rights and security (Dev et al., 2003). In addition, Quandat et al. (2017) revealed that the higher the number of tree species in agroforestry planted by the community, the higher the increment of economic capital. Moreover, the

---

* Reduced Impact Logging is one of the sustainable forest management strategies or techniques used in tree harvesting to reduce the negative impact of land degradation, residual damage and soils and increase forest conservation (Edwards et al., 2012).

TABLE 5.1

Key Elements in Assessing the Benefit of Tropical Tree Planting Based on Sustainable Livelihood Approach (SLA)

Attributes	Dimension	Indicators
Natural capital	Forest products and farm	Farm holding; size of farmland; collect forest products; agroforestry; distance to forest
Social capital	Membership, community involvement	Involvement in forest management; involvement in tropical tree planting; membership status; rights in decision-making;
Human capital	Knowledge and skills, labor capacity	Labor availability; training attendance; received information; level of contribution
Physical capital	Infrastructure facilities, tools and technology	Distance from farm to house; level of contribution (e.g. fertilizers, pesticides, seedlings); housing status; facility conditions
Financial or economic capital	Cash income; selling from forest products, agroforestry, and wages	Paying job; selling forest products and agroforestry; income from selling forest products and agroforestry; wages from tropical tree planting

distance to forest was taken into account to measure the benefit of natural resources to the local community. This characteristic has also been considered by many scholars, including Phompila et al. (2017) and Dehghani Pour and Milad (2018). Due to some limitations in evaluating environmental services, the topic of environmental services has been excluded in this chapter. This study also examined social capital in terms of the involvement in forest management and tree planting and status of rights. Dehghani Pour and Milad (2018) also included the level of cooperation in community forestry as one of the indicators in measuring social capital.

Moving on, indicators such as availability of labor, training attendance, and information reception were tested to measure human capital. Studies done by Batoa (2018) and Chinangwa et al. (2016) demonstrated that human capital significantly increases a community's employment and income. Physical capital such as distance from farm to house, level of contribution (raw materials), facility condition, and housing status were also taken into account. Finally, financial or economic capital, such as a paying job, wages from tree planting, the status of selling forest products, and income were assessed. It is worth noting that a study done by Tang et al. (2013) showed that reducing income from subsidies can help to sustain community development.

5.3.3 METHODS USED FOR DATA COLLECTION

The research was executed using triangulation methods involving household surveys, informant interviews, and observation. The validity of the response from the questionnaire and interview were proofed through discussion with the forestry department and some community groups. In addition, the household surveys were collected randomly using other triangulation sampling methods involving convenient, purposive, and snowball samplings. There were 425 households selected from four case studies of Mangkuwagu Forest Reserve, Gana-Lingkabau Forest Reserve, Tamparuli Forest Reserve, and Deramakot Forest Reserve. The closed and open-ended questionnaires were collected on different field visits from May 2014 to August 2018. Furthermore, face-to-face semi-structured interviews were conducted with 50 informants, both individually and with small groups of people, including forestry department officers, village heads, village committees, and women leaders. The observation was conducted at an agroforestry farm, rubber tree farm, forest area, home-garden, and housing area.

Moreover, the questions and observations on tropical tree planting were concerned with native forest tree species and fruit trees. Invasive fast-growing species or agricultural crops were excluded. On top of that, community involvement in tropical tree planting was considered based on their involvement in the consultation, planning, discussion, site-preparation, planting, monitoring, and harvesting stages. Heterogenous community background information, such as gender, age, education level, membership status, and income level, was also evaluated to gain more information and different views on a tree planting. Data on socio-economic profiles, livelihood assets, sources of income, and tropical tree-planting activity were also collected.

5.3.4 DATA SYNTHESIS

The results were presented qualitatively and quantitatively into two parts: socio-demographic profiles and SLA analysis. The data were compiled and analyzed using SPSS version 21. For households' socio-demographic features, the frequencies and percentages of the households' responses were presented. Furthermore, the pattern of the households' agreement on the benefit of tree planting based on SLA was analyzed using non-parametric tests with Kendall's Coefficient of Concordance *W*.

5.3.5 LIMITATIONS OF THE STUDY

Due to the limited time available, this study only presented tropical tree-planting activity based on the community perspective. The analysis was based on the community's perceptions and feedback from available selected respondents; in other words, whole households were not interviewed. Nevertheless, the households were considered to represent the case studies due to the high number of responses. The selection of the households was based on randomized sampling with an equal probability of being selected. The findings were also considered to be free from bias, in terms of selection sampling and methods of data collection.

5.4 IMPACTS OF TROPICAL TREE PLANTING ON COMMUNITY LIVELIHOODS

5.4.1 PREFERENCES FROM A SOCIO-DEMOGRAPHIC PERSPECTIVE

In this study, about 425 households, with members between the age of 17 and 96 years old and of heterogenous features such as gender, social status, and membership status in four community forestry management areas, were interviewed. Based on the results shown in Table 5.2, the majority of the households involved in this study were male. For instance, 70% of Mangkuwagu Forest Reserve respondents and 64% of Deramakot Forest Reserve were male. The percentage of women involved in tropical tree planting was lower compared to men, particularly in Deramakot Forest Reserve (38%) and Mangkuwagu Forest Reserve (32%). Nevertheless, in Tamparuli Forest Reserve and Gana-Lingkabau Forest Reserve, the percentage of women involved in tropical tree planting activity was considered higher than that in other forests (45% and 55%, respectively). This shows that both genders equally participate in tropical tree planting activities. Most women in these case studies are farmers, small-scale business owners, and housewives. Women are mostly involved in non-paid and subsistence activities such as the collection of non-wood forest products and agroforestry (Whiteman et al., 2015). Nevertheless, women's involvement in tree planting and forest management, especially in decision-making and benefit-sharing, is still low and limited (Coleman and Mwangi, 2013; Evans et al., 2017).

The average age of the households in Mangkuwagu Forest Reserve was 48 years old, followed by Tamparuli Forest Reserve (42 years old), Deramakot Forest Reserve (40 years old), and Gana-Lingkabau Forest Reserve (39 years old). Overall, 48% of the households were between 21 and 35 years old, and 35% were 56 years old and above. Most of the elders gave responsibility to the young people to replace them in agriculture, farm, and tropical tree planting activities due to deteriorating health. The households' age is one of the key indicators determining their involvement in forest management (Musyoki et al., 2013). Negi et al. (2018) mentioned that the increase in households' age can increase the household's involvement.

On the other hand, the results for educational background between the four case studies did not differ as much. The majority of the households in Mangkuwagu, Tamparuli, and Deramakot Forest Reserves had a secondary educational background. Nonetheless, about half of households in Gana-Lingkabau Forest Reserve had not received a formal education. The literacy rate among the households in these case studies were considered to be high, considering the limited education facilities and awareness of the importance of education. Apart from that, the huge distance between the secondary school and the village could also be responsible for the lower educational attainment of the households. Educational status has been used in many studies to determine households' involvement in forest management (Mutune and Lund, 2016; Paudel, 2018; Permadi et al., 2018).

Meanwhile, most households in three of the case studies were farmers, except for Deramakot Forest Reserve, where 42% of the households work with private companies in oil palm plantations and rubber tree plantations. In Deramakot Forest Reserve, most of the villages that were located outside the forest reserve depended on the oil palm plantation for their employment. In other case studies, the households were still dependent on agriculture as their main source of income and employment. Similar findings from other studies revealed that farming is the main occupation among households in community forestry areas (Moktan et al., 2016; Yadav et al., 2015).

TABLE 5.2

Socio-Demographic Features of the Households in Four Different Case Studies

Features		Mangkuwagu FR (n = 174)	Tamparuli FR (n = 53)	Deramakot FR (n = 59)	Gana-Lingkabau FR (n = 139)
Gender	Male	121 (70%)	28 (53%)	38 (64%)	86 (62%)
	Female	51 (30%)	25 (47%)	21 (36%)	53 (38%)
Age (in years)	Up to 20	3 (2%)	N/A	2 (3%)	4 (3%)
	21–35	39 (22%)	20 (38%)	22 (37%)	67 (48%)
	36–45	40 (23%)	15 (28%)	12 (20%)	28 (20%)
	46–55	32 (18%)	9 (17%)	14 (24%)	22 (16%)
	56 and above	60 (35%)	9 (17%)	9 (15%)	18 (13%)
Education level	No formal education	31 (18%)	10 (19%)	6 (10%)	75 (54%)
	Primary education	58 (33%)	7 (13%)	19 (32%)	29 (21%)
	Secondary education	81 (47%)	36 (68%)	32 (54%)	29 (21%)
	Higher education	4 (2%)	N/A	2 (3%)	6 (4%)
Type of occupation	Public sector	10 (6%)	1 (2%)	7 (12%)	7 (5%)
	Private sector	14 (8%)	12 (23%)	25 (42%)	1 (1%)
	Self-employment:				
	• Farmer	143 (82%)	37 (70%)	4 (7%)	90 (65%)
	• Small scale business	3 (2%)	1 (2%)	15 (25%)	25 (18%)
	• Fisherman	1 (1%)	N/A	2 (2%)	N/A
	Housewife	3 (2%)	1 (2%)	15 (25%)	25 (18%)
	Unemployed	N/A	N/A	2 (3%)	11 (7%)

Notes: In order to balance the unequal sample size for all case studies, weighted data were used in all analyses.

N/A represents that the data are not available and have no value.

The data present the number of respondents and the percentage in parentheses.

5.4.2 NATURAL CAPITAL ATTRIBUTES

Natural capital may play a pertinent role in the success of tree planting activity and deforestation mitigation (Narita et al., 2018; Ojha et al., 2016). For this study, natural capital was classified into five attributes including farm holding, size of farm, collection of forest products, and distance to forest (Table 5.3). Households with farm holding had higher percentage, including Gana-Lingkabau Forest Reserve (72%), Mangkuwagu Forest Reserve (70%), and Tamparuli Forest Reserve (68%). The households from Deramakot Forest Reserve stated that they had less farm holding, with only 32% of them having ownership. The farm holding included land allocated by the forestry department to plant tropical forest trees, rubber trees, and agroforestry trees. It also included land illegally owned by the households (the household assumed that they owned the land inherited from their ancestors). Apart from that, the households were also asked to determine the size of their farm. On average, households who had a farm holding mentioned that they had at farms of least 0.51–2.90 hectares. Households with large farm size had greater influence on economic and production growth (Gautam and Ahmed, 2018; Rada and Fuglie, 2018). Most of the households with individual farm land planted rubber trees, fruit trees, agricultural crops, and some forest trees on their land.

Moreover, the percentage of households that collected forest products was high for all the case studies, ranging between 72% and 89%. Most of the households collected edible plants such as ferns, wild mushroom, bamboo shoots, rattan shoots, young leaves, fuelwood, and other medicinal plants. The collection of forest products, especially non-wood forest products, was still practiced by the households to support their daily needs and subsistence (Coulibaly-Lingani et al., 2009; Harbi et al., 2018). When the households were asked whether they produce agroforestry products or not, most of them admitted to having agroforestry products such as rubber trees,

TABLE 5.3

Comparison of Natural Capital Attributes between the Four Case Studies

Natural Capital Attributes	Mangkuwagu FR (n = 174)	Tamparuli FR (n = 53)	Deramakot FR (n = 59)	Gana-Lingkabau FR (n = 139)
Owned farm holding (% of households)	122 (70%)	36 (68%)	19 (32%)	100 (72%)
Size of farm/hectare (mean±std. dev.)	0.46 ± 0.51	0.63 ± 0.94	0.91 ± 1.02	1.78 ± 2.90
Collected forest products (% of households)	126 (72%)	42 (79%)	32 (74%)	123 (89%)
Distance to forest/ kilometers (mean ± std. dev.)	0.57 ± 1.03	0.12 ± 0.36	2.51 ± 1.82	3.56 ± 1.85

Note: In order to balance the unequal sample size for all case studies, weighted data were used in all analyses.

Std. dev represents the standard deviation.

fruit trees (durian, rambutan, and langsat), and other crops (pineapple, banana, and vegetables). Only 37% of the households from Deramakot Forest Reserve stated that they had agroforestry products. The households were asked about the distance from the house to the forest to determine the distance that needed to be traveled to collect forest products. On average, the households travelled between 0.12 and 1.85 kilometers to access the forest area to collect forest products. Studies done by Engida and Mengistu (2013) and Musyoki et al. (2013) revealed that the distance to the forest can influence households' involvement in forest management activity.

5.4.3 SOCIAL CAPITAL ATTRIBUTES

The social capital with each key indicator is summarized in Table 5.4. The households from Mangkuwagu Forest Reserve (97%), Tamparuli Forest Reserve (94%), Deramakot Forest Reserve (78%), and Gana-Lingkabau Forest Reserve (58%) stated that they were involved in forest management through various activities initiated by the forestry department. The involvement in forest management includes participation in discussion, tropical tree planting, meetings, and other activities related to forest management. Moreover, the number of households involved in tropical tree planting was higher than those who were not involved. The majority of the households in tropical tree planting reported that they were involved in tree planting activity – Mangkuwagu Forest Reserve (93%), Tamparuli Forest Reserve (79%), Gana-Lingkabau Forest Reserve (59%), and Deramakot Forest Reserve (53%). The type of involvements can vary from voluntary, individual, and group selection to paid or non-paid work. Tree planting activity can assist in reforestation, enhance the

TABLE 5.4
Social Capital Attributes for Four Case Studies

Social Capital Attributes	Mangkuwagu FR (n = 174)	Tamparuli FR (n = 53)	Deramakot FR (n = 59)	Gana-Lingkabau FR (n = 139)
Involved in forest management (% of households)	169 (97%)	50 (94%)	46 (78%)	82 (59%)
Involved in tropical tree planting (% of households)	162 (93%)	42 (79%)	31 (53%)	82 (59%)
Owned membership (% of households)	93 (53%)	16 (30%)	57 (41%)	33 (56%)
Involved in decision-making (% of households)	39 (22%)	8 (15%)	16 (27%)	27 (19%)

Note: In order to balance the unequal sample size for all case studies, weighted data were used in all analyses.

Involvement in forest management refers to the households' involvement in all forest-related activities, including talks, meetings, training, decision-making, etc.

biodiversity of tree plants, sustain the ecosystem (Nolan et al., 2018), and promote community land tenure and property rights (Coulibaly-Lingani et al., 2009; Legesse et al., 2018).

Furthermore, the households were asked about their membership status in forest management. The results demonstrated that 56% of Gana-Lingkabau Forest Reserve villagers and 53% of Mangkuwagu Forest Reserve villagers engaged in forest management. The membership was slightly lower for Tamparuli Forest Reserve (30%) villagers and Deramakot Forest Reserve (41%) villagers. This is not to say that the extent of villagers' engagement in forest management depended on the type of membership itself. For example, participation in forest management in Tamparuli Forest Reserve was voluntary. In contrast, villagers who participated in forest management in the Mangkuwagu Forest Reserve and Gana-Lingkabau Forest Reserve were selected to execute the program. Meanwhile, in Deramakot Forest Reserve, the households were involved through casual tasks and on a group contract basis.

Membership status could influence the households' involvement in forest management activity. Following this, disadvantaged groups (those with no membership status) should be given equal opportunities to get involved in forest management. This will increase forest management efficiency (Anderson et al., 2015; Yadav et al., 2015). Furthermore, only 19%–27% of the households within the forest reserves were involved in the tropical tree planting decision-making process. Previous studies agreed with this, showing that in Sabah, the community did not participate actively in decision-making processes (Paimin et al., 2014; Yahya et al., 2015). Creating opportunities and increasing awareness on decision-making processes may improve a community's involvement in forest management (Chirenje et al., 2013). Batoa (2018) supported this, saying that social capital may influence the empowerment of a social community.

5.4.4   Human Capital Attributes

Moving on, labor availability has a significant impact on households' involvement in forest management activity (Gatiso, 2017). In this study, the households were asked about their labor availability, their education level, their background related to tree planting, their reception of information, and the level of information contributed by them (Table 5.5). From the table, it can be seen that labor availability was the highest in Mangkuwagu Forest Reserve (96%), followed by Tamparuli Forest Reserve (93%) and Gana-Lingkabau Forest Reserve (63%). The percentage of labor availability was the lowest in Deramakot Forest Reserve (34%). From the results, it was deduced that the low percentage of labor availability in Deramakot Forest Reserve was due to the fact that most households work for oil palm plantations and other private companies. Oil palm plantations create positive employment and increase household income (Jelsma et al., 2017; Obidzinski et al., 2012) by providing attractive salaries to support household's monthly expenses and needs.

Moreover, only a few households reported that they had attended training related to tropical tree planting. For instance, only 37% of the households from Deramakot Forest Reserve, 16% from Mangkuwagu Forest Reserve, and 9% from Tamparuli Forest Reserve attended any training. On the other hand, 42% of the

TABLE 5.5
Assessment of Human Capital Attributes for Four Case Studies

Human Capital Attributes	Mangkuwagu FR (n = 174)	Tamparuli FR (n = 53)	Deramakot FR (n = 59)	Gana-Lingkabau FR (n = 139)
Have labor availability (% of households)	167 (96%)	49 (93%)	20 (34%)	87 (63%)
Attending training (% of households)	28 (16%)	5 (9%)	22 (37%)	59 (42%)
Receive information (% of households)	146 (84%)	43 (81%)	20 (33%)	136 (98%)
Level of information contribution (median)	1 (55%)	1 (77%)	4 (48%)	1 (73%)

Notes: In order to balance the unequal sample size for all case studies, weighted data were used in all analyses.

The median point is based on a five-point Likert scale from 1(low) to 5(high) indicating the level of information contribution related to tropical tree planting.

The information contribution is based on the households' perception and included the contribution of the households to decision-making, consultation, training, meetings, etc.

households in Gana-Lingkabau Forest Reserve had participated in training. The figures demonstrated that there was a lack of involvement in training, and most of the households failed to see the benefit of the training to them. In addition, these households would also prefer activities related to agricultural crops than tree planting (Dinh et al., 2017).

When households were asked about the status of the information that they received, the majority of the households mentioned that they received information related to tropical tree planting from the forestry department officer, the head of the village, the village representative, or other villagers. It is worth noting that the result from Deramakot Forest Reserve was not consistent with the other three case studies. Only 33% of the households from Deramakot Forest Reserve stated that they had received information related to tropical tree planting. Drawing from this, there needs to be an improvement in information relay in the future. Households' knowledge about forest resource management and their involvement in training are key to sustainable forest management and the success of community forestry (Hermudananto et al., 2018; Jafari et al., 2018).

Moreover, the level of information contributed was significantly low for each case study, except for Deramakot Forest Reserve. On average, the median for information contribution ranged from 1 to 4. The five-point Likert scale ranging from 1 (low) to 5 (high) was used to determine the local community perception. As low as it was, the data were not surprising. As demonstrated before, most households did not actively participate in decision-making processes and even less in contributing information. This, combined with a lack of information, led to a low level of contribution by the local community (Sapkota et al., 20118). However, the households in Deramakot Forest Reserve stated that they had actively contributed information.

5.4.5   PHYSICAL CAPITAL ATTRIBUTES

Physical capital is the most important asset in forest management strategy (Dehghani Pour and Milad, 2018). The households were asked to determine their physical capital attributes in terms of facilities, distance from farm to house, level of physical contribution, and housing status (Table 5.6). The table shows that 89% of the households from Gana-Lingkabau Forest Reserve stated that they received facilities provided by the Sabah Forestry Department, followed by 68% of Deramakot Forest Reserve households and 53% of Tamparuli Forest Reserve households. The facilities include houses, road access, schools, clinics, and water supply. Nonetheless, only 46% of the households in Mangkuwagu Forest Reserve stated that they had received facilities. In general, the distribution of facilities to households within forest reserves is the responsibility of the forestry department. The regulations state that the establishment of settlements is restricted and they are not allowed to be built inside a forest reserve area. Despite these regulations, the Sabah Forestry Department tried their best to provide the communities with basic facilities and infrastructures, such as road access in Gana-Lingkabau Forest Reserve (Figure 5.5).

The survey also asked households about the distance between their farm and the community settlement. Based on the households' response, the average distance ranged between 0.21 and 2.31 kilometers. According to studies by Poudel et al. (2015) and Rasolofoson et al. (2015), distance can significantly influence the ability of households to access the forest or farm areas.

Moving on, households' contribution of raw materials such as fertilizers, pesticides, seedlings, and tools varied among the four case studies. The local community used a five-point Likert scale ranging from 1 (low) to 5 (high) based on

---

TABLE 5.6

Comparison of Physical Capital Attributes between Four Case Studies

Physical Capital Attributes	Mangkuwagu FR (n = 174)	Tamparuli FR (n = 53)	Deramakot FR (n = 59)	Gana-Lingkabau FR (n = 139)
Received facilities (% of households)	80 (46%)	28 (53%)	40 (68%)	124 (89%)
Distance from farm to house/ kilometers (mean ± std.dev.)	0.54 ± 1.05	0.21 ± 0.26	2.06 ± 1.05	2.31 ± 1.03
Level of contribution on raw materials (median)	2 (48%)	2 (47%)	4 (36%)	1 (83%)
Owned housing facility (% of households)	12 (7%)	2 (4%)	125 (90%)	58 (98%)

Notes: In order to balance the unequal sample size for all case studies, weighted data were used in all analyses.

The contribution on raw materials refers to contributions in fertilizers, pesticides, seedlings, and tools.

The median point is based on a five-point Likert scale from 1(low) to 5(high) indicating the level of contribution related to tropical tree planting.

Std. dev represents the standard deviation.

FIGURE 5.5 Resettlement housing view from Gana-Lingkabau Forest Reserve.

their perception of their contribution of raw materials. The households in Deramakot Forest Reserve stated that they contributed a significant amount of raw materials for tropical tree planting. Meanwhile, the households in Mangkuwagu and Tamparuli Forest Reserves showed somewhat low contributions of raw materials. The households in Gana-Lingkabau Forest Reserve demonstrated an even lower contribution of raw materials. One reason for this situation is that most of the raw materials are provided by the forestry department as part of their subsidies and incentives. In addition, most households assumed that the raw materials that they contributed cost less, as many of them cultivate seedlings themselves. In many cases, households produced seedlings from fruits given to them by their neighbors or family (Cemansky, 2015; Coomes et al., 2015).

5.4.6 FINANCIAL OR ECONOMIC CAPITAL ATTRIBUTES

Financial or economic capital is a significant mechanism that can help to generate funding for forest management activity (Dev et al., 2003; Feurer et al., 2018). Four key items were used in assessing financial or economic capital attributes (Table 5.7). First, the households' paying jobs were determined, in which they needed to explain if the households had been involved in tropical tree planting. Approximately 58% of the households from Deramakot Forest Reserve mentioned that they had a sufficient salaried job working for oil palm plantations and private companies. On the other hand, other households from the other three case studies reported few salaried jobs – Tamparuli Forest Reserve (28%), Mangkuwagu Forest Reserve (16%), and Gana-Lingkabau Forest Reserve (11%). It should be noted that a salaried job was

TABLE 5.7

Assessment of Financial or Economic Capital Attributes from Four Case Studies

Financial or Economic Capital Attributes	Mangkuwagu FR (n = 174)	Tamparuli FR (n = 53)	Deramakot FR (n = 59)	Gana-Lingkabau FR (n = 139)
Salaried job (% of households)	27 (16%)	15 (28%)	34 (58%)	15 (11%)
Selling forest products and agroforestry (% of households)	27 (16%)	49 (93)	55 (40%)	3 (5%)
Annual income from selling forest products and agroforestry (mean per capita)	RM4,914	RM12,861	RM1,476	RM333
Annual income from wages related to tropical tree planting (mean per capita)	RM1,581	RM2,993	RM374	RM28

Notes: In order to balance the unequal sample size for all case studies, weighted data were used in all analyses.

Income represents the raw value of income and excluded the total cost.

The exchange rate for RM1 is equivalent to USD0.24.

The wages income from tropical tree planting included site-preparation, seedling planting, and maintenance (e.g. fertilizing, weeding) activities.

defined as an occupation paid either by the government or private companies or on a casual-labor basis. Labor income was observed to significantly benefit the households in supporting their income and needs (Humphries et al., 2012).

Furthermore, selling forest products and agroforestry can create job opportunities and support the community's income (Endamana et al., 2016). Interestingly, the sales of forest products and agroforestry activity were higher among the households from Tamparuli Forest Reserve (93%). Most of the households sold agroforestry products such as latex sheets, rubber-tree bowls, and tropical fruit trees such as *durian*, *rambutan*, *langsat*, and *chempedak*. Apart from that, other forest products sold by the households included rattan, bamboo culms, rattan shoots, wild vegetables, and edible mushrooms. Other households from the other three case studies did not participate much in this activity; only 40% of households in Deramakot Forest Reserve participate in sale activities, followed by Mangkuwagu Forest Reserve (16%) and Gana-Lingkabau Forest Reserve (5%).

The results of this study demonstrated that forest-related income contributed less to the households' income compared to agricultural income, but that farming showed promising potential to increase households' income (Bakkegaard et al., 2017). In Tamparuli Forest Reserve, the mean per-capita annual income from forests and agroforestry products can be up to RM12,860 per year. Most households in Tamparuli sold rubber-tree products, including rubber sheets and latex bowls, as well as other tropical fruits. Moreover, the households from Mangkuwagu Forest Reserve can receive up to RM4,914 mean per-capita annual income from selling forest products, followed by Gana-Lingkabau Forest Reserve (RM1,476). Forest and agroforestry income made up 55% of the mean annual total income for households in Gana-Lingkabau Forest Reserve and 54% of that in Tamparuli Forest Reserve. Meanwhile, the households from Deramakot Forest Reserve received at least RM333 mean per-capita annual income from forest and agroforestry products. These numbers correspond with the findings of Oli et al. (2016), who observed that the income from forests can contribute up to 17% of household income.

Finally, the wages produced from forest-related employment create various benefits for the local community, including access to forest resources (Sapkota et al., 2018) and employment (Pokharel et al., 2015; Teitelbaum, 2014). The households of Tamparuli Forest Reserve reported that they can earn RM2,993 mean per-capita annual income from their wages related to tropical tree planting activity. Similarly, the households of Mangkuwagu Forest Reserve stated that they can earn up to RM1,581 mean per-capita annual income from their wages from tropical tree planting. On the other hand, the wages produced from tropical tree planting activity were slightly lower in Deramakot Forest Reserve (RM374) and Gana-Lingkabau Forest Reserve (RM28). The wages earned from tropical tree planting activity can be earned individually or through group contract. Most of this work involves site preparation, planting, and maintenance (e.g. fertilizing and weeding). The involvement of the community in forest management activity such as protection, weeding, and planting is considered a community right and responsibility (Afroz et al., 2016; Anderson et al., 2015).

5.4.7 BENEFITS OF TROPICAL TREE PLANTING ACTIVITY

The findings of this research regarding the benefits of tropical tree planting activity were presented in a spider diagram, and the mean ranks from five capital assets (SLA) among four case studies were compared. In Figure 5.6, the five capital assets among the households from the four case studies were compared. The level of benefit was determined according to a five-point rating ranging from 1 (low) to 5 (high) based on the household's perception. From the results, the highest mean rank for social capital was 4.12 (somewhat high) for Tamparuli Forest Reserve, followed by the Mangkuwagu Forest Reserve (mean rank = 3.73), the Deramakot Forest Reserve (mean rank = 3.45), and the Gana-Lingkabau Forest Reserve (mean rank = 3.15) with moderate level. Furthermore, in terms of financial or economic capital, moderate benefit was accorded to the households from Mangkuwagu Forest Reserve (mean rank = 3.31), followed by Deramakot Forest Reserve (mean rank = 3.21), Tamparuli Forest Reserve (mean rank = 3.10), and Gana-Lingkabau Forest Reserve (mean rank = 3.03). These figures demonstrated that social capital provided more positive benefits to the community compared to other capital assets. This pattern can be influenced by the increment of households' involvement in forest management and tropical tree planting activities organized by the forestry department. Pour et al. (2018) contradicted this finding when they demonstrated that financial assets are

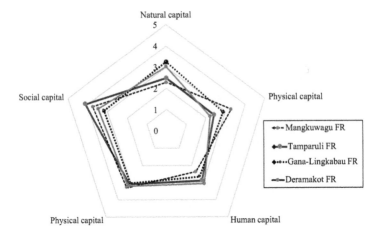

FIGURE 5.6 Level of benefit from tropical tree planting activity based on sustainable livelihood approach (SLA).

most preferred in improving community livelihood compared to other capital. In contrast, Harbi et al. (2018) reported that the social capital provided the highest contribution to the local community livelihood and forest conservation. Moreover, Sapkota et al. (2018) argued that the social capital is critical to forest management. Equal opportunities, accessibility, and the ability of the households need to be considered to motivate participation in tropical tree planting activity.

Furthermore, Kendall's W coefficient of concordance was used to determine the agreement pattern between the households from the four case studies on the level of benefits in terms of SLA that they could receive from tropical tree planting activity (Table 5.8). The assessment was based on the households' agreement in terms of the five capital assets mentioned in the previous figure. The Kendall's W can be

TABLE 5.8
Results of the Households' Agreement Pattern Using Kendall's W test

Case Studies	Kendall's W	X^2	p
Mangkuwagu Forest Reserve	0.228	158.368	0.0005
Tamparuli Forest Reserve	0.263	111.708	0.0005
Gana-Lingkabau Forest Reserve	0.034	55.921	0.001
Deramakot Forest Reserve	0.161	152.076	0.005

Note: In order to balance the unequal sample size for all case studies, weighted data were used in all analyses.
The Kendall's W is 0, which determines that there is no overall trend of agreement among the respondents.
The significant level is $p<0.05$.

used to measure the agreement or judgement pattern of people in different subjects or situations (Gearhart et al., 2013). The finding stated that most households from the Mangkuwagu Forest Reserve agreed with their statement. Nonetheless, the level of agreement was significantly very low; $W = 0.228$, $p < 0.0005$. Moreover, 26% of the agreement from the households of Tamparuli Forest Reserve had similar perceptions with the households in Mangkuwagu Forest Reserve, where most of the households stated somewhat low to somewhat high agreement on the benefit of tropical tree planting; $W = 0.263$, $p < 0.0005$. Similarly, the findings from the Deramakot Forest Reserve ($W = 0.161$, $p < 0.0005$) and Gana-Lingkabau Forest Reserve ($W = 0.034$, $p < 0.0005$) showed that there was no overall trend of agreement among the households. The pattern of agreement demonstrated that the benefits that the households received from tropical tree planting was significant, but the assessment was very low. Therefore, the benefits of tropical tree planting are still up for debate. Poudel et al. (2014) reveal that forest management can offer benefits to the forest-dependent community. Moreover, Watkins et al. (2018) conceded that most of the households had different perceptions of the benefits of tree planting to their social livelihood.

Moving on, Table 5.9 presents the list of tropical tree species planted by the local community in the four case studies. This is considered to be reforestation activity, where most of the tropical forest trees were planted on the degraded forest land. Fruit tree species including *Artocarpus integar (Cempedak)*, *Artocarpus odoratissimus (Tarap)*, *Durio spp. (Durian)*, *Mangifera spp.* (wild mango), and *Nephelium lappaceum (Rambutan)* were commonly planted by the households in the four case studies. Note that the *Dipterocapus spp.* (Dipterocarpus) was the most preferred commercial native tropical forest tree species planted. Other native tropical fast-growing tree species such as *Neolamarckia cadamba* (Laran) and *Octomeles sumatrana* (Binuang) were among the most common tree species to be planted in forest reserve areas. Encouraging the local community to integrate tropical food crops, native forest trees, and fruit tree planting in the forest area can help to sustain the ecological conservation and improve income and livelihood in Malaysia (Yacob et al., 2012). Lampela et al. (2017) mention that most of the tree species that are selected for reforestation are chosen based on scientific research and that the local community preferred multipurpose tree species. Figure 5.7 shows the signboard of a forest restoration project at Deramakot Forest Reserve, where the local community were involved in tropical tree planting work. Furthermore, the donations made by private companies to reforestation programs in Deramakot Forest Reserve have increased the employment rate among the local community and enhanced the forest composition (Sabah Forestry Department, 2008).

5.5 CONCLUSIONS AND RECOMMENDATIONS

This study explored the benefits of tropical tree planting in the context of community forestry management and community livelihood in Sabah. From the empirical findings, the extent of the community's involvement in tropical tree planting from four case studies of community forestry areas in Sabah were determined. Households' involvement in tropical tree planting was determined through observations of

TABLE 5.9

Sample List of Tropical Tree Species Planted by the Households in Four Case Studies

Tropical Tree Species and Local Name	Mangkuwagu FR	Tamparuli FR	Deramakot FR	Gana-Lingkabau FR
		Case Study Area		
Artocarpus integer (Chempedak)	x	x	x	x
Artocarpus odoratissimus (Tarap)	x	x	x	x
Aquilaria maleccensis (Gaharu)		x		
Baccaurea macrocarpa (Tampoi)				x
Calophyllum spp. (Bintangor)			x	
Dipterocarpus spp. (Dipterocarp)	x	x	x	x
Durio spp. (Durian)	x	x	x	x
Eugneia spp. (Obah)				x
Hevea brasiliensis (Rubber)	x	x		x
Mangifera spp. (Wild manggo)	x	x	x	x
Nauclea subdita (Pulai Bangkal Kuning)			x	
Neolamarckia cadamba (Laran)		x	x	x
Nephelium lappaceum (Rambutan)	x	x	x	x
Octomeles sumatrana (Binuang)		x	x	
Pangium edule (Kepayang)				x
Peronema canescens (Sungkai)		x		
Sandoricum koetjape (Sentul)			x	
Scaphium macropodum (Kembang Semangkok)	x			x

Note: The tropical tree species refer to the local names of plants and could be different in other locations.

socio-demographic status based on SLA. Based on 425 household responses, this study showed that the most promising benefit of tropical tree planting based on SLA was social capital. Two indicators from the social capital element were reported to have been given more attention by the households: their involvement in forest management and tropical tree planting activity. The results showed that the level of involvement in tree planting was high among the households from the four case studies. In addition, women's involvement in tropical tree planting was low compared to male households, but the impact of their involvement was still relevant. Moreover, physical assets were also mentioned by the households, with most of them stating that they received benefits from the facilities provided by the forestry department. Through the findings, the positive effects of community involvement in tropical tree

FIGURE 5.7 Forest restoration project at Deramakot Forest Reserve that involved local community.

planting activity can be promoted by strengthening community livelihood assets such as social, financial, human, natural, and physical capital. On top of that, promoting local community involvement in tropical tree planting should be enhanced through various strategies including forestry extension, support, training, and other relevant activities.

ACKNOWLEDGEMENTS

The author would like to acknowledge the contributions of colleagues and the editor (Dr. Kang Chiang Liew) for their edits and feedbacks. The contributions from the Sabah Forestry Department and the local villagers are highly appreciated.

REFERENCES

Afroz, Sharmin, Rob Cramb, and Clemens Grünbühel. 2016. Ideals and Institutions: Systemic Reasons for the Failure of a Social Forestry Program in South-West Bangladesh. *Geoforum* 77:161–173. doi:10.1016/j.geoforum.2016.11.001.
Anderson, Jon, Shreya Mehta, Edna Epelu, and Brian Cohen. 2015. Managing Leftovers: Does Community Forestry Increase Secure and Equitable Access to Valuable Resources for the Rural Poor? *Forest Policy and Economics* 58:47–55. doi:10.1016/j. forpol.2014.12.004.
Angelsen, Arild, Pamela Jagger, Ronnie Babigumira, Brian Belcher, Nicholas J. Hogarth, Simone Bauch, Jan Börner, Carsten Smith-Hall, and Sven Wunder. 2014. Environmental Income and Rural Livelihoods: A Global-Comparative Analysis. *World Development* 64:S12–S28. doi:10.1016/j.worlddev.2014.03.006.

Ayine, Dominic, M. 2008. *Social Responsibility Agreements in Ghana's Forestry Sector.* Developing Legal Tools for Citizen Empowerment Series. IIED, London.

Azima, A.M., Novel Lyndon, and Mohd Shafiq Akmal. 2015. Understanding of the Meaning of Native Customary Land (NCL) Boundaries and Ownership by the Bidayuh Community in Sarawak, Malaysia. *Mediterranean Journal of Social Sciences* 6(5): 342–348. doi:10.5901/mjss.2015.v6n5s1p342.

Bakkegaard, Riyong Kim, Nicholas J. Hogarth, Indah Waty Bong, Aske S. Bosselmann, and Sven Wunder. 2017. Measuring Forest and Wild Product Contributions to Household Welfare: Testing a Scalable Household Survey Instrument in Indonesia. *Forest Policy and Economics* 84:20–28. doi:10.1016/j.forpol.2016.10.005.

Barnes, Clare, Rachel Claus, Peter Driessen, Maria Joao Ferreira Dos Santos, Mary Ann George, and Frank Van Laerhoven. 2017. Uniting Forest and Livelihood Outcomes? Analyzing External Actor Interventions in Sustainable Livelihoods in a Community Forest Management Context. *International Journal of the Commons* 11(1):532–571. doi:10.18352/ijc.750.

Batoa, Hartina, La Rianda Weka Widayati, Dasmin Sidu, Nur Rahmah, and Outu Arimbawa. 2018. The Effect of Human Capital, Social Capital, and Competency on the Empowerment of Bajo Ethnic Communtiy in the Regency of Muna: A Gender Perspective. *Agriculture, Forestry and Fisheries* 7(1):6–10. doi:10.11648/j.aff.20180701.12.

Baynes, Jack, John Herbohn, Carl Smith, Robert Fisher, and David Bray. 2015. Key Factors Which Influence the Success of Community Forestry in Developing Countries. *Global Environmental Change* 35:226–236. doi:10.1016/j.gloenvcha.2015.09.011.

Bernama. 2012. State's Customary Land Right Sabah State Government's Priority. https://www.theborneopost.com/2012/10/22/states-customary-land-right-sabah-state-governments-priority-new.html. (Accessed October 22, 2018).

Cemansky, Rachel. 2015. Africa's Indigenous Fruit Trees: A Blessing in Decline. *Environmental Health Perspectives* 123(12):A291–A296. doi:10.1289/ehp.123-A291.

Chinangwa, Linda, Andrew S. Pullin, and Neal Hockley. 2016. Livelihoods and Welfare Impacts of Forest Comanagement. *International Journal of Forestry Research* 2016:1–12. doi:10.1155/2016/5847068.

Chirenje, Leonard I., Richard A. Giliba, and Emmanuel B. Musamba. 2013. Local Communities' Participation in Decision-Making Processes through Planning and Budgeting in African Countries. *Chinese Journal of Population, Resources and Environment* 11(1):10–16. doi:10.1080/10042857.2013.777198.

Coleman, Eric A., and Esther Mwangi. 2013. Women's Participation in Forest Management: A Cross-Country Analysis. *Global Environmental Change* 23(1):193–205. doi:10.1016/j.gloenvcha.2012.10.005.

Colombo, Emanuela, Francesco Romeo, Lorenzo Mattarolo, Jacopo Barbieri, and Mariano Morazzo. 2018. An Impact Evaluation Framework Based on Sustainable Livelihoods for Energy Development Projects: An Application to Ethiopia. *Energy Research and Social Science* 39:78–92. doi:10.1016/j.erss.2017.10.048.

Coomes, Oliver T., Shawn J. McGuire, Eric Garine, Sophie Caillon, Doyle McKey, Elise Demeulenaere, Devra Jarvis, et al. 2015. Farmer Seed Networks Make a Limited Contribution to Agriculture? Four Common Misconceptions. *Food Policy* 56:41–50. doi:10.1016/j.foodpol.2015.07.008.

Coulibaly-Lingani, Pascaline, Mulualem Tigabu, Patrice Savadogo, Per Christer Oden, and Jean Marie Ouadba. 2009. Determinants of Access to Forest Products in Southern Burkina Faso. *Forest Policy and Economics* 11(7):516–524. doi:10.1016/j.forpol.2009.06.002.

Dev, Om Prakash, Nagendra Prasad Yadav, Oliver Springate-Baginski, and John Soussan. 2003. Impacts of Community Forestry on Livelihoods in the Middle Hills of Nepal. *Journal of Forest and Livelihood* 3(1):64–77. doi:10.1002/ldr.438.

Dhruba Bijaya, G.C., S. Cheng, Z. Xu, J. Bhandari, L. Wang, and X. Liu. 2016. Community Forestry and Livelihood in Nepal: A Review. *Journal of Animal and Plant Sciences* 26(1):1–12. doi:10.1029/2005GB002620.

Dinh, Hoang Huu, Trung Thanh Nguyen, Hoang Viet Ngu, and Clevo Wilson. 2017. Economic Incentive and Factors Affecting Tree Planting of Rural Households: Evidence from the Central Highlands of Vietnam. *Journal of Forest Economics* 29(Part A):14–24. doi:10.1016/j.jfe.2017.08.001.

Edwards, David P., Paul Woodcock, Felicity A. Edwards, Trond H. Larsen, Wayne W. Hsu, Suzan Benedick, and David S. Wilcove. 2012. Reduced-Impact Logging and Biodiversity Conservation: A Case Study from Borneo. *Ecological Applications* 22(2):561–571. doi:10.1890/11-1362.1.

Emanuel, Maria, and Tumpe Ndimbwa. 2013. Traditional Mechanisms of Resolving Conflicts Over Land Resource: A Case of Gorowa Community in Northern Tanzania. *International Journal of Academic Research in Business and Social Sciences* 3(11):214–224. doi:10.6007/IJARBSS/v3-i11/334.

Endamana, D., K.A. Angu, G.N. Akwah, G. Shepherd, and B.C. Ntumwel. 2016. Contribution of Non-Timber Forest Products to Cash and Non-Cash Income of Remote Forest Communities in Central Africa. *International Forestry Review* 18(3):280–295. doi: 10.1505/146554816819501682.

Engida, Tadesse, and Abay Tafere Mengistu. 2013. Explaining the Determinant of Community Based Forest Management: Evidence from Alamata, Ethiopia. *International Journal of Community Development* 1(2):63–70. doi:11634/233028791301431.

Evans, Kristen, Selmira Flores, Anne M. Larson, Roberto Marchena, Pilar Müller, and Alejandro Pikitle. 2017. Challenges for Women's Participation in Communal Forests: Experience from Nicaragua's Indigenous Territories. *Women's Studies International Forum* 65:37–46. doi:10.1016/j.wsif.2016.08.004.

FAO. 2004. *Selecting Tree Species on the Basis of Community Needs*. Community Forestry Field Manual 5. Food and Agriculture Organization of the United Nations, Rome, Italy.

FAO. 2012. *FRA 2015 Terms and Definitions*. Forest Resources Assessment Working Paper 180. Food and Agriculture Organization of the United Nations, Rome, Italy.

Feurer, Melanie, David Gritten, and Maung Than. 2018. Community Forestry For Livelihoods: Benefiting from Myanmar's Mangroves. *Forests* 9(150):2–15. doi:10.3390/f9030150.

Gatiso, Tsegaye T. 2017. Households' Dependence on Community Forest and Their Contribution to Participatory Forest Management: Evidence from Rural Ethiopia. *Environment, Development and Sustainability* 1:17. doi:10.1007/s10668-017-0029-3.

Gautam, Madhur, and Mansur Ahmed. 2018. Too Small to Be Beautiful? The Farm Size and Productivity Relationship in Bangladesh. *Food Policy* 2018:2–11. doi:10.1016/j.foodpol.2018.03.013.

Gearhart Amanda, D. Terrance Booth, Kevin Sedivec, and Christopher Schauer. 2013. Use of Kendall's Coefficient of Concordance to Assess Agreement among Observers of Very High Resolution Imagery. *Geocarto International* 28(6):517–526. doi:10.1080/101060 49.2012.725775.

Gebreegziabher, Zenebe, and G. Cornelis van Kooten. 2013. Does Community and Household Tree Planting Imply Increased Use of Wood for Fuel? Evidence from Ethiopia. *Forest Policy and Economics* 34:30–40. doi:10.1016/j.forpol.2013.03.003.

Hajjar, Reem, Johan A. Oldekop, Peter Cronkleton, Emily Etue, Peter Newton, Aaron J.M. Russel, Januarti Sinarra Tjajadi, Wen Zhou, and Arun Agrawal. 2016. The Data Not Collected on Community Forestry. *Conservation Biology* 30(6):1357–1362. doi:10.1111/cobi.12732.

Harbi, Jun, James Thomas Erbaugh, Mohammad Sidiq, Berthold Haasler, and Dodik Ridho Nurrochmat. 2018. Making a Bridge between Livelihoods and Forest Conservation: Lessons from Non Timber Forest Products' Utilization in South Sumatera, Indonesia. *Forest Policy and Economics* 94:1–10. doi:10.1016/j.forpol.2018.05.011.

He, Jun, Myong Hyok Ho, and Jianchu Xu. 2015. Participatory Selection of Tree Species for Agroforestry on Sloping Land in North Korea. *Mountain Research and Development* 35(4):318–327. doi:10.1659/MRD-JOURNAL-D-15-00046.1.

Hermudananto, Claudia Romero, Ruslandi, and Francis E. Putz. 2018. Analysis of Corrective Action Requests from Forest Stewardship Council Audits of Natural Forest Management in Indonesia. *Forest Policy and Economics* 96:28–37. doi:10.1016/j.forpol.2018.07.012.

Humphries, Shoana, Thomas P. Holmes, Karen Kainer, Carlos Gabriel Gonçalves Koury, Edson Cruz, and Rosana de Miranda Rocha. 2012. Are Community-Based Forest Enterprises in the Tropics Financially Viable? Case Studies from the Brazilian Amazon. *Ecological Economics* 77:62–73. doi:10.1016/j.ecolecon.2011.10.018.

Jafari, Ali, Hamdollah Sadeghi Kaji, Hossein Azadi, Kindeya Gebrehiwot, Fateme Aghamir, and Steven Van Passel. 2018. Assessing the Sustainability of Community Forest Management: A Case Study from Iran. *Forest Policy and Economics* 96:1–8. doi:10.1016/j.forpol.2018.08.001.

Jelsma, Idsert, G.C. Schoneveld, A. Zoomers, and A.C.M. van Westen. 2017. Unpacking Indonesia's Independent Oil Palm Smallholders: An Actor-Disaggregated Approach to Identifying Environmental and Social Performance Challenges. *Land Use Policy* 69:281–297. doi:10.1016/j.landusepol.2017.08.012.

Kaskoyo, Hari, Abrar Juhar Mohammed, and Makoto Inoue. 2017. Impact of Community Forest Program in Protection Forest on Livelihood Outcomes: A Case Study of Lampung Province, Indonesia. *Journal of Sustainable Forestry* 36(3):250–263. doi:1 0.1080/10549811.2017.1296774.

Kibria, Abu S.M.G., Robert Costanza, Colin Groves, and Alison M. Behie. 2018. The Interactions between Livelihood Capitals and Access of Local Communities to the Forest Provisioning Services of the Sundarbans Mangrove Forest, Bangladesh. *Ecosystem Services* 32:41–49. doi:10.1016/j.ecoser.2018.05.003.

Knutsson, Per. 2006. The Sustainable Livelihoods Approach: A Framework for Knowledge Integration Assessment. *Human Ecology Review* 13(1):90–99. doi:10.1080/13657300903156092.

Kulindwa, Yusuph J., and Erik O. Ahlgren. 2018. Evaluation of the Impact of Fuelwood Tree Planting Programmes in Tanzania. *Energy Procedia* 147:154–161. doi:10.1016/j.egypro.2018.07.047.

Lampela, Maija, Jyrki Jauhiainen, Sakari Sarkkola, and Harri Vasander. 2017. Promising Native Tree Species for Reforestation of Degraded Tropical Peatlands. *Forest Ecology and Management* 394:52–63. doi:10.1016/j.foreco.2016.12.004.

Lee, Yohan, Indri Puji Rianti, and Mi Sun Park. 2017. Measuring Social Capital in Indonesian Community Forest Management. *Forestry Science and Technology* 13(3):133–141. doi: 10.1080/21580103.2017.1355335.

Legesse, Befikadu A., Kenrett Jefferson-Moore, and Terrence Thomas. 2018. Impacts of Land Tenure and Property Rights on Reforestation Intervention in Ethiopia. *Land Use Policy* 70:494–499. doi:10.1016/j.landusepol.2017.11.018.

Lunkapis, Gaim James. 2015. Secure Land Tenure as Prerequisite Towards Sustainable Living: A Case Study of Native Communities in Mantob Village, Sabah, Malaysia. *SpringerPlus* 4:549. doi:10.1186/s40064-015-1329-4.

Malkamäki, Arttu, Dalia D'Amato, Nicholas J. Hogarth, Markku Kanninen, Romain Pirard, Anne Toppinen, and Wen Zhou. 2018. A Systematic Review of the Socio-Economic Impacts of Large-Scale Tree Plantations, Worldwide. *Global Environmental Change* 53:90–103. doi:10.1016/j.gloenvcha.2018.09.001.

Maryudi, Ahmad, Rosan R. Devkota, Carsten Schusser, Cornelius Yufanyi, Manjola Salla, Helene Aurenhammer, Ratchananth Rotchanaphatharawit, and Max Krott. 2012. Back to Basics: Considerations in Evaluating the Outcomes of Community Forestry. *Forest Policy and Economics* 14(1):1–5. doi:10.1016/j.forpol.2011.07.017.

Massoud, May A., Sarah Issa, Mutasem El Fadel, and Ibrahim Jamali. 2016. Sustainable Livelihood Approach Towards Enhanced Management of Rural Resources. *International Journal of Sustainable Society.* doi:10.1504/IJSSOC.2016.074947.

Moktan, Mani Ram, Lungten Norbu, and Kunzang Choden. 2016. Can Community Forestry Contribute to Household Income and Sustainable Forestry Practices in Rural Area? A Case Study from Tshapey and Zariphensum in Bhutan. *Forest Policy and Economics* 62:149–157. doi:10.1016/j.forpol.2015.08.011.

Musyoki, Josephine Kamene, Jayne Mugwe, Kennedy Mutundu, and Mbae Muchiri. 2013. Determinants of Household Decision to Join Community Forest Associations: A Case Study of Kenya. *ISRN Forestry* 2013:1–10. doi:10.1155/2013/902325.

Mutune, Jane Mutheu, and Jens Friss Lund. 2016. Unpacking the Impacts of 'Participatory' Forestry Policies: Evidence from Kenya. *Forest Policy and Economics* 69:45–52. doi:10.1016/j.forpol.2016.03.004.

Narita, Daiju, Mulugeta Lemenih, Yukimi Shimoda, and Alemayehu N. Ayana. 2018. Economic Accounting of Ethiopian Forests: A Natural Capital Approach. *Forest Policy and Economics* 97:198–200. doi:10.1016/j.forpol.2018.10.002.

Negi, Swati, Thu Thuy Pham, Bhaskar Karky, and Claude Garcia. 2018. Role of Community and User Attributes in Collective Action: Case Study of Community-Based Forest Management in Nepal. *Forests* 9(3):1–20. doi:10.3390/f9030136.

Nolan, Rachael H., David M. Drew, Anthony P. O'Grady, Elizabeth A. Pinkard, Keryn Paul, Stephen H. Roxburgh, Patrick J. Mitchell, Jody Bruce, Michael Battaglia, and Daniel Ramp. 2018. Safeguarding Reforestation Efforts against Changes in Climate and Disturbance Regimes. *Forest Ecology and Management* 424:458–467. doi:10.1016/j.foreco.2018.05.025.

Obidzinski, Krystof, Rubeta Andriani, Heru Komarudin, and Agus Andrianto. 2012. Environmental and Social Impacts of Oil Palm Plantations and Their Implications for Biofuel Production in Indonesia. *Ecology and Society* 17(1):25. doi:10.5751/ES-04775-170125.

Obodai, Jacob, Lawrencia Pkuah Siaw, Foster Frempong, and James Boafo. 2014. Participatory Forestry Intervention: Assessing the Contribution of the Expended Plantation Program to Community Livelihood Sustainability and Poverty Reduction. *International Research Journal of Arts and Social Science* 3(4):104–108. doi:10.14303/irjass.2014.056.

Ojha, Hemant R., Rebecca Ford, Rodney J. Keenan, Digby Race, Dora CariasVega, Himlal Baral, and Prativa Sapkota. 2016. Delocalizing Communities: Changing Forms of Community Engagement in Natural Resources Governance. *World Development* 87:274–290. doi:10.1016/j.worlddev.2016.06.017.

Oli, Bishwa Nath, Thorsten Treue, and Carsten Smith-Hall. 2016. The Relative Importance of Community Forests, Government Forests, and Private Forests for Household-Level Incomes in the Middle Hills of Nepal. *Forest Policy and Economics* 70:155–163. doi:10.1016/j.forpol.2016.06.026.

Ostrom, Elinor, and Harini Nagendra. 2006. Insights on Linking Forests, Trees, and People from the Air, on the Ground, and in the Laboratory. *Proceedings of the National Academy of Sciences of the United States of America* 103(51):19224–19231. doi:10.1073/pnas.0607962103.

Paimin, N.F. Velnisa, S. Modilih, S.H. Mogindol, C. Johnny, and J.A. Thamburaj. 2014. Community Participation and Barriers in Rural Tourism: A Case Study in Kiulu, Sabah. *SHS Web of Conferences* 12:1–7. doi:10.1051/shsconf/20141201003.

Paudel, Jayash. 2018. Community-Managed Forests, Household Fuelwood Use and Food Consumption. *Ecological Economics* 147:62–73. doi:10.1016/j.ecolecon.2018.01.003.

Permadi, Dwiko B., Michael Burton, Ram Pandit, Digby Race, and Iain Walker. 2018. Local Community's Preferences for Accepting a Forestry Partnership Contract to Grow Pulpwood in Indonesia: A Choice Experiment Study. *Forest Policy and Economics* 91:73–83. doi:10.1016/j.forpol.2017.11.008.

Phompila, Chittana, Megan Lewis, Bertram Ostendorf, and Kenneth Clarke. 2017. Forest Cover Changes in Lao Tropical Forests: Physical and Socio-Economic Factors Are the Most Important Drivers. *Land* 6(23):3–14. doi:10.3390/land6020023.

Poudel, Narayan Raj, Nobuhiko Fuwa, and Keijiro Otsuka. 2015. The Impacts of a Community Forestry Program on Forest Conditions, Management Intensity and Revenue Generation in the Dang District of Nepal. *Environment and Development Economics* 13–24. doi:10.1017/S1355770X14000473.

Pour, Milad Dehghani, Ali Akbar Barati, Hossein Azadi, and Jürgen Scheffran. 2018. Revealing the Role of Livelihood Assets in Livelihood Strategies: Towards Enhancing Conservation and Livelihood Development in the Hara Biosphere Reserve, Iran. *Ecological Indicators* 11:516–524. doi:10.1016/j.ecolind.2018.05.074.

Quandt, Amy, Henry Neufeldt, and J. Terrence McCabe. 2017. The Role of Agroforestry in Building Livelihood Resilience to Floods and Drought in Semiarid Kenya. *Ecology and Society* 22(3):10. doi:10.5751/ES-09461-220310.

Rada, Nicholas E., and Keith O. Fuglie. 2018. New Perspectives on Farm Size and Productivity. *Food Policy* 1–6. doi:10.1016/j.foodpol.2018.03.015.

Raintree, John B. 1991. *Socioeconomic Attributes of Trees and Tree Planting Practices.* FAO Community Forestry Note 9:115. Food and Agriculture Organization of the United Nations, Rome, Italy.

Rasolofoson, Ranaivo A., Paul J. Ferraro, Clinton N. Jenkins, and Julia P.G. Jones. 2015. Effectiveness of Community Forest Management at Reducing Deforestation in Madagascar. *Biological Conservation* 184:271–277. doi:10.1016/j.biocon.2015.01.027.

Rollan, Catherine Denise, Richard Li, Jayne Lois San Juan, Liezel Dizon, and Karl Benedict Ong. 2018. A Planning Tool for Tree Species Selection and Planting Schedule in Forestation Projects Considering Environmental and Socio-Economic Benefits. *Journal of Environmental Management* 206:319–329. doi:10.1016/j.jenvman.2017.10.044.

Sabah Forestry Department. 2006. Annual Report 2006. http://www.forest.sabah.gov.my/download/2006/ar2006.htm. (Accessed October 23, 2018).

Sabah Forestry Department. 2008. Annual Report 2008. http://www.forest.sabah.gov.my/download/2008/index.htm. (Accessed October 15, 2018).

Sabah Forestry Department. 2010. Annual Report 2010. http://www.forest.sabah.gov.my/pdf/ar2010/index.htm. (Accessed October 15, 2018).

Sabah Forestry Department. 2013. Annual Report 2013. http://www.forest.sabah.gov.my/annual-reports/2013. (Accessed October 9, 2018).

Sabah Forestry Department. 2014. Annual Report 2014. http://www.forest.sabah.gov.my/annual-reports/2014. (Accessed October 15, 2018).

Sabah Forestry Department. 2015. Annual Report 2015. http://www.forest.sabah.gov.my/annual-reports/2015. (Accessed October 15, 2018).

Sabah Forestry Department. 2016. Annual Report 2016. Sabah Forestry Department. http://www.forest.sabah.gov.my/publication/annual-reports/130-publications/annual-report/686-annual-report-2016. (Accessed October 23, 2018).

Sapkota, Prativa, Rodney J. Keenan, and Hemant R. Ojha. 2018. Community Institutions, Social Marginalization and the Adaptive Capacity: A Case Study of a Community Forestry User Group in the Nepal Himalayas. *Forest Policy and Economics* 29:55–64. doi:10.1016/j.forpol.2018.04.001.

Schroth, Götz, and François Ruf. 2014. Farmer Strategies for Tree Crop Diversification in the Humid Tropics. A Review. *Agronomy for Sustainable Development* 34(1):139–154. doi:10.1007/s13593-013-0175-4.

Scoones, Ian. 1998. Sustainable Rural Livelihoods a Framework for Analysis. *Institute Development Studies Series* 72:1–22. doi:10.1057/palgrave.development.1110037.

Serrat, Olivier. 2017. The Sustainable Livelihoods Approach. In *Knowledge Solutions*, pp. 21–26. doi:10.1007/978-981-10-0983-9_5.

Sunderlin, William D. 2006. Poverty Alleviation through Community Forestry in Cambodia, Laos, and Vietnam: An Assessment of the Potential. *Forest Policy and Economics* 8(4):386–396. doi:10.1016/j.forpol.2005.08.008.

Tang, Qing, Sean J. Bennett, Yong Xu, and Yang Li. 2013. Agricultural Practices and Sustainable Livelihoods: Rural Transformation within the Loess Plateau, China. *Applied Geography* 41:15–23. doi:10.1016/j.apgeog.2013.03.007.

Teitelbaum, Sara. 2014. Criteria and Indicators for the Assessment of Community Forestry Outcomes: A Comparative Analysis from Canada. *Journal of Environmental Management* 132:257–267. doi:10.1016/j.jenvman.2013.11.013.

Timberwell Berhad. 2015. Map of Gana-Lingkabau Forest Reserve. https://timwell.com.my. (Accessed October 23, 2018).

Toh, Su Mei, and Kevin T. Grace. 2006. Case Study: Sabah Forest Ownership. In *Understanding Forest Tenure in South and Southeast Asia edited by Food and Agriculture Organization of the United Nations, 253-280*. Forestry Policy and Institutions Working Paper: FAO.

Valmassoi, Arianna, Salem Gharbia, Santa Stibe, Silvana Di Sabatino, and Francesco Pilla. 2017. Future Impacts of the Reforestation Policy on the Atmospheric Parameters: A Sensitivity Study over Ireland. *Procedia Computer Science* 109C:367–375. doi:10.1016/j.procs.2017.05.403.

Watkins, Shannon Lea, Jess Vogt, Sarah K. Mincey, Burnell C. Fischer, Rachael A. Bergmann, Sarah E. Widney, Lynne M. Westphal, and Sean Sweeney. 2018. Does Collaborative Tree Planting between Nonprofits and Neighborhood Groups Improve Neighborhood Community Capacity? *Cities* 78:83–99. doi:10.1016/j.cities.2017.11.006.

Whiteman, Adrian, Anoja Wickramasinghe, and Leticia Piña. 2015. Global Trends in Forest Ownership, Public Income and Expenditure on Forestry and Forestry Employment. *Forest Ecology and Management* 352:99–108. doi:10.1016/j.foreco.2015.04.011.

Yacob, Ahmad Nawawi, Tan Luck Lee, and Hasan Bin Ibrahim. 2012. Conservation, Consolidation and Economic Generation of Indigenous Community Agriculture Sustainable Food Yielding Reforestation. *Procedia – Social and Behavioral Sciences* 68:319–329. doi:10.1016/j.sbspro.2012.12.230.

Yadav, Bhagwan Dutta, Hugh Bigsby, and Ian MacDonald. 2015. How Can Poor and Disadvantaged Households Get an Opportunity to Become a Leader in Community Forestry in Nepal? *Forest Policy and Economics* 52:27–35. doi:10.1016/j.forpol.2014.11.010.

Yahya, Hardawati, Roszehan Mohd Idrus, Hamimah Talib, and Eunice Fong. 2012. Perspective on Forest Conservation : A Case Study of Community at Gana Resettlement and Integrated Development Project (GRID), Sabah, Malaysia. *Journal of Forest Science* 28(3):185–193. doi:10.7747/JFS.2012.28.3.185.

6 Response of *Shorea* Species to Drought Stress

Rhema D. Maripa, Rahila Dahlan
David and Nor Hayati Daud

CONTENTS

6.1 INTRODUCTION

Shorea is a genus from the family of *Dipterocarpaceae* and the subfamily of *Dipterocarpoideae* which is highly distributed and native to Southeast Asia (Appanah and Turnbull, 1998), from the Philippines, Indonesia and Malaysia to Northern India (Wikipedia, 2018). The mature trees from this family are usually gigantic and offer a good quality of wood. Most of the commercialized *Shorea* species are categorized as heavy hardwood with a remarkable timber strength and a very fine and even wood texture (Shorea Organization, 2018). Due to the high density of the timber, *Shorea* is considered one of the most popular woods in the world (Shorea Organization, 2018), its uses ranging from light construction material to home-based furniture. *Shorea laevis* Ridl. and *Shorea maxwelliana* King (Balau) are examples of the popular Group A wood (Malaysian Timber Council, 2018).

Other *Shorea* species from the Meranti group such as Dark and Light Red Meranti, White Meranti and Yellow Meranti have their own niche in the wood timber industry. Therefore, the *Shorea* species are very attractive for large-scale plantation in the tropical region. However, they are not immune to environmental factors such as drought stress, which could jeopardize the planting effort. There are about 196 *Shorea* species in nature that had been currently identified and described (Ashton, 2004; Wikipedia, 2018). Based on reports published since the 1980s relating to tropical species' response to drought stress, *Shorea* is probably the most studied genus among dipterocarps.

Drought stress can be defined as an unfavorable influence on plants due to lack of water, and it is usually measured in relation to growth, plant survival, or the primary assimilation of carbon dioxide (CO_2) and mineral uptake (Taiz and Zeiger, 2006). Drought stress develops when there is no rainfall for a period of time, which then causes a depletion of soil moisture that impacts the plant (Larcher, 2003; Pallardy, 2008). In the humid tropical climate regions where *Shorea* naturally grows, drought stress can develop when the annual total precipitation is lower than the annual evaporation (Larcher, 2003). According to Walsh (1996), several months of rain below 100 mm caused drought in tropical forests. Drought stress can also develop more randomly depending on soil type and its capability to hold moisture, type of vegetation cover, level of radiation and rate of evapotranspiration (Pallardy, 2008). Unlike other stresses faced by plants, drought stress develops gradually and intensifies over time (Larcher, 2003). How well the plant species respond to the stress could determine their survivability.

The physiological responses of plants under drought stress can be described using the stress concept originally developed by Selye (1936) and the model refined by Lichtenthaler (1998). A response is a reaction of an organism to a change in its environment (Cambridge Academic Content Dictionary, 2018). In the model described by Selye (1936) and Lichtenthaler (1998), the sequence of stress syndromes endured by plants is divided into four phases: the alarm, resistance, exhaustion and regeneration phases (Lichtenthaler, 1998). At the start of the stress, the plant will start to respond once an alarm reaction has been activated and the plant begins to experience structural changes to resist the stress. If stress continues, it can lead to a resource exhaustion which eventually cause acute damage to the plant. As the stress continues, the plant enters the resistance stage, where it begins to adjust, repair and harden itself until it reaches the optimum stage of resistance. However, when the stress intensifies or is prolonged, the affected plant can be exhausted and decline towards chronic damage or cell death. If the stressor is removed, the plant can regenerate itself and recover with renewed physiological vigor.

This chapter concerns the prospects for and utilization of *Shorea* as a tropical plantation tree by exploring the history of failures and successes in planting this genus across different regions. In view of more frequent drought events in the future that could hamper efforts to plant more *Shorea*, we discuss here a case study and methodology to evaluate the level of drought-tolerance characteristics of underutilized species. This information hopefully will benefit plantation owners, managers and researchers who are interested in finding new tropical tree species for their plantation projects.

6.2 THE POTENTIAL OF *SHOREA* AS A PLANTATION SPECIES

Most of the *Shorea* supply to the wood industry in the 1980s and 1990s was sourced from natural tropical forests in regions such as Malaysia, Indonesia, the Philippines and Northern India (Shorea Organization, 2018). The logs from naturally extracted trees were usually large and of premium quality. This is because *Shorea* are mostly slow-growing emergent species that take many decades to achieve maturity. In their natural environment, the seeds, which are mostly winged, are distributed through wind dispersal not far from their mother trees. As the recalcitrant seeds fall onto the moist forest floor, germination will start to occur within 30–40 days. The wildings will continue to develop underneath forest canopies with low light conditions until they reach a young age, when competition for resources such as growing space, light, water and nutrients limit their growth. The saplings will eventually dominate the forest gaps which soon create layers within the forest. As the natural forests are diminished over time, so is the supply of good-quality timber species such as *Shorea*. Therefore, growing *Shorea* as a plantation species is deemed important to ensure a continuous supply in the future.

Shorea species are among dipterocarps that have been widely used as forest rehabilitation species throughout the Southeast Asia region (Utsugi et al., 2009; Widiyatno et al., 2014). As *Shorea* trees require low sunlight exposure during their early establishment, most of them are planted in a disturbed or degraded forest as part of a reforestation process rather than in an open monoculture plantation. In the state of Sarawak, Malaysia, *Shorea* has contributed more than 50 % (Table 6.1) of the total timber species planted in former shifting cultivation and degraded land since 1979 (Perumal, 2012). Follow-up research on the growth of *Shorea macrophylla* in the Sarawak plantation reported a range of 5–34.7 cm diameter at breast height (DBH) for a 13-year-old stand (321 trees) and 18.3–66 cm DBH for a 23-year-old stand (125 trees) (Nibu and Duju, 2011). The species has a longer rotation period and therefore are mixed with other local or exotic fast-growing timber species (FAO, 2002). However, some *Shorea* species have been reported to be able to grow well with minimum shade trees or without shade trees entirely. According to Tata et al. (2010), *Shorea selanica* and *Shorea lamellata* grow well within 1- and 10-year-old rubber agroforests in Sumatra and Kalimantan, Indonesia. They listed *Shorea javanica*, *Shorea johorensis*, *Shorea lamellate*, *Shorea leprosula*, *Shorea macroptera*, *Shorea macropylla*, *Shorea parvifolia*, *Shorea pinanga* and *Shorea stenoptera* as suitable species to be planted in smallholder rubber agroforests in their region.

Shorea species does have high prospects as a plantation species due to several factors. There is an existing local and international market for *Shorea* timber species. The timber of selected species such as Balau, Selangan Batu and Meranti from Southeast Asian countries, including *Shorea robusta* (Sal) from Northern India, have already been recognized on the international market, with secure demand not only from European countries such as the United Kingdom and the Commonwealth, but also from giant eastern consumers such as China, Japan, Korea, Taiwan, Thailand and India (Wong et al., 2005). The supply of high demand species from natural tropical forests is becoming less available due to decades of extensive logging activities and land degradation. Scarcity in supply can increase the price of good quality logs.

TABLE 6.1

List of *Shorea* Species Planted in Malaysia, Indonesia and the Philippines

No.	*Shorea* Timber Species	Vernacular Name	Area (ha)	Region	Reference
1.	*S. macrophylla*	Engkabang jantong	7,314.4	Sarawak, Malaysia	Perumal (2012)
2.	*S. pinanga*	Engkabang langgai bukit	996.7	Sarawak, Malaysia	Perumal (2012)
3.	Other 6 *Shorea* spp. including *S. parvifolia*	Not available	143.5	Sarawak, Malaysia	Perumal (2012)
4.	*S. stenoptera*	Light Red Meranti	Not available	Indonesia	Ecosia Knowledge Base (2018)
5.	*S. agsaboensis*	Tiaong	Not available	Philippines	Lantican (2015)
	S. almon	Almon			
	S. astylosa	Yakal			
	S. gisok	Gisok			
	S. guiso	Guijo			
	S. kalunti	Kalunti			
	S. negrosessis	Red lauan			
	S. philippinensis	Manggasinoro			
	S. plagata	Malaguijo			
	S. polita	Malaanonang			
	S. polysperma	Tangile			
	S. squamata	Mayapis			

Although plantation trees do not grow bigger and taller than naturally found trees, they promise a good return for future investment.

6.3 DROUGHT AS A MAJOR THREAT IN *SHOREA* SURVIVAL

Drought may be one of the main factors influencing the survival of any plant species in natural or plantation forests. Drought caused by El Niño-related events is expected to become more frequent and to intensify in the future (Shamala, 2015; Bebber et al., 2004; Malhi and Wright, 2004). A study by Newbery and Lingenfelder (2009) of the primarily lowland dipterocarp forest in Danum Valley Conservation Area (DVCA), Sabah, reported that dipterocarps, which include *Shorea* species, were affected by the worst drought ever recorded in Sabah history in 1997–1998. The regeneration of the affected area was overtaken by fast-growing, light-demanding and drought-tolerant understory plants, which were likely a pioneer species. On the other hand, *Shorea* are mostly slow-growing and shade-demanding species in comparison to non-dipterocarps. In their early establishment, they are sensitive to open and dry growing conditions, which are common in most forest plantations in the tropics. By taking into consideration the natural characteristics of young seedlings of *Shorea*, planting and providing them with nursing trees may be a way to tackle unfavorable conditions on forest floors.

Another aspect that can contribute to the success of *Shorea* plantation is the utilization of native species which are local to the plantation site. Our recent study revealed that dipterocarps are site specific even within the same forest. Some species thrive on ridges and in mountainous areas, while others can only be found in valleys and low-lying areas (Maripa, 2018). Different species were found to dominate different forests in a dry inland region and a wet coastal forest (Maripa, 2018). The existing composition of dipterocarp forests in the tropic regions is in fact believed to be the result of their tolerance to long-term and persistent drought stress environments over millennia.

The utilization of commercially high-value indigenous species in enrichment plantings and rehabilitation programs has been implemented for quite some time in the Philippines (Dierick and Holscher, 2009), Malaysia and Indonesia (Kettle, 2010). For logged forests, enrichment plantings of selected native seedlings are planted between remaining stands, while for degraded land such as mining areas, fast growing non-dipterocarp species are first established to provide shade for dipterocarps that will be planted underneath nursing trees at a later stage (Kettle, 2010). Since enrichment and restoration plantings are quite expensive, the survival of every individual tree is vital. In this case, species selection to match site conditions either in natural or plantation forests is crucial, and extensive knowledge is required on species characteristics and ability to adapt to the planting environment (Norisada and Kojima, 2005; Tolentino and Camacho, 2006; Utsugi et al., 2009).

6.4 DETERMINING DROUGHT TOLERANCE AMONG *SHOREA* SPECIES

Every plant species has a different level of drought tolerance to water stress depending on their natural characteristics. Many local authorities and organizations involved in the reforestation or rehabilitation of degraded land recognize the need to select and replant native species into their environment. However, previous studies have shown that there is a niche within a site that influences the success of reintroducing local species back into their ecosystems (Maripa, 2018). Therefore, a series of drought-tolerance tests can be performed prior to species selection. Here, we present a case study on three selected *Shorea* species as an example. The experiment was carried out for five consecutive weeks in February and March 2016 at the greenhouse of the Forestry Complex, Faculty of Science and Natural Resources, Universiti Malaysia Sabah.

6.4.1 PREPARING SEEDLINGS FOR DROUGHT STRESS TEST

Seedlings used in any drought test needed to be nurtured and stabilized prior to treatment. In this case, seeds of *Shorea argentifolia*, *Shorea leprosula* and *Shorea seminis* were collected from dipterocarp forests and raised in the nursery before being transplanted into plastic pots in the greenhouse (Figure 6.1). The ecology and mature leaf traits of three selected *Shorea* are presented in Table 6.2. The seedlings were watered daily for the next 3 months and protected from pest and diseases. The preparation and stabilization methods were adapted from

FIGURE 6.1 (A). From left to right; leaves of *Shorea argentifolia*, *Shorea leprosula* and *Shorea seminis*. (B). A typical environment in the greenhouse where seedlings are arranged in blocks.

Bertolli et al. (2012). After about 3 months, only the healthy and vigorous seed-lings were selected for study. The selection process is important to eliminate undersize or oversize seedlings. Drought stress treatment of 40% was given to treated seedlings based on their estimated daily transpiration rate. The amount of water that evaporates through the surface of the pot plus the water transpired by the plant can be obtained by measuring its daily pre-watering and post-watering weight. Ten empty plastic pots without seedlings were used to estimate evaporation in the greenhouse.

6.4.2 COMPARING GROWTH RATE OF STRESSED AND CONTROLLED SEEDLINGS

Measuring the growth of the seedlings over time is one of the easiest ways to deter-mine the impact of drought stress on the plant growth. Plants generally grow in diameter of stem and in height. For seedlings, diameter increments can be measured approximately 2 cm above the root collar, while the height can be measured directly

TABLE 6.2

Description of Ecology and Mature Leaf Traits of Three Selected *Shorea* Species

Species	Ecology[a,b]	Mature leaf traits[a,b]
Shorea argentifolia Symington	Mixed dipterocarp forest on clay rich alluvium soils, on hills, slopes and sometimes ridges at altitude up to 900 m above sea level.	Thinly coriaceous, 6 to 11 cm in length, small leaves.
Shorea leprosula Miquel	Lowland Dipterocarp forest, lower hill slopes and valleys in hill Dipterocarp forest	Thinly coriaceous leaves, 8 to 14 cm in length, moderate size leaves.
Shorea seminis (de Vriese) Slooten	Common on clay and silt alluvium river banks along lowland rivers which is at altitude up to 300 m above sea level.	9 to 18 cm in length, big leaves.

Notes: [a] (Soepadmo et al., 2004) [b] (Newman et al., 2000).

from the soil surface to the tip of shoot. Their growing condition and survival can be monitored throughout the experiment. The number of leaves was excluded in this case study due to the destructive harvesting required for certain parameters to obtain its data.

6.4.3 MEASURING WATER LEVEL OF WHOLE PLANTS

Pre-dawn water potential (Ψ_{PD}) is usually assumed to represent the water available in the soil that is available for the plants to uptake from the roots (Williams and Araujo, 2002). This is because photosynthesis does not occur at night and the stomata remain closed (Hopkins and Hüner, 2008). Therefore, during the night, the water within the plant equilibrates with the source of water in the soil. The measurement of Ψ_{PD} is usually done during the early morning between 4:30 and 6:30 am, before sunrise. This is because during this time, the amount of water present in the leaf is the same as the amount of water in the soil (Tinus, 1996). A measurement taken at noon is called *midday water potential* (Ψ_{MD}); at this point the plant is likely to be under the most drought stress, around 11:30 am to 1:30 pm. The instrument used to measure pre-dawn water potential is the plant moisture stress system (PMS) also known as *pressure bomb* (Model 600 Pressure Chamber Instrument, 2009). Both types of sampling are destructive, and therefore large numbers of treated seedlings are required.

A leaf sample is taken from the sample seedlings and placed into the chamber in the PMS. It is sealed tight in the chamber with its lid and rubber gasket. This chamber will be pressurized by switching on the PMS to allow nitrogen gas from the compressed gas cylinder into the chamber. The leaf is observed through a magnifying glass from outside the chamber while it is being pressurized. When a droplet of water comes out of the petiole of the leaf, it indicates that the pressure has reached balance level; the switch is then immediately turned off and the measurement from the pressure gauge is recorded and tabulated in a table. The equation to calculate the pre-dawn water potential is adapted from the research of Williams and Araujo (2002) in their research paper on water potentials as shown in Equation 6.1.

$$\Psi_{PD} = \Psi_P + \Psi_S \tag{6.1}$$

whereby, Ψ_P is the pressure applied by the chamber in PMS and Ψ_S is the effect of solute in the xylem sap. The pre-dawn water potential calculated using this equation will be tabulated for further analysis.

6.4.4 WEIGHTING WATER CONTENT CHANGES IN PLANT CELLS

The relative water content (RWC) is the measurement of water content in the leaves of any plants that are fully hydrated. The method used to measure this RWC is a destructive type. This method is adapted from the research of Arndt et al. (2015). In this case study, a leaf sample was taken from the seedlings, sealed in a plastic and placed in a cooler box at around 10°C–15°C. The leaves were brought to the lab and their fresh weight was measured immediately using a digital weighing machine. In the lab, the leaves were placed in a beaker filled with distilled water to be hydrated to full turgidity for 3–4 hours at room light and temperature. After hydration, the samples were taken out of the water and dried with tissues and immediately weighed to obtain their turgid weight. Then these leaves were placed in the oven at a temperature of 80°C for a minimum of 3 days. The leaves were inspected after 3 days and taken out if they were completely dry; if they were not, the leaves were put back in the oven for another day or two until they were dry. Once the leaves had completely dried, the dry weight was measured after they had being cooled down in a desiccator. The RWC could then be calculated with all the measured values. The equation used to calculate the percentage of RWC was adapted from the research of Arndt et al. (2015) in their research paper on introducing the right way to measure RWC as shown in Equation 6.2.

$$\mathrm{RWC}\ (\%) = \left[(W - DW)/(TW - DW) \right] \times 100\% \tag{6.2}$$

whereby, W is the sample fresh weight, TW is the sample turgid weight and DW is the sample dry weight. With this equation the calculated relative water content was tabulated in a table for further analysis.

6.4.5 CALCULATING WATER LOSS THROUGH TRANSPIRATION

The TPS-2 Portable Photosynthesis System (PP Systems TPS-2 Portable Photosynthesis System Operator's Manual Version 2.01, 2007) is a high-precision instrument for measuring water loss through transpiration. The transpiration rate is determined by the flow of air entering and leaving the leaf cuvette (Bertolli et al., 2012). The transpiration rate is not directly measured through this instrument but is obtained by measuring a few important units to calculate and provide the photosynthesis rate. A leaf is sealed into a chamber in this instrument called the *leaf cuvette*. A measured flow of air is passed to this cuvette. The equation that is used to calculate the transpiration rate in this instrument is adapted from the official PP

Systems TPS-2 Portable Photosynthesis System Operator's Manual version 2.01 as shown in Equation 6.3.

$$E = \left[W \times \left(e_{\text{out}} - e_{\text{in}} \right) \right] / \left(P - e_{\text{out}} \right) \tag{6.3}$$

whereby, W is the mass flow of air per unit leaf area entering the cuvette, e_{in} is the water vapor pressure of the air entering the cuvette, e_{out} is the water vapor pressure of the air leaving cuvette, and P is the atmospheric pressure. The transpiration rate (E) is now calculated from the water vapor pressure of the air entering e_{in} and leaving the cuvette e_{out} by considering the atmospheric pressure, P and the mass flow of air per unit leaf area entering the cuvette, W.

6.4.6 CALCULATING PHOTOSYNTHESIS RATE IN LEAVES

The TPS-2 Portable Photosynthesis System can also measure the assimilation of carbon dioxide (CO_2), which is the gas exchanged in the photosynthesis and respiration process. The principle this instrument operates by using the open system Bertolli et al. (2012). The TPS-2 uses the infrared gas analysis technique to determine CO_2 concentration in the leaves. This instrument uses a non-destructive method to obtain the photosynthesis rate.

The photosynthesis rate is not directly measured through this instrument. It is obtained by measuring a few important units in order to calculate and provide the photosynthesis rate. The TPS-2 uses a gas-switching technique, and it first samples the CO_2 and water (H_2O) of the air going to the cuvette, which in the instrument is referred to as *reference*; and then the air leaving the cuvette, which is referred as *analysis*. Based on this flow, the CO_2 assimilation can be determined, and thus, the photosynthesis rate can be determined too. The equation that is used to calculate the photosynthesis rate in this instrument is adapted from the official PP Systems TPS-2 Portable Photosynthesis System Operator's Manual version 2.01 as shown in Equation 6.4.

$$A = C_{\text{in}} \times W - C_{\text{out}} \times \left(W + E \right)$$

$$A = - \left[W \times \left(C_{\text{out}} - C_{\text{in}} + C_{\text{out}} \times E \right) \right] \tag{6.4}$$

whereby, C_{in} is the CO_2 concentration that is entering the leaf in the cuvette, C_{out} is the CO_2 concentration that is leaving the leaf in the cuvette, W is the mass flow of air per unit leaf area entering the cuvette and E is the transpiration rate.

The photosynthesis rate (A) is calculated from the difference in the CO_2 concentration entering (C_{in}), that leaving (C_{out}), and the flow rate through the cuvette. The CO_2 readings are assumed to be corrected for water vapor, temperature and atmospheric pressure. Also, the addition of water vapor by transpiration in the leaf cuvette dilutes the outgoing air, and this must be compensated for in the calculation as shown above. This instrument offers an optional LED light unit feature for use with the cuvette for manual control of the light intensity in the cuvette. The photosynthesis rate can be obtained under normal sunlight conditions and under LED light conditions. The LED light intensity used in this research is set at the maximum level of 9 in the PPS which has the range of 0–2500 $\mu mol\ m^{-2}\ s^{-1}$.

6.5 THE GROWTH RATE OF DROUGHT-STRESSED SEEDLINGS

In general, the size of a seedling should increase as time passes, and the same goes for the sample seedlings in this case study. However, the growth of the drought-stressed seedlings was lesser than the control seedlings. In term of diameter increment, the drought stress caused a decrease of 30%–50% of the total diameter of the control seedlings (Table 6.3) while the height increment was lowered to 23%–50% of that of the unstressed seedlings. In terms of diameter, *Shorea argentifolia* was better than *Shorea leprosula* and *Shorea seminis*. However, in terms of height increment, *Shorea leprosula* was better than *Shorea seminis* and *Shorea argentifolia*. This shows that the growth rate slowed down under gradual water stress. Throughout the study, the air temperature ranged from 35°C to 40°C with a relative humidity of 49%–62%. The greenhouse weather during data collection was very sunny and hot as there was a drought attack during the period of data collection for this study.

TABLE 6.3
Descriptive Analysis of Growth Measurement of Three Selected *Shorea* Species

Parameter/ Species	Block	Pre-Treatment	Post-Treatment	Changes	Sig. (2-tailed)	Treatment Differences (%)
			(cm)			
		Mean of Diameter				
S. argentifolia	Control	5.5	6.7	+1.2	***	66.7
	Treatment	4.8	5.6	+0.8	***	
S. leprosula	Control	5.8	6.9	+1.1	***	54.5
	Treatment	5.7	6.3	+0.6	***	
S. seminis	Control	5.6	6.8	+1.2	***	50.0
	Treatment	5.8	6.4	+0.6	***	
		Mean of Height				
S. argentifolia	Control	82.5	83.9	+1.4	**	50.0
	Treatment	81.6	82.3	+0.7	***	
S. leprosula	Control	79.4	81.1	+1.7	*	76.5
	Treatment	80	81.3	+1.3	**	
S. seminis	Control	67.5	69.7	+2.2	*	59.1
	Treatment	67.9	69.2	+1.3	**	

Significant differences between pre- and post-treatment are shown in the Sig. (2-tailed) column: ***$p < 0.0005$, **$p < 0.001$ and *$p < 0.05$. The rightmost column indicates the differences of changes in mean of diameter or height of treated seedlings, in comparison to control.

Based on an observation of physical changes, the control and treatment of *Shorea argentifolia* looked fresh and healthy with lots of leaves in W_0. In W_5, the control of *Shorea argentifolia* looked fresh and healthy and still had lots of leaves. The control and treatment of *Shorea leprosula* looked fresh and healthy with lots of leaves in W_0, but it had less leaves than *Shorea argentifolia*. In W_5, the control of *Shorea leprosula* looked fresh and healthy and still had lots of leaves. The control and treatment of *Shorea seminis* looked fresh and healthy with lots of leaves in W_0, but it had less leaves than *Shorea argentifolia* and *Shorea leprosula*. In W_5, the control of *Shorea seminis* looked fresh and healthy and still had lots of leaves. The physical observation of these control and treatment blocks shows there was a physical, observable change. (Figure 6.2).

6.6 THE WATER POTENTIAL OF DROUGHT-STRESSED SEEDLINGS

Stressed plants usually displayed more negative or lower water potential (Ψ) value than unstressed plants. According to Tinus (1996), the unstressed pre-dawn water potential (ψ_{PD}) should be from around −0.2 to −0.4 MPa, the moderately stressed Ψ_{PD} from −0.5 to −0.8 MPa and the very stressed Ψ_{PD} more than −0.8 MPa. In this study, the Ψ_{PD} ranged from −0.25 ± 0.07 MPa to −0.40 ± 0.08 MPa, which means no stress was detected among the studied seedlings. In the meanwhile, the midday water potential (Ψ_{MD}) ranged from −0.53 ± 0.24 MPa to −0.73 ± 0.38 MPa (Table 6.4). Among the species, *Shorea leprosula* displayed the lowest Ψ_{MD} among the species.

Table 6.5 shows a range of *Shorea* water potentials measured during rainy or dry seasons. These *Shorea* were either in a seedling, sapling or mature state and found in either undisturbed or partially disturbed land, and at higher elevations. As shown in Table 6.5, it was reported that during the dry season, *Shorea multiflora* in the Brunei heath forest experienced moderate stress while *Shorea robusta* in Central Himalaya suffered serious drought stress in the dry season.

6.7 THE RELATIVE WATER CONTENT OF DROUGHT-STRESSED SEEDLINGS

Relative water content (RWC) is probably the least affected parameter during drought tests and only shows significant changes in cases of severe stress. In most laboratory and greenhouse studies related to the response of plants to drought stress, the RWC of leaves often displayed a slight reduction depending on the amount of water supplied on treated seedlings. According to Stanton and Mickelbart (2014), healthy leaves have a RWC of 90%–98%, while wilting leaves have a RWC of 60%–70% and severely desiccated and senescing leaves a RWC of 40%. In the study case, the treated seedlings displayed a mild form of drought stress, with *Shorea argentifolia* having a RWC of less than 90% (Table 6.6). Prolonged drought-stress tests should be able to detect more response from the species.

Based on the estimated marginal means of the percentage of RWC of leaves for treatment, *Shorea seminis* has the highest mean value for both treatment levels followed by *Shorea leprosula* and *Shorea argentifolia*. However, the mean value is not less than 88% even for treatment, which shows that gradual water stress in this experiment does not affect the RWC of the leaves significantly. During the experiment,

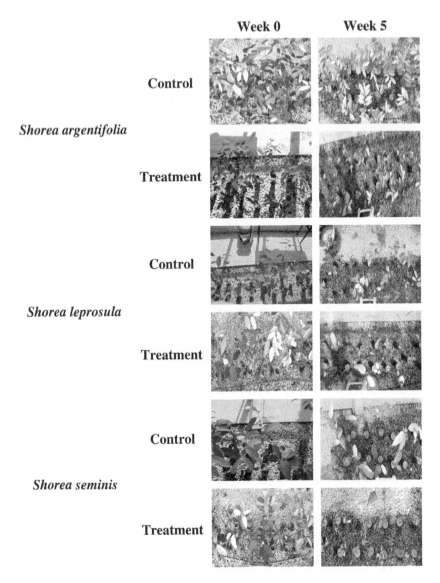

FIGURE 6.2 Physical observation of three *Shorea* species, pre- and post-treatment at the greenhouse.

the air temperature was warm and humid at the beginning of February and it started to increase from the end of February into March. The air temperature ranged from 24.5°C to 38°C. The relative humidity ranged from 64.5% to 92%.

6.8 THE TRANSPIRATION RATE OF DROUGHT-STRESSED SEEDLINGS

The transpiration rate was measured simultaneously while measuring the photosynthesis rate under sunlight using the same instrument. A problem was identified wherein the

TABLE 6.4

Mean of Predawn (Ψ_{PD}) and Midday Water Potential (Ψ_{MD}) of Three Selected *Shorea*

Treatment	Species	Ψ_{PD}		Ψ_{MD}		
		Mean (MPa)	Std. Deviation	Mean (MPa)	Std. Deviation	N
100%	*Shorea argentifolia*	−0.30	0.08	−0.64	0.27	72
	Shorea leprosula	−0.35	0.07	−0.59	0.29	72
	Shorea seminis	−0.25	0.07	−0.54	0.25	72
60%	*Shorea argentifolia*	−0.35	0.07	−0.64	0.30	72
	Shorea leprosula	−0.40	0.08	−0.73	0.38	72
	Shorea seminis	−0.30	0.07	−0.53	0.24	72

air temperature of the greenhouse in the afternoon was very hot, from 36°C to 40°C. The high temperature not only affects the photosynthesis rate to decrease or stop, it also causes the stomata to close. Stomata are very important for a plant as they regulate the gaseous exchange and transpiration processes (Hopkins and Hüner, 2008). When the temperature is too hot, the stomata get sensitive, and they close to avoid any excessive loss of water from the plant through the transpiration process (Koyama and Takemoto, 2014). This is another reason why the data is collected very early in the morning from 7.00 to 8.00 am. The air temperature during data collection ranges from 31.0 to 35.5°C. The relative humidity during data collection ranges from 56% to 79%.

Theoretically, the transpiration rate must decrease when the plant is under water stress. Based on these descriptive statistics, it can be seen that only *Shorea argentifolia* showed an increase in transpiration rate at 60% treatment level as compared to 100% treatment level. These mean values of photosynthesis rate range from 0.57 mmol m^{-2} s^{-1} to 0.92 mmol m^{-2} s^{-1}, which is in the range of the dipterocarp tree species *Shorea seminis* transpiration rate as recorded in the research of Eschenbach *et al.* (1998) on lowland dipterocarp rainforest species at Sabah, with a range of 2.5 mmol m^{-2} s^{-1} on average. Table 6.7 displayed a range of transpiration for seedlings and mature trees.

In the case study, it is shown that the transpiration rate for *Shorea seminis* is the highest, followed by *Shorea leprosula* and then by *Shorea argentifolia* (Figure 6.3) This result suggested that *Shorea argentifolia* was most affected, followed by *Shorea leprosula*, while *Shorea seminis* is not affected. This is because the larger number of leaves means more stomata, and gradual water stress hastens the loss of water when there are more stomata as well as when the surrounding temperature is hot and dry (Lawson and Blatt, 2014). All three species react differently to the treatment imposed. *Shorea argentifolia* shows an increase in transpiration rate at a treatment level of 60%. *Shorea leprosula* shows a slight decrease in transpiration rate at a treatment level of 60%, and *Shorea seminis* shows a very drastic drop in transpiration rate at a treatment level of 60%. The results show that not all species are affected by gradual water stress.

TABLE 6.5
A Summary of Water Potentials (Ψ) of Dipterocarps during Rainy and Dry Season

Species	Location	Rainy Season[a]		Dry Season[a]		Reference
		Ψ_{PD} (MPa)	Ψ_{MD} (MPa)	Ψ_{PD} (MPa)	Ψ_{MD} (MPa)	
Seedling						
Shorea johorensis	Undisturbed	−0.06				Clearwater et al. (1999)
Shorea johorensis	Partially open	−0.08				Clearwater et al. (1999)
Shorea johorensis	Fully open 14 days	−0.14				Clearwater et al. (1999)
Shorea johorensis	Fully open 90 days	−0.08				Clearwater et al. (1999)
Shorea multiflora	Brunei heath forest	−0.22	−0.2	−0.5	−1.2	Cao (2000)
Shorea robusta	Central Himalaya	−0.28	−0.9	−0.98	−1.84	Garkoti et al. (2003)
Sapling/Juvenile/Young Tree						
Cotylelobium buckii	Brunei heath forest	−0.23	−0.18	−0.6	−0.6	Cao (2000)
Dipterocarpus borneensis	Brunei heath forest	0	0	−0.45	−0.5	Cao (2000)
Hopea pentanervia	Brunei heath forest	−0.25	−0.43	−1.45	−2.2	Cao (2000)
Parashorea chinensis	Plantation	−0.23	−1.03	−0.56	−1.87	Zhang et al. (2009)
Shorea robusta	1860 m elevation	−0.21	−0.27	−0.54	−1.15	Singh et al. (2006)
Mature Tree						
Dipterocarpus pachyphyllus[b]	Emergent	−0.57 to −1.18				Hiromi et al. (2012)
Dryobalanops aromatica[b]	Emergent	−0.50 to −0.92				Hiromi et al. (2012)
Shorea parvifolia[b]	Emergent	−0.64 to −1.98				Hiromi et al. (2012)
Shorea smithiana[b]	Emergent	−0.59 to −1.91				Hiromi et al. (2012)
Dipterocarpus retusus	Plantation	−0.33	−0.44	−0.5	−1.38	Zhang et al. (2009)
Vatica xishuangbannaensis	Plantation	−0.38	−0.63	−0.48	−1.15	Zhang et al. (2009)
Hopea hainanensis	Plantation	−0.63	−0.63	−0.48	−2.43	Zhang et al. (2009)
Parashorea chinensis	Plantation	−0.23	−1.03	−0.56	−1.87	Zhang et al. (2009)
Mean of 17 Dipterocarps[c]	Plantation		−0.61			Zhang and Cao (2009)
Shorea robusta	500 m elevation	−0.17	−0.36	−0.6	−1.37	Tewari (1999) (approx. values)
Shorea robusta	Plain			−0.55	−1.35	Mainali et al. (2006)
Shorea robusta	Hill			−1.15	−1.74	Mainali et al. (2006)

[a] categorized according to interest of current study.

[b] measured between 0600 to 2000.

[c] *Anisoptera costata, Dipterocarpus alatus, D. intricatus, D. retusus, D. tuberculatus, D. turbinatus, Hopea chinensis, H. hainanensis, H. hongayensis, H. mollissima, Parashorea hinensis, Shorea assamica, S. robusta, S. Spp., Vatica guangxiensis, V. angachapoi, V. xishuangbannaensis.*

TABLE 6.6
Mean of Relative Water Content (RWC) of Three Selected *Shorea*

Treatment	Species	RWC (%)		
		Mean (MPa)	Std. Deviation	N
100 %	*Shorea argentifolia*	92	1.65	72
	Shorea leprosula	95	1.92	72
	Shorea seminis	97	1.88	72
60 %	*Shorea argentifolia*	88	1.72	72
	Shorea leprosula	91	1.65	72
	Shorea seminis	92	2.10	72

6.9 THE PHOTOSYNTHESIS RATE OF DROUGHT-STRESSED SEEDLINGS

According to Pietragalla and Pask (2012), the ideal time to measure the photosynthesis rate is close to solar noon, which is typically from 11:00am to 2:00pm. However, during the trial period of this research, the photosynthesis rate was measured at this time of day, and the photosynthesis rate was in a negative value ranging from −0.8 to −66.8 μmol m^{-2} s^{-1}, which is too small a value to be taken as reliable data. This occurs due to the fact that the seedlings are highly sensitive in this hot environment. A problem was identified where the air temperature of the greenhouse was very hot, ranging from 36 to 40°C. This hot air temperature in the greenhouse caused an increase in the leaf surface temperature from 36 to 38°C during the trial week. The hot air temperature in the greenhouse occurred due to the drought attack during the months of February and March 2016, which caused a reduction in rainfall of up to 60% on the average rainfall (Malaysian Metrological Department, 2016).

Theoretically, the rate of photosynthesis is supposed to be high at noon, but if the temperature is too hot for that particular plant, the photosynthesis rate will decrease or even stop because the enzymes involved in photosynthesis, such as ribulose biphosphate carboxylase (RubisCo), are sensitive and a high temperature will reduce RubisCo activation for most plants, including dipterocarps (Koyama and Takemoto, 2014). This RubisCo is important enzyme in carbon fixation during photosynthesis (Koyama and Takemoto, 2014). Therefore, the measurement of the photosynthesis rate was taken at 7am–8am at which temperatures are lower than 36°C, ranging from 31.0 to 35.5°C. The relative humidity during data collection ranged from 56% to 79%. Although the humidity was above 70%, the leaf surface was dry during data collection and not moist due to the very high temperature.

On the other hand, Figure 6.4 shows the mean photosynthesis rate under sunlight and LED light. *Shorea seminis* have the highest mean value followed by *Shorea leprosula* and *Shorea argentifolia* for treatment of both 100% and 60%. These mean values of the photosynthesis rate range from 4.9 μmol m^{-2} s^{-1} to 12.8 μmol m^{-2} s^{-1} which is in the range of dipterocarp tree species *Shorea seminis*' photosynthesis rate

TABLE 6.7
Mean of Photosynthesis and Transpiration Rate of Selected *Shorea*

Species	Location	A_{net} µmol m^{-2} s^{-1}	E mmol m^{-2} s^{-1}	Authors
		Seedling		
Anisoptera thurifera	Nursery		0.979	Tolentino and Camacho (2006)
Dipterocarpus chartaceus	Degraded	10.1		Norisada and Kojima (2005)
Dipterocarpus cornulus	Nursery	4.89		Maruyama, et al. (1997)
Dipterocarpus obtusifolius	Degraded	20.9		Norisada and Kojima (2005)
Hopea odorata	Degraded	13.5		Norisada and Kojima (2005)
Hopea odorata	Nursery	15.89		Maruyama et al. (1997a); Maruyama et al. (1997b)
Hopea plagata	Nursery		0.880*	Tolentino and Camacho (2006)
Neobalanocarpus heimii	Nursery	3.93		Maruyama et al. (1997a); Maruyama et al. (1997b)
Parashorea malaanonan	Nursery		0.700*	Tolentino and Camacho (2006)
Shorea acuminata	Nursery	7.46		Maruyama et al. (1997b)
Shorea almon	Nursery		0.790*	Tolentino and Camacho (2006)
Shorea assamica	Nursery	6.25		Maruyama et al. (1997a); Maruyama et al. (1997b)
Shorea contorta	Nursery		0.380*	Tolentino and Camacho (2006)
Shorea curtisii	Nursery	5.49		Maruyama et al. (1997b)
Shorea leprosula	Nursery	7.85		Maruyama et al. (1997b)
Shorea macroptera	Nursery	3.58		Maruyama et al. (1997b)
Shorea ovalis	Nursery	4.97		Maruyama et al. (1997b)
Shorea parvifolia	Nursery	7.54		Maruyama et al. (1997b)
Shorea pauciflora	Nursery	7.31		Maruyama et al. (1997b)
		Mature Tree		
Shorea robusta		0.0318 to 0.0439+		Roa et al. (2008)
Mean of 17 Dipterocarps	Plantation			Zhang and Cao (2009)
Dipterocarpus pachyphyllus	Field	6.06		Hiromi et al. (2012)
Dryobalanops aromatic	Field	4.09		Hiromi et al. (2012)
Shorea parvifolia	Field	9.79		Hiromi et al. (2012)
Shores smithiana	Field	7.96		Hiromi et al. (2012)

Notes: E = Transpiration; * = approximate value derived from figure; A_{net} = value in mol m^{-2} s^{-1}; + = value in g g^{-1} d^{-1}.

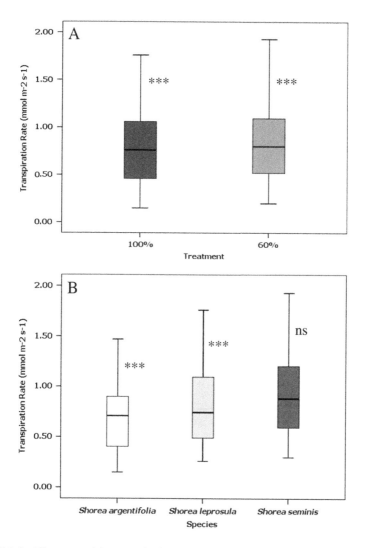

FIGURE 6.3 The mean of the transpiration rate according to treatment (A) and species (B). Significant differences between treatment and control; ***$p < 0.0005$, **$p < 0.001$, *$p < 0.05$ and ns = non-significant.

as recorded in the research of Eschenbach et al. (1998) on lowland dipterocarp rainforest species at Sabah with the similar range 5.96 µmol m^{-2} s^{-1} as average.

The result also shows that *Shorea argentifolia* and *Shorea seminis* are affected by gradual water stress on the photosynthesis rate under sunlight. This is because *Shorea argentifolia* is distributed on hills, slopes and ridges in highlands with a surrounding temperature of 26°C, whereas *Shorea seminis* is distributed on riverbanks in lowlands with surrounding temperatures of 28–31°C (Appanah and Turnbull, 1998). The higher surrounding temperature in the greenhouse caused a reduction in RubisCo activity, and the gradual water stress caused the closure of stomata,

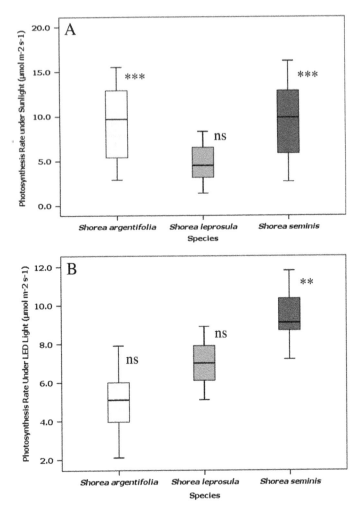

FIGURE 6.4 Mean of photosynthesis rate under sunlight (A) and LED light (B) of three selected *Shorea*. Significant differences between treatment and control; ***$p < 0.0005$, **$p < 0.001$, *$p < 0.05$ and ns = non-significant.

which affected the photosynthesis rate under sunlight for *Shorea argentifolia* and *Shorea seminis* (Koyama and Takemoto, 2014). Only *Shorea seminis* was affected by gradual water stress on the photosynthesis rate under LED light. This is because the ecology conditions wherein *Shorea seminis* is usually found are very moist as it is near riverbanks, which are made up of clay and silt alluvium soil (Appanah and Turnbull, 1998). Thus, gradual water stress affected the photosynthesis rate of *Shorea seminis* although LED light was used. *Shorea leprosula* was not affected by gradual water stress for the photosynthesis rate under either sunlight or LED light because it is distributed on lowland, low-lying hills, slopes and valley with sandy soils and a surrounding temperature of 31°C. Its natural ecology of dry land and sandy soils means faster drainage, and the surrounding temperature is higher than the ecology of both the other species (Appanah and Turnbull, 1998). Therefore, it can withstand gradual water stress. *Shorea argentifolia* is also not affected by gradual water stress on the photosynthesis rate under LED light as the LED light promotes the species to carry out the photosynthesis process even with water stressed condition.

6.10 CONCLUSION

This case study was set up to investigate the effects of gradual water stress on the photosynthesis rate and transpiration rate of three selected *Shorea* species: *Shorea argentifolia*, *Shorea leprosula* and *Shorea seminis*. The aim of this chapter is achieved by obtaining the results that show the interactions between independent and dependent variables that is involved in this study. These interactions show how gradual water stress can affect the photosynthesis rate and transpiration rate as well as the pre-dawn water potential of leaves. In general, gradual water stress affects the photosynthesis rate and the transpiration rate under sunlight and under LED light. However, it does not affect the pre-dawn water potential of leaves and the relative water content of leaves.

It can be observed that for the photosynthesis rate under sunlight, *Shorea argentifolia* is most affected by gradual stress, followed by *Shorea seminis*, and that the photosynthesis rate under sunlight of *Shorea leprosula* is not affected. It can be observed that for the photosynthesis rate under LED light, *Shorea seminis* is most affected by gradual stress but both *Shorea argentifolia* and *Shorea leprosula* are not affected, where *Shorea leprosula* is the least affected. It can be observed that for the transpiration rate, *Shorea argentifolia* is most affected by gradual stress followed by *Shorea leprosula*, while the photosynthesis rate under sunlight of *Shorea seminis* is not affected. Based on these trends, the conclusion is that *Shorea leprosula* is best able to withstand gradual water stress followed by *Shorea seminis* and then *Shorea argentifolia* as the least able to withstand gradual water stress.

REFERENCES

Appanah, S., and Turnbull, J. 1998. *A Review of Dipterocarps Taxonomy, Ecology and Silviculture.* Center for International Forestry Research: Jakarta.

Arndt, S. K., Irawan, A., and Sanders, G. J. 2015. Apoplastic water fraction and rehydration techniques introduce significant errors in measurements of relative water content and osmotic potential in plant leaves. *Physiologia Plantarum*, 155(4): 355–368.

Ashton, P. S. 2004. Dipterocarpaceae. In: *Tree Flora of Sabah and Sarawak*, Volume 5. Soepadmo, E., Saw, L. G., and Chung, R. C. K., (eds.). Government of Malaysia: Kuala Lumpur, Malaysia.

Bebber, D. P., Brown, N. D., and Speight, M. R. 2004. Dipterocarp seedling population dynamics in Bornean primary lowland forest during the 1997–8 El Niño Southern Oscillation. *Journal of Tropical Ecology*, 20(1): 11–19.

Bertolli, S., Rapchan, G., and Souza, G. 2012. Photosynthetic limitations caused by different rates of water-deficit induction in Glycine max and Vigna unguiculata. *Photosynthetica*, 50(3): 329–336.

Cambridge Academic Content Dictionary 2018. The meaning of response in English Dictionary. https://dictionary.cambridge.org/dictionary/english/response. Accessed on 14 November 2018.

Cao, K. F. 2000. Water relations and gas exchange of tropical saplings during a prolonged drought in a Bornean heath forest, with reference to root architecture. *Journal of Tropical Ecology*, 16(1, January): 101–116.

Clearwater, M. J., Susilawaty, R., Effendi, R., and van Gardingen, P. R. 1999. Rapid photosynthetic acclimation of Shorea johorensis seedlings after logging disturbance in Central Kalimantan. *Oecologia*, 121(4): 478–488.

Dierick, D., and Holscher, D. 2009. Species-specific tree water use characteristics in reforestation stands in the Philippines. *Agricultural and Forest Meteorology*, 149(8): 1317–1326.

Ecosia Knowledge Base 2018. What kinds of trees are planted in Indonesia? https://ecosia. zendesk.com/hc/en-us/articles/115002389385-What-kinds-of-trees-are-planted-in-Indonesia-. Accessed on 14 November 2018.

Eschenbach, C., Glauner, R., Kleine, M., and Kappen, L. 1998. Photosynthesis rates of selected tree species in lowland Dipterocarp rainforest of Sabah, Malaysia. *Journal of Ecology*, 12(6): 356–365.

FAO 2002. Case study of tropical forest plantations in Malaysia by D.B.A Krishnapillay. Forest Plantations Working Paper 23. Forest Resources Development Service, Forest Resources Division. FAO: Rome. Unpublished.

Garkoti, S. C., Zobel, D. B., and Singh, S. P. 2003. Variation in drought response of Sal (*Shorea robusta*) seedlings. *Tree Physiology*, 23(5): 1021–1030.

Hiromi, T., Ichie, T., Kenzo, T., and Ninomiya, I. 2012. Interspecific variation in leaf water use associated with drought tolerance in four emergent dipterocarp species of a tropical rain forest in Borneo. *The Journal of Forest Resources*, 17(4): 369–377.

Hopkins, W., and Hüner, N. 2008. *Introduction to Plant Physiology* (4th ed.). John Wiley and Sons, Inc: Hoboken, NJ.

Kettle, C. 2010. Ecological considerations for using dipterocarps for restoration of lowland rainforest in Southeast Asia. *Biodiversity and Conservation*, 19(4): 1137–1151.

Koyama, K., and Takemoto, S. 2014. Morning reduction of photosynthetic capacity before midday depression. *Scientific Reports*, 4.

Lantican, C. B. 2015. Philippine native trees—What to plant in different provinces. file:///C:/ Users/Now%20or%20Never/Downloads/Philippine%20Native%20Trees.pdf-DrCBLantican-30jan2015.pdf. Accessed on 14 November 2018.

Larcher, W. 2003. *Physiological Plant Ecology: Ecophysiology and Stress Physiology of Functional Groups* (4th ed.). Springer-Verlag: Berlin and Heidelberg.

Lawson, T., and Blatt, M. R. 2014. Stomatal size, speed and responsiveness impact on photosynthesis and water use efficiency. *Plant Physiology*, 164(4): 1556–1570.

Lichtenthaler, H. K. 1998. The stress concept in plants: An introduction. *Annals of the New York Academy of Sciences*, 851: 187–198.

Mainali, K. P., Tripathi, R., Jha, P. K., and Zobel, D. B. 2006. Water relations of *Shorea robusta* in mixed Sal forest. *International Journal of Ecology and Environmental*, 32(2): 143–152.

Malaysian Meteorological Department 2016. Bulletin of monthly weather February and March 2016. http://www.met.gov.my/web/metmalaysia/publications/bulletinpreview/ monthlyweather?p_p_id=122_INSTANCE_phshFS8dJslQandp_p_lifecycle=0andp_p_ state=normalandp_p_mode=viewandp_p_col_id=118_INSTANCE_F96odbUpU62g_ column-2andp_p_col_count=1andp_r_p_564233524_resetCur=trueandp_r_p_ 564233524_categoryId=30094. Accessed on 1 June 2016.

Malaysian Timber Council 2018. MTC wood wizard. http://mtc.com.my/wizards/mtc_tud/ items/report(95).php. Accessed on 14 November 2018.

Malhi, Y., and Wright, J. 2004. Spatial patterns and recent trends in t0he climate of tropical rainforest regions. *Philosophical Transactions of the Royal Society Series B: Biological Sciences*, 359(1443): 311–329.

Maripa, R. D. 2018. Adaptation of Dipterocarp species to drought stress in Sabah, Malaysia. PhD Thesis. University of Melbourne: Australia. Unpublished.

Maruyama, Y., Matsumoto, Y., Morikawa, Y., Ang, L. H., and Yap, S. K. 1997a. Leaf water relations of some dipterocarps. *Journal of Tropical Forest Science*, 19(2): 249–255.

Maruyama, Y., Toma, T., Ishida, A., Matsumoto, Y., Morikawa, Y., Ang, L. H., and Iwasa, M. 1997b. Photosynthesis and water use efficiency of 19 tropical tree species. *Journal of Tropical Forest Science*, 9(3): 434–438.

Model 600 Pressure Chamber Instrument 2009. http://www.pmsinstrument.com/products/model-600-pressure-chamber-instrument. Accessed on 14 November 2018.

Newberry, D. M., and Lingenfelder, M. 2009. Plurality of tree species responses to drought perturbation in Bornean tropical rain forest. *Plant Ecology*, 201(1): 147–167.

Newman, M., Burgess, P., and Whitmore, T. 2000. *Manuals of Dipterocarps for Foresters: Borneo Island Light Hardwoods*. CIFOR and Royal Botanical Garden Edinburgh: Edinburgh.

Nibu, A. N., and Duju, A. 2011. Sampling and collection. In: *Properties of Shorea macrophylla (Engkabang Jantong) Planted in Sarawak*. Lim, N. P. T., Tan, Y. E., Gan, K. S. and Lim, S. C. (eds.). Forest Research Institute Malaysia: Selangor.

Norisada, M., and Kojima, K. 2005. Photosynthetic characteristics of dipterocarp species planted on degraded sandy soils in southern Thailand. *Photosynthetica*, 43(4): 491–499.

Pallardy, S. G. 2008. *Physiology of Woody Plants* (3rd ed.). Elsevier Inc: New York, NY.

Perumal, M. 2012. Growth performance and survival rate of planted *Shorea macrophylla* at various age stands in Sampadi Forest Reserve. Bachelor of Science with Honours (Plant Resource Science and Management Thesis. Universiti Malaysia Sarawak. Unpublished.

Pietragalla, J., and Pask, A. 2012. *Stomatal Conductance. Physiological Breeding II: A Field Guide to Wheatphenotyping*. CIMMYT: Mexico, pp. 15–17.

PP Systems TPS-2 Portable Photosynthesis System Operator's Manual Version 2.01 2007. PP Systems Incorporation: Hertfordshire.

Roa, P. B., Kaur, A., and Tewari, A. 2008. Drought resistance in seedlings of five important tree species in Tarai region of Uttarakhand. *Tropical Ecology*, 49(1): 43–52.

Selye, H. 1936. A syndrome produced by diverse nocuous agents. *Nature*, 138(32): 32–34.

Shamala, S. 2015. Expect moderate El-Niño from September. Daily Express Newspaper Online, Sabah, Malaysia. http://www.dailyexpress.com.my/news.cfm?NewsID=101049. Accessed on 15 October 2015.

Shorea Organization 2018. What makes *Shorea* so valuable? http://www.shorea.org/. Accessed on 18 October 2018.

Singh, S. P., Zobel, D. B., Garkoti, S. C., Tewari, A., and Negi, C. M. S. 2006. Patterns in water relations of central Himalayan trees. *Tropical Ecology*, 47(2): 159–182.

Soepadmo, E., Saw, L. G., and Chung, R. C. K. 2004. *Tree Flora of Sabah and Sarawak*, Volume 5. Forestry Department: Sabah.

Stanton, K., and Mickelbart, M. V. 2014. Maintenance of water uptake and reduced water lost contribute to water stress tolerance of *Spirea alba* Du Roi and *Spirea tomentosa* L. *Horticulture Research*, 14033.

Taiz, L., and Zeiger, E. 2006. *Plant Physiology* (4th ed.). Sinauer Associates, Inc., Publishers: Sunderland, MA.

Tata, H. L., Wibawa, G., and Joshi, L. 2010. *Enrichment Planting with Dipterocarpaceae Species in Rubber Agroforests: Manual*. World Agroforestry Centre (ICRAF): Bongor, p. 23. http://www.worldagroforestry. Org/sea/publication?do=view:pub_detailandpub_no=MN0047-11. Accessed on 14 November 2018.

Tewari, A. 1999. Tree water relations study in Sal (*Shorea robusta* Gaertn.) forest in Kumaon Central Himalaya. *Journal of Environmental Biology*, 20(4): 353–357.

Tinus, R. W. 1996. Cold hardiness testing to time lifting and packing of container stock: A case history. *Tree Planters' Notes*, 47: 62–67.

Tolentino, J. E. L., and Camacho, L. D. 2006. Dipterocarps as forest restoration species: Environmental and economic implications of ecophysiological studies in the nursery. Paper presented at the 8th Round-Table Conference on Dipterocarps, Ho Chi Minh City, Vietnam. http://vaf.gov.vn/en/2006/09/dipterocarps-as-forest-restoration-species-environmental-and-economic-implications-of-ecophysiological-studies-in-the-nursery/.

Utsugi, H., Okuda, S., Luna, A. C., and Gascon, A. F. 2009. Differences in growth and photosynthesis performance of two dipterocarp species planted in Laguna, the Philippines. *Japan Agricultural Research Quarterly: JARQ*, 43(1): 45–53.

Walsh, R. P. D. 1996. Drought frequency changes in Sabah and adjacent parts of northern Borneo since the late nineteenth century and possible implications for tropical rain forest dynamics. *Journal of Tropical Ecology*, 12(3): 385–407.

Widiyantno, W., Soekotjo, S., Naiem, M., Purnomo, S., and Setiyanto, P. E. 2014. Early performance of 23 dipterocarp species planted in logged-over rainforest. *Journal of Tropical Forest Science*, 26(2): 259–266.

Wikipedia 2018. Shorea. https://en.wikipedia.org/wiki/Shorea. Wikipedia Foundation, Inc. Accessed on 17 September 2018.

Williams, L., and Araujo, F. 2002. Correlations among predawn leaf, midday leaf, and midday stem water potential and their correlations with other measures of soil and plant water status in *Vitis vinifera*. *Journal of the American Society for Horticultural Science*, 127(3): 448–454.

Wong, A. H. H., Kim, Y. S., Singh, A. P., and Ling, W. C. 2005. Natural durability of tropical species with emphasis on Malaysian Hardwoods—Variations and prospects. file:///C:/Users/Now%20or%20Never/Downloads/Natural_Durability_of_Tropical_Species_with_Emphas.pdf. Accessed on 14 November 2018.

Zhang, J. L., and Cao, K. F. 2009. Stem hydraulics mediates leaf water status, carbon gain, nutrient use efficiencies and plant growth rates across dipterocarp species. *Functional Ecology*, 23(4): 658–667.

Zhang, J. L., Meng., L. Z., and Cao, K. F. 2009. Sustained diurnal photosynthetic depression in uppermost-canopy leaves of four dipterocarp species in the rainy and dry seasons: Does photorespiration play a role in photoprotection? *Tree Physiology*, 29(2): 217–228.

7 Wood Anatomical Structure of *Aquilaria malaccensis* Associated with Agarwood Occurrence and the Uses of Agarwood

Shirley Marylinda Bakansing,
Mohd Hamami Sahri and Kang Chiang Liew

CONTENTS

7.1 INTRODUCTION

Aquilaria malaccensis is a tree in Thymelaeaceae highly valued and sought for its resinous scented wood, known widely as *agarwood, aloeswood, eaglewood* and *gaharu*. The tree produces a unique aromatic resin when it is wounded or in a stress condition. The wound allows pathogenic infection, triggering the wood cells to secrete resin as a defensive chemical. The secretion and deposit of resin in the wood cells transforms the wood into agarwood. It is suggested that the ability of *A. malaccensis* to produce aromatic resin is due to its unique anatomical structure; its physiological process, along with endophyte activity in the wood, produces chemicals as a defense mechanism against fungi and bacteria when the tree is wounded.

A. *malaccensis* wood is composed of secondary xylem and interxylary phloem. The secondary xylem consists of fibres, vessel elements and parenchyma cells. There are abundant axial and ray parenchyma cells in the stem and branches of *A. malaccensis*, which are used for carbohydrate storage, food and chemical transportation and extractive secretion. The interxylary phloem is composed of sieve elements, companion cells and axial and ray parenchyma cells. The physiological activity of the xylem parenchyma cells and interxylary phloem could be the source of resin formation. In infected or inoculated trees, brown or dark brown resin is found accumulated in the interxylary phloem and xylem parenchyma cells. In highly impregnated resinous wood, resin can also be found accumulated in the fibre wall and vessel lumens. The formation of agarwood is followed by the development of an aromatic scent in the wood, a change in colour, an increase in weight and density and an increase in hardness.

Agarwood has many uses, mainly perfumery, frankincense, pharmaceuticals, and medicinal and ritual practice. The high market demand for agarwood has resulted in the species becoming scarce due to overharvesting, and because of this it has been listed in CITES (Appendix II in the Convention on International Trade in Endangered Species of Wild Fauna and Flora) as a threatened species since 1995 (CITES, 2005). Nowadays, *A. malaccensis* has become a plantation species planted in small farms or large plantations, and it has become a domestic tree, planted in garden yards in villages for its agarwood and other products.

Hence, this article aims to elaborate on the anatomical structure of *A. malaccensis* wood in relation to agarwood resin formation. This article also gives a brief review of the uses of agarwood.

7.1.1 Synonyms for *Aquilaria malaccensis* Lamk. and Local Names

Aquilaria malaccensis is a species in Thymelaeaceae that produces important and unique scented oleoresin in the wood, known widely as agarwood. *Aquilaria malaccensis* is synonymous with *Agallochum malaccense* (Lamk.) O.K., *Agallochum malaicense* Rumph., *Agallochum secundarium coinamense* Rumph., *Aquilaria agallocha* Roxb., *Aquilaria agallocha* Roxb. ex DC., *Aloexylum agallochum* Lour., *Aquilaria agallochum* Roxb., *Aquilaria moluccensis* Oken, *Aquilaria ovata* Cav., *Aquilaria secundaria* DC, *Aquilariella malaccensis* (Lamk.) van Tiegh and *Aquilariella maccens* (Slik, 2009) and *A. grandiflora* (National Parks Board, Singapore (NPBS)).

A. malaccensis can be found in evergreen tropical rainforest and sub-tropical forests in Southeast Asia, viz., in Bangladesh, Bhutan, Cambodia, India, Indonesia, Laos, Malaysia, Myanmar, Philippines, Singapore, Thailand, Vietnam and Papua New Guinea (Barden et al., 2000). *A. malaccensis* is also grown in China, Hong Kong and Papua New Guinea (Burkill, 1966). Barden et al. (2000) stated that out of 27 *Aquilaria* species worldwide, 24 *Aquilaria* species can be found growing naturally in Southeast Asian countries. About 17 *Aquilaria* species grow as sub-canopy trees in the hardwood hill forests of tropical rainforests in Southeast Asia (Chung and Purwaningsih, 1999). There have been reports that the best quality agarwood has been found in Bhutan and on Borneo Island (Yamada, 1995).

The *A. malaccensis* tree is known by different local names in different countries. The local names also vary in different states. Slik (2009) stated that in Borneo Island, this tree species and its wood are known variously as *Alas, Calambac, Ching Karas, Gaharu, Galoop, Garu, Gharu, Karas, Kayu gaharu, Kekaras, Kepang, Laroo, Mengkaras,* Ngalas, *Sigi-sigi, Tabak, Taras gharu* and *Tengkaras*. In Malaysia, the *A. malacccensis* tree is commonly known as *Karas* and *Tengkaras* and *Gaharu*. In Indonesia, *A. malaccensis* is known as *Mengkaras* (Sumatra), *Ki Karas* (Sundanese) and *Gaharu*. In Northeast India, it is called the *Sanchi* plant (Nath and Saikia, 2002).

7.1.2 BOTANICAL AND GROWTH CHARACTERISTICS OF *AQUILARIA MALACCENSIS*

The botanical and characteristics of *A. malaccensis* are shown in Table 7.1 and Figure 7.1.

The fruiting, seedling and sapling characteristics of *A. malaccensis* have been studied to gather information about its regeneration. Chua (2008) in her report cited that *A.malaccensis* produces fruit after 7–9 years. The fruit viability is about 1 week and the seed germinates between 16 and 63 days (Ng, 1992). However, after transplanting of seedlings to the farm, some seedlings die. Lok et al. (1999) found that in the *Aquilaria* farm, the survival rate was 40.3%. Soehartono and Newton (2001) found that seedlings were distributed under the mother tree canopy a few meters from the tree. At this stage, light is important for their growth. Seedlings can also grow under the mother tree, around the tree base, if the tree crown is less dense, allowing 60–70% of light to reach the forest floor. Paoli et al. (2001) reported that seedlings and saplings grow better if they are away from the mother tree. The seedling will grow better when the mature tree has been cut down. Another study by Paoli et al. (2001) in Kalimantan revealed that the abundance of seedlings is not influenced by the size of mother trees, and they reported that the mean seedlings' density within 3–7 m of *Aquilaria* is more than 2 trees/m^2 and reduced rapidly to <0.01 trees/m^2 beyond 15 m.

The growth rate and timber volume of *A. malaccensis* in the natural forest and on the farm were studied for yield estimation. *A. malaccensis* has a mean height of 26.7 m, a mean diameter breast height of 38.2 cm and a a mean clear bole height of 15.7 m (Lok and Zuhaidi, 1996). A report by the Forestry Department Peninsular Malaysia stated that in the natural forest, it was estimated that there were 0.62 stems per hectare for size classes more than 10 cm dbh (diameter breast height) for all tree species of *Aquilaria*. These stems produce an estimated 0.311 m^3/ha of timber volume. It was also estimated that trees with 10–14.9 cm dbh contributed 48% of the

TABLE 7.1

Botanical Characteristics of *Aquilaria malaccensis*

Part of Tree	Characteristics	Source
Tree height	Up to 20 m (or 49 m)	Donovan and Puri, (2004); Slik (2009)
Bole diameter	Up to 60 cm	Slik, (2009); Chakrabarty et al. (1994)
Bole shape	Usually straight, sometimes fluted or with thick (10 cm) buttresses up to 2 m high	Slik (2009)
Bark	Smooth, whitish	Slik (2009)
Branchlets	Slender, pale brown, pubescent, glabrescent.	Slik (2009)
Leaves	Simple, alternate; petiole 4–6 mm long; blade elliptical-oblong to oblong-lanceolate, 7.5–12 cm × 2.5–5.5 cm, chartaceous to subcoriaceous, glabrous, sometimes pubescent and glabrescent beneath, shiny on both surfaces, papery texture, base acute, attenuate or obtuse, apex acuminate, acumen up to 2 cm long; veins in 12–16 pairs, rather irregular, often branched, elevated and distinct beneath, curving upward to the margin, plane or obscure above.	Slik (2009); Lee and Mohamed (2016)
Inflorescence	A terminal, axillary or supra-axillary, sometimes internodal umbel, usually branched into 2–3 umbels, each with about 10 flowers; peduncle 5–15 mm long; pedicel slender, 3–6 mm long; flowers 5-merous, campanulate, 5–6 mm long, green or dirty-yellow, scattered puberulous outside; floral tube nearly glabrous inside, distinctly 10-ribbed, persistent in fruit; calyx lobes 5, ovate-oblong, 2–3 mm long, almost as long as the tube, reflexed, densely puberulous within; petaloid appendages 10, inserted at the throat of the tube, oblong or slightly ovate-oblong, about 1 mm long, slightly incurved, densely pilose; stamens 10, free, emerging from the throat of the tube, filamentous, 1.2–2 mm long, episepalous ones longer than the others; anthers linear, obtuse; pistil included; ovary ovoid, 1–1.5 mm long, 2-celled, densely pubescent; style obscure, stigma capitate.	Slik (2009); Lee and Mohamed (2016)
Fruit	A loculicidal capsule, obovoid or obovoid-cylindrical, 3–4 cm × 2.5 cm, usually compressed, pubescent, glabrescent, base cuneate, apex rounded, pericarp woody; seed ovoid, 10 mm × 6 mm including a beak 4 mm long, densely red-haired, bearing from the base a twisted, tail-like, pubescent appendage as long as the seed; seedling with epigeal germination.	Slik (2009); Lee and Mohamed (2016)

FIGURE 7.1 *A. malaccensis.* A. A wild *A. malaccensis* tree. B. Planted sapling, photo shows leaf shape and arrangement. C. Young bark. D. Matured bark. E. Leaf veins. F. Dried leaf, fruit and seed.

total stem number, while trees with 15–44.9 cm dbh contributed 67% of the total volume (Anon, 2008). Asgarin (2004), in a study of 2 ha permanent forest plots in West Kalimantan, revealed that the growth rate of *A. malaccensis* over 6 years is 0.5 cm/year. *A. malaccensis* planted in Bengkulu, Sumatra, meanwhile, showed an annual growth of 2 cm/year in the early stages. Wulffraat (2006) in his study on small plantation trials related to soil, drainage and light demands (outside and inside the forest) revealed that the increase in the height and diameter of *A. malaccensis* is fastest when they are planted outside the forest.

7.1.3 Names of Agarwood

Agarwood is defined as an aromatic resinous wood from the trees of *Aquilaria* genus. Lim and Anak (2010) in their CITES Report stated that in Malaysia, the names *agarwood* and *gaharu* refer only to resin-impregnated pieces of wood (Grade C and above) that have been carved or separated from the non-impregnated wood, whereas the name *Karas* is the standard name for the timber of *Aquilaria* trees. Agarwood has various names according to different countries and languages. Agarwood is also called different names based on its grades. In Malaysia and Borneo Island, the resinous wood of *A. malaccensis* or from other *Aquilaria* species are called *Gaharu*, *Kayu Gaharu*, *Alas*, *Calambac*, *Ching karas*, *Galoop*, *Garu*, *Gharu*, *Karas*, *Kekaras*, *Kepang*, *Laroo*, *Mengkaras*, *Ngalas*, *Sigi-sigi*, *Tabak*, *Taras gharu*, *Tengkaras* (Slik, 2009), *kayu karas*, *gaharu* (Indonesia), calabac, *karas*, *kekaras*, *mengkaras* (Dayak), *galoop* (Melayu), *halim* (Lampung), *alim* (Batak) and *kareh* (Minang) (Hou, 1960). Lopez-Sampson and Page (2018) cited that in English, it is commonly known as *agarwood*, *aloeswood*, *Malayan aloeswood*, *Malayan eaglewood* or *eaglewood*. In other parts of Europe, it is called *Lignum aquila*, *Agilawood*, *Lignum aloes* or *aloeswood*. In French, it is called *Bois d'aigle*, *calambac* and *calambour*. In Myanmar, it is called *agar*. In Sanskrit, it is called *agāru*, and *aguru* means non-floating wood. In Greek it is called *agallochum*. In Hebrew, it is called *ahāloth*, and in Arabic it is called *alūwwa*, *ūd*, *oudh* or *oud*. In Chinese, it is called *Chan Hsiang*, *Chi Ku Hsiang*, *Huang Shu Hsiang*, *chén xiāng* or *kilam*; *bac*, mean 'Sinking incense', and *Cham Heong* in Cantonese. In Japanese, it is called *jinkoh*, meaning 'Sinking incense'. In Portugese, it is called *aguila* or *pao d'aguila*. The name is *Chim-Hyuang* in Korea, *Kritsana noi* and *Mai Kritsana* in Thailand, Mai Ketsana in Laos, *Tram Huong* in Vietnam, *oud* in the Middle East, *agor* in Bangladesh, *Agar* or *Aguru* in Bengali, Agar in Pakistan, *Agaru* in Tibetan, *Ghara* or *eaglewood* in Papua New Guinea, *adlerholz* in Germany, *kikaras* in Sundanese, *alim*, *halim* or *karek* in Sumatra, *agaru* or *sasi* and *Akil* (in Tamil) in India and *kanakoh* and *Tram huong* in Vietnam (Ng, et al., 1997; Yamada, 1995; Chakrabarty et al., 1994).

Agarwood is also called 'black gold', 'wood of the Gods', 'sinking wood', 'fragrant wood', 'scented wood' and 'aromatic resinous wood', while the aromatic odour is called 'scent from heaven' (Akter et al., 2013).

7.2 *AQUILARIA MALACCENSIS* WOOD MACROSCOPIC AND MICROSCOPIC CHARACTERISTICS

A. malaccensis is a unique hardwood which can be identified by the existence of xylem and interxylary phloem or includes phloem tissues in the stem, branches and roots. The xylem of *A. malaccensis* wood is composed of fibres, vessel elements and axial and ray parenchyma cells. The interxylary phloem tissues occurs consistently among the xylem tissues in the wood stem, branches and roots (Figure 7.2). The interxylary phloem tissues exist as islands arranged tangentially in elongated shape. *A. malaccensis* wood is capable of producing resin; however, epithelial cells, resin canal or resin ductus do not exist in the wood.

7.2.1 Tree Rings

Tree rings exist in the *A. malaccensis* stem but are indistinct (Figure 7.3). The tree rings are marked by dark and light colours due to changes in fibre diameter. Rings of a light colour formed by fibres of wide diameter are similar to 'earlywood'; the darker

FIGURE 7.2   Cross-section of twig wood of *A. malaccensis*. 40×. f = fibre, v = vessel elements, rp = ray parenchyma, ip = interxylary phloem. Bar = 5 μm.

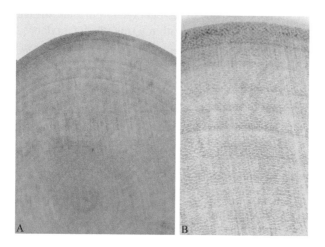

FIGURE 7.3   Transverse section of *A. malaccensis* stem showing tree rings. A. Irregular tree rings marked by colour changes in fibres and interxylary phloems. B. Distribution and arrangement of interxylary phloems contribute to tree-ring occurrence. Abundant interxylary phloem islands marked by elongated whitish spots.

coloured parts are formed by narrow-diameter fibres and are similar to 'latewood' in growth rings. Changes in fibre diameter take place abruptly in the transition to a new tree ring formation (Figure 7.4). The occurrence of tree rings in *A. malaccensis* stem is also due to the varying size, distribution and arrangement of the interxylary phloems (Figure 7.3B). The sizes, particularly the tangential diameters, vary along the radial direction and vary among the rings. The distribution of interxylary phloems varies within a ring and varies among the rings. In Figure 7.3B, the loose distribution of interxylary phloem causes the 'earlywood' to appear darker, which indicates

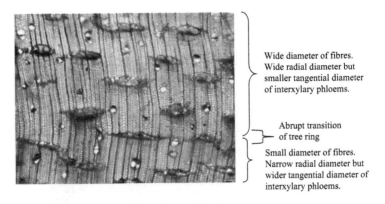

Wide diameter of fibres.
Wide radial diameter but
smaller tangential diameter
of interxylary phloems.

Abrupt transition
of tree ring

Small diameter of fibres.
Narrow radial diameter but
wider tangential diameter of
interxylary phloems.

FIGURE 7.4 Tree rings anatomical structure of *A. malaccensis* wood. Tree rings are formed by changes in fibre diameters and interxylary phloem sizes.

the presence of more fibres in this part. Meanwhile, the congested 'latewood' with smaller interxylary phloems cause the 'latewood' to look lighter in colour. The occurrence of tree rings is also due to the interxylary phloems' arrangement within a ring and along the radial direction. Within a ring, the interxylary phloem can be arranged in diffuse-alternate arrangement at the 'earlywood' and changes to a diffuse-opposite arrangement in the 'latewood', or the interxylary phloem could be entirely in an alternate arrangement but congested in the 'latewood'. Also within a ring, the interxylary can be found arranged in diffuse-alternate or diffuse-opposite in the 'earlywood' and can change to a diagonal arrangement in the 'latewood'. The diagonal pattern is might be due to combination effect of smaller sizes, congestion and alternate arrangement of interxylary phloems in the 'latewood'.

The number of tree rings in an *A. malaccensis* tree stem is inconsistent. In 5-year-old trees, stems with a diameter of 13 to 16 cm diameter have 11–16 tree rings. The tree rings within a stem showed irregular widths, ranging from 2.4 to 13.4 cm wide.

7.2.2 WOOD COLOUR AND ODOUR

Naturally, the *A. malaccensis* wood is soft, light, white or yellowish in colour and non-aromatic or odourless. The wood colour is white or pale yellowish (Figure 7.5A). In mature trees, the sapwood colour can be differentiated by the pale yellowish-brown colour of the heartwood. In young trees, the colour of the sapwood and heartwood is not differentiated (Figure 7.6A). Infected or inoculated stems will develop resinous wood, changing the light colour to a dark brown (Figure 7.6B). According to Ogata et al. (2008) and Richter and Dallwitz (2000), the sapwood and heartwood are not differentiated by colour, which is a pale yellowish grey to pale yellowish brown, without lustre. In infected wood of *A. malaccensis*, the wood is dark brown with white stripes or black with brown stripes. The colour is usually irregular, from light brown to dark brown or blackish dark brown (Figure 7.5B–D). A change in colour is followed by the development of an aromatic or sweet odour. The odour of agarwood varies considerably, and therefore the colour and aromatic odour are usually

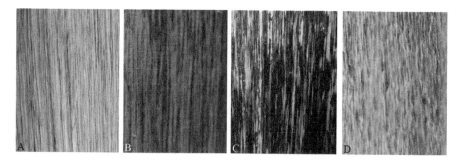

FIGURE 7.5   Colour of *A. malaccensis* wood. A. Original white colour of *A. malaccensis* wood. Brown stripes are a sign of agarwood formation. B. White colour changed to dark and brown stripes, indicates more formation of agarwood. The wood sample was obtained from natural forest. C. Dark brown blackish colour of agarwood. D. Brown reddish colour of agarwood. Agarwood developed after inoculation.

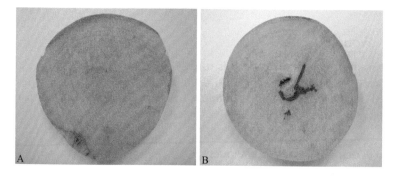

FIGURE 7.6   Cross-section of young *A. malaccensis* stem. A. Similar colour of heartwood and sapwood. B. Dark areas (dark brown) indicate developed agarwood in the heartwood.

used to determine agarwood quality. A dark colour and aromatic scent in the wood are caused by the deposition of resin mainly inside the parenchyma and interxylary phloem tissues, and the impregnation of resin into the fibre wall. A darker colour and strong aromatic scent are usually related to a high resin content in agarwood, and thus these two characteristics are used to determine its quality and grade. Liu et al. (2017) in their review stated that agarwood from different countries shows wide variations in resin colour, such as dark green, green, red, gold, black, red, brown, yellow and white. Agarwood pieces with a darker colour are considered to have a high resin content, and thus are classified as high-grade agarwood. However, agarwood pieces may be mistakenly low graded and priced cheaply if the non-resinous wood is not scraped off the agarwood.

Liu et al. (2017) described the aromatic odour of agarwood as pleasing and complex. The aroma from agarwood when it is burnt is widely used in the art of incense in China and Japan. The aroma is the main factor used by customers to select agarwood when buying it. Usually, customers will burn an agarwood piece and smell the aroma. Agarwood producing a softer aroma is considered to be a higher grade agarwood compared to an agarwood emitting a strong aroma. This type of agarwood is more favorable and expensive. Low-quality agarwood pieces produce a woody odour that irritates the nose and eyes. In high-quality agarwood, the pieces will burn evenly, gradually releasing the aroma, and the aroma lingers in the room for a longer time.

Liu et al. (2017) further explained in their review that in Japan, the aroma and taste of agarwood, or *jinkoh*, are used to classify the agarwood for *kodo* (Way of Fragrance), the art of appreciating Japanese incense. The agarwood is classified as *kyara, rakoku, manaban, manaka, sasora* and *sumatora*. For example, *kyara* (from the Sanskrit word *kara*, meaning 'black') is the highest quality agarwood, having five flavours of taste and being valued for its elegant aroma.

Jung (2011) cited that professional perfumers describe the agarwood fragrance as a 'highly complex accord' because of various chemical contents, and it can be distinguished by an aromatic mixture of 'oriental-woody' and 'very soft fruity-floral'. The incense smoke is described as 'sweet-balsamic' and 'shades of vanilla and musk' aroma and ambergris.

7.2.3   WOOD GRAIN AND TEXTURE

The grain of *A. malaccensis* wood is usually straight (Figure 7.7A). The grain is occasionally interlocked (Figure 7.7B). In infected wood, the wood grain can be distorted (Figure 7.7C). The wood texture is fine, and can be uneven and coarse due to existence of interxylary phloem tissues among the xylem cells.

7.2.4   WOOD DENSITY AND HARDNESS

The weight and density of *A. malaccensis* wood increase when the wood cells are impregnated with resin. Chakrabarty et al. (1994) and Rao and Dayal (1994) found that the wood density of *A. malaccensis* is 0.4 g/cm^3. Karlinasari et al. (2016) reported that wood density was higher in inoculated *A. malaccensis* trees compared to untreated trees. The average wood densities in untreated trees were 0.55 g/cm^3 (green condition) and 0.27 g/cm^3 (air-dried condition), whereas the wood densities

FIGURE 7.7   *A. malaccensis* wood grain. A. Straight grain. B. Interlocked grain. C. Distorted grain due to infection. Black spots are agarwood developed in the wood.

in treated trees were 0.59 g/cm³ (green condition) and 0.30 g/cm³ (air-dried condition). Ogata et al. (2008) found that the air-dry specific gravity (SG) for normal or uninfected wood is 0.33–0.60, and commonly the SG is 0.35–0.50. The resin-impregnated wood is higher in SG, with an air-dry SG of more than 0.8. Based on this knowledge, traders and buyers of agarwood believe that the heavier the wood is, the higher the resin content of the agarwood and therefore the higher the grade. Liu et al. (2017) explain that the weight and density of agarwood is used to determine its quality. In China, agarwood quality is determined by sinking it in water. Agarwood pieces will be tested in a glass bowl or glass beaker filled with water to determine its sinkage properties and is classified based on three grades. The pieces that sink to the bottom are classified as *sinkage agarwood*, which is the highest grade. The half-sinkage or half-floating pieces are in second grade, classified as *stack incense*. The full-floating is classified as *half-mature incense* grade. However, the determination of high-grade agarwood does not necessarily follow this grading system. Hainan agarwood (originating from Hainan, China) and and *tagara* agarwood (top-grade agarwood) are in half-sinkage and full-floating grades. In the Chinese pharmaceuticals market, the grading is also based on four levels of resin content determined by weight ratio to agarwood weight: 80%, 60%, 40% and 25%.

The weight of agarwood is related to the resin content, which can be used to estimate the agarwood oil production. Antonopoulou et al. (2010), the United Arab Emirates (UAE) CITES management uses an official conversion formula established by a leading agarwood trading company in UAE. The formula explains that to produce 6 toulas of oil, 10 kg of agarwood is needed. Therefore, to produce 1 litre of oil, 143.6 kg of agarwood will be used. It was estimated that 1 kg/1l contains approximately 86 toulas, with 1 toula being calculated as equal to 11.6 g. Other traders estimate that 1 kg of agarwood will yield from 3 to 10 g of oil. Thus, between 100 and 333 kg of agarwood are needed to produce 1l g of oil. However, the yield depends on the efficiency of the oil distillation technology.

The hardness of *A. malaccensis* wood increases when the wood is impregnated with resin. The hardness can be easily determined just by touching or pressing the wood with a fingernail. It is obvious that resinous wood is harder than the non-resinous wood. Karlina et al. (2017) stated that the average wood hardness of inoculated trees was 146.08 g·cm⁻²; however, the study was not carried out on healthy trees for comparison.

7.3 *AQUILARIA MALACCENSIS* XYLEM MICROSCOPIC ANATOMICAL STRUCTURE

7.3.1 FIBRES

Fibre cells are the main tissues in the *A. malaccensis* wood stem, branches and roots (Figure 7.8). The fibres are xylem cells that function as the main mechanical support and water conduction mechanism for the tree growth. The fibres in *A. malaccensis* vary in structure and dimensions in different parts of the tree. The fibres are fibre tracheid with bordered pit of cell wall (Figure 7.8B). Generally, the fibres of *A. malaccensis* are slender, non-septate, short in length and small in diameter

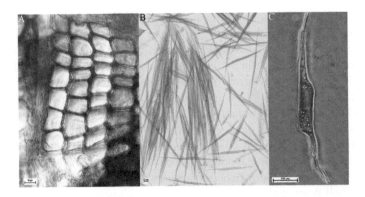

FIGURE 7.8 A. Cross-section of fibre cells of *A. malaccensis* wood. 200×. Bar = 10 μm. B. Fibre cells of *A. malaccensis*. 40×. Bar = 50 μm. C. Crooked shape and pointed end of fibre cell. 100×. Bar = 100 μm.

(Figure 7.8). The fibres can be crooked in shape. The fibre ends are pointed and blunt and sometimes have a forked end (Figure 7.9). Observation of the cross-section of *A. malaccensis* wood shows that the fibres are arranged in radial rows with varying diameters and irregular cell wall thickness, in a net-like arrangement (Figure 7.8A). The fibre length is 341–1217 μm and the fibre diameter is 19–55 μm. However, these dimensions vary in the stem and branch, and among trees from the same or different locations. Ogata et al. (2008) found that the fibres are non-septate, 1.1 (0.7–1.6) mm in length, 18–25 μm in diameter and with a cell wall thickness of 2–3 μm. The fibres have a tapering end with abundant pits on the radial walls. Sathyanathan et al. (2012) found that fibres are thin walled, spindle shaped, tapering, wide or narrow lumened and without lateral wall pits. The fibre lengths vary between 350–500 μm for wide fibres and and 500–600 μm for narrow ones.

7.3.2 Vessel Elements

Vessel elements are the xylem cells for water conduction in the tree. The vessel element cells form a pipe-line like system from the roots to the branches to conduct

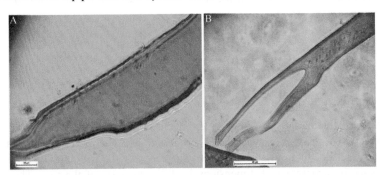

FIGURE 7.9 A. Blunt end of fibre cell. 400×. Bar = 20μm. B. Bordered pit and fork end of fibre cell. 400×. Bar = 50 μm.

water to the crown for the photosynthesis process (Figure 7.10A). The vessel element occurrence in *A. malaccensis* indicates that the wood is a hardwood type. They are diffuse porous. The vessel elements of *A. malaccensis* are solitary and multiple pores. Most of the vessel elements are arranged in radial multiple pores with 2 to 5 (up to 7) pores in group (Figure 7.10B). The intervessel pits and vessel-ray pits are alternate, and the walls in contact with the rays are congested with pits of the bordered pit type (Figure 7.11A). The vessel elements have simple perforation plates at both ends, oblique or horizontal (Figure 7.12 A–G). The vessel lumens do not contain tyloses; however, resin would be deposited in the vessel lumen when resin is produced (Figure 7.10B). Similar findings by Ogata et al. (2008) stated that vessel elements are 2–4 (–10) pores in a radial direction, about 8–14/mm². The tangential diameter of a solitary pore is 110–160 μm and multiple pores is 160–240 μm. The perforation plate is simple. The intervessel pits are alternate, with diameter of ca. 5 μm; vestured and vessel-ray pits are alternate. Tyloses were not observed in the vessel lumen.

The vessel elements in *A. malaccensis* show variations in structure, shape and cell morphology. The shape and structure may change due to the growth rate, location in

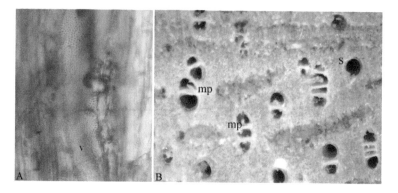

FIGURE 7.10 Vessel elements. A. Vessel elements in longitudinal series. 200×. B. Cross-section of *A. malaccensis* wood showing vessel elements types. 50×. v = vessel, s = solitary pore, mp = multiple pores.

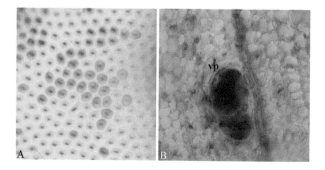

FIGURE 7.11 Vessel elements. A. Bordered intervessel pits of vessel wall. 1000x. B. Vessel elements with resin deposited in lumen. 200×. vp = vessel pores.

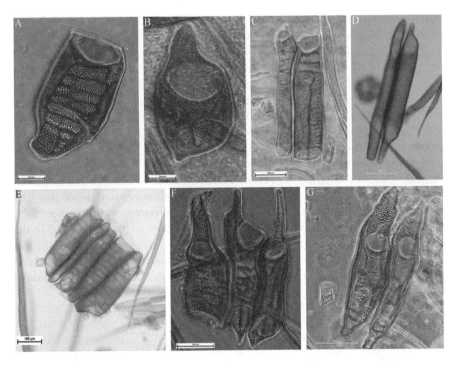

FIGURE 7.12 Various anatomical structures of vessel elements in *A. malaccensis* wood. pp = perforation plate; ip = intervessel pit; irp = intervessel-ray parenchyma pits. A. 400×, Bar = 50 μm. B. 400×, Bar = 50 μm. C. 200×, Bar = 100 μm. D. 200×, Bar = 100 μm. E. 100×, Bar = 100 μm. F. 100×, Bar = 100 μm. G. 200×, Bar = 100 μm.

the tree and tree-ring formation. The variations can be observed based on cell morphology, wall pitting, perforation plates and the presence of an extended cell wall/ tail on one or both ends. The cell shape may be (1) a barrel-like shape without tails or extended cell end with varying diameter; (2) a wide vessel element with one or both cell ends extended; (3) a slender vessel element with or without extended cell end; (4) a slender vessel element with long extended cell end; (5) a short vessel element; (6) a vessel element with side perforation plates or a wider extended cell end; (7) a vessel element with a crooked cell wall; (8) a vessel element with a crooked cell wall with side perforation plates (Figure 7.12A-G).

7.3.3 AXIAL PARENCHYMA

Xylem axial parenchyma in *A. malaccensis* is paratracheal scanty around the vessel elements. The axial parenchyma around the vessel elements are scarce. The paratracheal axial parenchyma cells also exist as unilateral and vasicentric associated with the interxylary phloem islands. The unilateral axial parenchyma also occur as a tangential band (Figure 7.13A) connecting the interxylary phloem islands. In Figure 7.13A, the axial parenchyma adjacent to the interxylary phloems has developed resin that contributes to the brown colour of the cells. Ogata et al. (2008) and

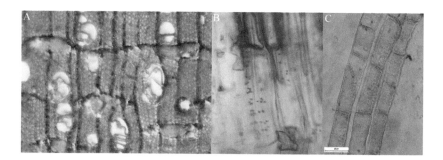

FIGURE 7.13 A. Cross-section of *A. malaccensis* showing xylem axial parenchyma arranged tangentially as bands associated with the interxylary phloem. Parenchyma cells are deposited with resin of a brown colour. 40×. B. Xylem axial parenchyma at longitudinal-tangential section of *A. malaccensis* wood. 200×. C. Xylem axial parenchyma cells. 200×.

Adams et al. (2014) found that the axial parenchyma is paratracheal scanty and found tissues around the interxylary phloem. According to Richter and Dallwitz (2000) and Mohamad et al. (2013), the axial parenchyma strands occur as paratracheal scanty with an average of two cells per strand, and as incomplete sheaths adjacent to vessels.

Xylem axial parenchyma cells in *A. malaccensis* wood are isodiametric, long, narrow and thin walled with simple pits (Figure 7.13B and Figure 7.13C). A similar result was found by Sathyanathan et al. (2012). Tapering axial parenchyma cells are sometimes present, and two tapering axial parenchyma are connected by simple pits on the cell wall (Figure 7.13B). Silica grain was found in the *A. malaccensis* wood, although Richter and Dallwitz (2000) did not find silica in *Aquilaria* wood. Crystals or mineral inclusions were not observed in the axial parenchyma.

Basically, xylem axial parenchyma cells in *A. malaccensis* are used for food (starch) and nutrient storage, food transportation, secretion and wound healing. The axial parenchym cells also store water and ions. The axial parenchyma cells in *A. malaccensis* do not form epithelial cells to secrete resin but it is suggested that all the parenchyma cells, together with the interxylary phloem tissues activities, are capable to produce resin as defense to pathological invasion. The resin produced is accumulated in the axial parenchyma cells and can be seen as a brownish substance in the cell (Figure 7.13A).

7.3.4 RAY PARENCHYMA

Xylem ray parenchyma cells are abundant in *A. malaccensis* stems, and they have a similar function to xylem axial parenchyma. The ray parenchyma cells are arranged radially in groups as uniseriate and biseriate, as can be seen in cross-section and tangential section (Figure 7.14). The xylem rays are in a non-storied arrangement, as seen in tangential section. The xylem rays are in an isodiametric shape, and are short with irregular lengths. The xylem rays are upright and procumbent, mostly of the upright type. Ogata et al. (2008) found that the xylem rays of *A. malaccensis* are ray uniseriate or biseriate, with a ray height measurement of 440–720 µm. In healthy wood, the xylem rays are without coloured content. According to Adams et al. (2016),

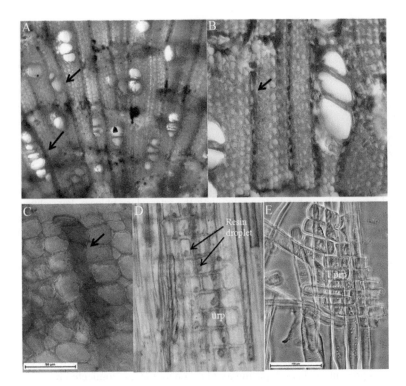

FIGURE 7.14 Xylem ray parenchyma in *A. malaccensis* wood. A. Uniseriate and biseriate xylem rays (arrows). 40×. B. Cross-section of *A. malaccensis* wood showing xylem rays (arrow). 200×. Rays are filled with brown-coloured resin. 400×. C. Cross-section of *A. malaccensis* wood showing uniseriate xylem rays (arrow). 400×. D. Radial section of *A. malaccensis* wood showing upright parenchyma cells. Each cell contains one brown resin droplet. 400×. urp = upright ray parenchyma. E. Procumbent ray parenchyma cells with irregular sizes. 400×. prp = procumbent ray parenchyma.

the rays are 1–3 layers thick and commonly single layered, seen in transverse section. In tangential section, the rays are arranged in a uniseriate and occasionally biseriate configuration. Rays are mostly homocellular but are occasionally heterocellular.

In infected or inoculated wood, the xylem rays secrete resin and accumulate it inside the cells, changing the whitish wood to a brown colour. The resin in the ray cells can be seen as resin droplets (Figure 7.14B).

7.4 INTERXYLARY PHLOEM

Interxylary phloem, or included phloem, tissues exist among the xylem cells in *A. malaccensis* wood. The tissues are islands among the xylem cells and exist consistently and in diffuse, alternate or diagonal arrangements along the radial of the stem, as can be observed in transverse section. The islands of interxylary phloem are tangentially elongated in shape and irregular in size. The islands appear whitish in colour (Figure 7.15A). The interxylary phloem tissues can be easily differentiated

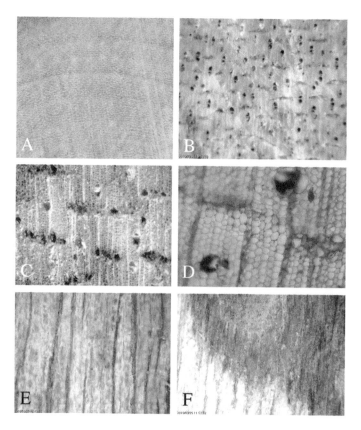

FIGURE 7.15 *A. malaccensis* wood. A. Cross-section of healthy *A. malaccensis* wood. Whitish elongated spots are interxylary phloem in diffuse, alternate and opposite arrangements. B. Impregnated wood. Resin is accumulated in interxylary phloems. 50×. C. Interxylary phloem and xylem cells impregnated with dark brown or blackish resin. 100×. D. An island of interxylary phloem consists of thin cell wall tissues and lignified cells. Brown resin is accumulated in interxylary phloem. 400×. E. Radial-longitudinal section showing resin-impregnated interxylary phloems (wide brown lines). 50×. F. Tangential-longitudinal section showing black-brown resin developed in the infected area. 50×.

from xylem cells by their shape and cell structure (Figure 7.15B and Figure 7.15C). The interxylary phloems are composed of several types of cells with a thin cell wall (Figure 7.15D). The tangential diameter is normally wider than the radial diameter (Figure 7.15C and Figure 7.15D). Ogata et al. (2008) stated that included phloem are scattered islands as seen in cross section, with a radial diameter of 120–200 µm and a tangential diameter of 320–380 µm.

An island of interxylary phloem consists of phloem fibres, or sieve elements, companion cells and phloem axial and ray parenchyma cells (Figure 7.16). The phloem fibres are elongated with wide diameter, thin cell wall and with simple wall pits. The sieve element is a long cell with wide lumen, thin cell wall, foraminate perforation plate and simple wall pits. The companion cells are smaller parenchyma cells associated with the sieve elements. The phloem axial parenchyma are cells surrounding

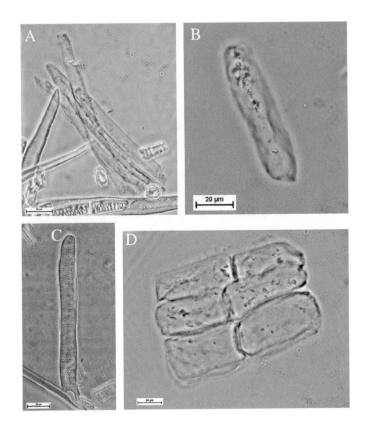

FIGURE 7.16   Interxylary phloem cells in *A. malaccensis*. A. Sieve elements or phloem fibres. 100×. Bar = 50 μm. B. Companion cells, 400×. Bar = 50 μm. C. Axial parenchyma. 400×. Bar = 50 μm. D. Ray parenchyma. 400×. Bar = 20 μm.

the island. The axial parenchyma cells have thin cell walls or may develop ligni-fied walls. The phloem ray cells are cells connected to the xylem rays, which make continuous rows of rays from pith to cambium. The xylem and phloem rays can be differentiated by the cell wall. The phloem rays have thin walls whereas the xylem rays have thicker walls (Figure 7.15D).

In infected wounded wood or inoculated trees, the xylem parenchyma and maybe the interxylary phloem cells will be triggered, and together they produce agarwood resin in them. The secretion and accumulation of resin will spread to the fibre wall, fibre lumens and vessel lumens. Accumulation and impregnation of resin changes the colour of the wood to dark brown or blackish brown (Figure 7.15E and Figure 7.15F).

7.5   A REVIEW OF AGARWOOD CHEMICAL CONTENT AND USES

7.5.1   Agarwood Source from Thymelaeaceae Trees

Agarwood is the resin produced from the trees of the Thymelaeaceae family. There are six genera in the family that are known to produce agarwood, including *Aquilaria*,

Aetoxylon, *Enkleia*, *Gonystylus*, *Gyrinops* and *Wikstroemia*. Researchers recorded 13–20 *Aquilaria* species that produce agarwood: *A. malaccensis/A. agallocha*, *A. subintegra*, *A. crassna*, *A. apiculata*, *A. acuminata*, *A. baillonii*, *A. banaensae*, *A. beccariana*, *A. brachyantha*, *A. cumingiana*, *A. filaria*, *A. grandiflora*, *A. hirta*, *A. khasiana*, *A. microcarpa*, *A. rostrata*, *A. parvifolia*, *A. sinensis* and *A. ophispermum*. Other *Aquilaria* trees that have been recorded are *A. pentandra*, *A. decemcostata*, *A. rugosa*, *A. urdanetensis*, *A. yunnanensis* and *A. Citrinicarpa*; from the genera *Gyrinops*: *G. walla, G. ledermannii, G. versteegii* and *G cumingiana*; from the genera *Aetoxylon*: *A. Sympethalum*; from the genera *Enkleia*: *E. Malaccensis*; from the genera *Gonystylus*: *G. banccanus* and *G. Macrophyllus*; from the genera *Wikstroemia*: *W. polyantha, W. androsaemofolia* and *W. tenuriamis* (Hashim et al., 2016; Hou, 1960; Lee and Mohamed, 2016; Ng et al., 1997).

These agarwood-producing trees are distributed in Asian countries, and these countries have become the main exporters of agarwood. These exporting countries include Malaysia, Indonesia, India, Singapore, Thailand, Laos, Vietnam, Cambodia and Myanmar.

7.5.2 Factors of Agarwood Resin Formation in *Aquilaria* Trees

Many studies on agarwood formation suggested that agarwood is formed in *Aquilaria* trees after they are wounded or infected as a response against pathogen attack or stress condition (Chakrabarty et al., 1994; Ng et al., 1997; Oldfield, 1998). Stress conditions occur from lightning strike and fires, whereas injury and invasion of fungii can occur from slashing, cutting, drilling, nailing, natural injury and animal grazing (Barden et al., 2000; Nobuchi and Hamami, 2008; Pojanagaroon and Kaewrak, 2005; Figueiredo et al., 2008; Yang et al., 2016). In natural forest, agarwood can be found in the stem of old trees, broken branches, knots and injured parts. According to Chakrabarty et al. (1994), infected *Aquilaria* trees develop resin from the age of 20 years old and onwards, with 50-year-old trees showing the highest concentration of yield (2–3 kg per tree). It has been estimated that only 7%–10% of trees contain agarwood (Gibson, 1977; Ng et al., 1997), and only 10%–20% of the whole stem can be processed into raw agarwood (in the form of chips and flakes); the other part can be processed into powder/dust for oil distillation (Barden et al., 2000). The low yield from a single tree and knowledge of agarwood formation has led researchers and planters to induce planted *Aquilaria* to produce agarwood at an early age to shorten the duration of yield harvesting. To induce agarwood formation, several techniques are applied, for example: (1) the inoculation of *Aquilaria* trees with organic or chemical inoculant at 5 years old and above; (2) a combination of inoculation and artificial wounding; and (3) artifical wounding. Inoculation methods that have been applied are (1) drilling holes and injecting inoculant into the holes using syringe; (2) drilling holes and dripping inoculant in using a hose tube attached to a bottle; (3) drilling holes and inserting inoculant sticks into the holes; and (4) drilling holes, inserting PVC pipes for aeration and injecting with inoculant. In the second and third methods, trees are artificially wounded by slashing, nailing with iron nails, drilling holes, and peeling off bark. Wounding will create a stress condition for the trees and allow fungus to infect the tree stems. The method of peeling off the bark followed by brushing inoculant on the exposed stem is another way to induce resin formation. These methods showed successful results; however, agarwood from farms is always considered to be of a lower grade compared to agarwood from natural forest (Akter et al., 2013; Blanchette and Heuveling van Beek, 2005; Chakrabarty et al., 1994).

Although agarwood can be induced successfully, studies are continuously being carried out to find out more, for instance, on factors related to agarwood formation. According to many studies, several factors influence the formation and quality of agarwood:

1. Infection of pathogens if the trees are wounded or injured (Chakrabarty et al., 1994; Ng et al., 1997)
2. Stress conditions, for example, lightning strike, fire, wounding or injury (Yang et al., 2016)
3. Tree species and genetic variation (Kaiser, 2006; Ng et al., 1997)
4. Age of tree (Chakrabarty et al., 1994; Kaiser, 2006)
5. Location, elevation at which the tree grows or is planted (Donovan and Puri, 2004)
6. Soil type and habitat (Donovan and Puri, 2004)
7. Microclimate (Donovan and Puri, 2004)
8. Seasonal variation (Ng et al., 1997)
9. Environmental variation (Ng et al., 1997)
10. Region of origin of tree (Kaiser, 2006)
11. Type of fungus and bacteria invading the trees (Donovan and Puri, 2004)
12. Physiology process in the trees as the immune system and defence mechanism produce essential oil (Nobuchi and Hamami, 2008; Pojanagaroon and Kaewrak, 2005; Figueiredo et al., 2008)
13. Endophytes living in the tree (Shoeb et al., 2010)
14. Location in the tree where agarwood is formed or taken from (Kaiser, 2006)
15. Length and intensity of light exposure in inoculated wood (Rozi et al., 2014).
16. Inoculation (Blanchette and Heuveling van Beek, 2005)

Burfield and Kirkham (2005) wrote that the formation of the scented resinous agarwood is special, formed due to pathological infection after wounding and/or other processes. The non-resinous wood is odourless, yellow-whitish in colour, soft, even-grained and low in density. Past studies have suggested that certain fungi cause an immune reaction in the wood that triggers the production of agarwood oleo-resin. In the infected area, irregular patches of streaks developed. Occasionally, dark fibres at the wood stem surface can be seen as a sign of the existence of agarwood. However, to confirm the occurrence of the dark and heavier resinous wood, the tree has to be cut down. In most cases, it is difficult to determine whether an *Aquilaria* tree has developed agarwood. According to Donovan and Puri (2004), the Penan Benalui tribe in Sarawak, Malaysia, look for holes made by borer insects in the *Aquilaria* or *Sekau* stem, or any deformed bole, and then slash it with a large bush knife to confirm agarwood in the stem. Chakrabarty et al. (1994) mentioned that some external features of *Aqularia* trees are used as an indication of agarwood occurence in the stem: (1) a poor crown, decayed branches and uneven bole; (2) swellings or depressions and cankers on the bole; (3) the appearance of hordes of ants in the fissures; and (4) an obvious yellowish brown colour of the stem under the bark.

Wounding followed by invasion of a pathogen or inoculation has been shown to have been successful in the formation of agarwood. From agarwood, several fungi have been isolated, including *Aspergillus* sp., *Alternaria* sp., *Botryodyplodia* sp., *Cladosporium* sp., *Diplodia* sp., *Fusarium bulbiferum*, *F. laterium*, *F. oxysporum*, *F. solani*, *Penicillium* sp., *Pythium* sp., *Cunninghamella*, *Curvularia*, *Trichoderma*, *Phaeoacremonium* and *Lasiodiplodia* (Barden et al., 2000; Chakrabarty et al., 1994; Premalatha and Karla, 2013; Mohamed et al., 2010). Several endophytic bacteria have also been isolated from *A. malaccensis*, including *Bacillus cereus*, *Burkholderia*, *Staphylococcus aureus*, *Escherichia coli*, *Shigella boydii*, *Pseudomonas aeruginosa*, *Salmonella typhi* and *Vibrio cholerae* (Ryan et al., 2008; Shoeb et al., 2010).

Previous studies have shown that agarwood resin is formed in *Aquilaria* trees after the trees have been infected or inoculated. The resin is the extractive produced in the *A. malaccensis* wood as a chemical defense against any pathogen. Siburian et al. (2013) noted that secondary metabolite compounds suppress and inhibit the development of pathogens. Reseachers found that xylem parenchyma cells and interxylary phloem turned starch granules into oleo-resin, which accumulated as droplets in the xylem parenchyma and interxylary phloems. The accumulation of oleo-resin is followed by an aromatic odour. The oleo-resin is the secondary metabolite product in the wood, produced as a defence against pathogens (e.g. fungi, bacteria, insects) after the trees have been naturally infected, put under stress, struck by lightning, injured, wounded or mechanically wounded (Adams et.al., 2016; Nobuchi and Siripatanadilok, 1991; Nobuchi and Hamami, 2008; Barden et al., 2000; Pojanagaroon and Kaewrak, 2005; Figueiredo et al., 2008; Rao and Dayal, 1992; Tabata et al., 2003). The secondary metabolites found in *A. malacccensis* agarwood contained terpenoids, phenolics, phenolic-terpenoid complexes, lipid droplets and alkaloids, mostly accumulated in the interxylary phloems and ray cells of the infected area (Adams et.al., 2016). The volatile chemicals in agarwood oil are mostly terpenoid substances (mono-, sesqui- and di-terpenes) and are among the most valuable compounds produced by plants, along with alkaloids and phenolic substances (Figueiredo et al., 2008). The sesquiterpenes and phenolic compounds (chromone derivatives) have an important function in the plant defense system and are also found in inoculated wood of *Aquilaria* trees. It has also been found that to produce these chemicals, the fungus disintergrated the cell wall including lignin (Adams et al., 2016; Novriyanti and Santosa, 2011). The phenolic content in agarwood formation is highly related to the infected area of *Aquilaria* trees after inoculation, where the amount of phenolic content decreased in the infected wood of these trees. Another possible factor in resin formation is the endophytes (fungus and bacteria) in *Aquilaria* trees, which live in all parts of the trees as symbiotic organisms to support healthy growth and the plant's defence system. A study by Shoeb et al. (2010) revealed that nine endophytic fungal strains and two types of bacteria (*Bacillus cereus* and *Staphylococcus aureus*) were isolated from *A. malaccensis*, and three chemical compounds, 1,7-dihydroxy-3-methoxyanthraquinone, propyl p-methoxy phenyl ether and 6-methoxy-7-O-(p-methoxyphenyl)- coumarin, were obtained from the agarwood. Ryan et al. (2008) stated that *Bacillus*, *Pseudomonas* and *Burkholderia* bacteria are endophytes in plants well known for their wide range of secondary metabolites, including chemicals that can potentially be used as antibiotics, anticancer compounds, volatile organic compounds, and antifungal, antiviral, insecticidal and immunosuppressant agents.

7.5.3 AGARWOOD GRADES RELATED TO ITS TYPE AND USES

The types and grading of agarwood are important in the market to determine the price and uses of the agarwood. In Table 7.2, Table 7.3 and Table 7.4, examples of the types and grades of agarwood used in Malaysia and other countries are shown. Generally, agarwood products are in the form of logs, wood blocks, chips, flakes, wood powder, chunks, carved agarwood and agarwood oil (Figure 7.17). The agarwood products are used as incense, fragrance, perfume, medicinal products and decorations and for ritual practice. The grades vary among countries and within countries. The grading of agarwood is based on its colour, odour and shape and the agarwood's weight or density. It is considered that dark, scented, heavy agarwood contains more resin, and the price is determine based on the grade. A large, dark-coloured, heavy agarwood chunk with an aromatic odour fetches a higher price than smaller agarwood chunks, chips and light-coloured agarwood. However, problems often occur in deciding the grade of agarwood due to inconsistent quality based on size, colour, odour and weight. Kaiser (2006) notes that agarwood naturally varies widely in character and quality. It is influenced by the region of origin, the tree species, the age of the tree and the different parts of agarwood taken from a tree. Thus, the agarwood pieces contain varying composition and amounts of resin. In Table 7.2 and Table 7.3, examples of agarwood type, grade and price are listed; however, prices increase from time to time due to a high demand for agarwood in the global market.

Chua (2008) stated that the terms *agarwood* and *gaharu* are used only for resin-impregnated pieces of wood (Grade C and above) that have been at least partially shaved of non-impregnated wood (the CITES terminology for these pieces is *wood chips*). Most forms of semi-processed or raw agarwood are about 10 cm in length and are referred to as *chips, fragments, shavings* and *splinters,* and the wood also comes in

FIGURE 7.17 A. Agarwood chunks with tiger stripes in a black and brown colour. Grade A agarwood. B. Agarwood chunks and chips of varying size. Colour ranges from whitish and light brown to dark brown. Grade B agarwood.

TABLE 7.2

Types and Grades of Agarwood in Malaysia

Types of Agarwood	Characteristics	Grade	Price (RM) per Kilogram
Wood chunk	Big chunk, dark colour, shiny, 100% impregnated with resin and strong aromatic odour. Weight 500 g–3 kg	Super King	RM20,000 and above /kg
Wood chunk	Big chunk, dark colour, shiny, 100% impregnated with resin and strong aromatic odour. Weight 200–500 g	Triple Super	RM20,000 and above /kg
Wood chunk	Smaller blocks, 90–100% impregnated with resin, dark colour, shiny and strong aromatic odour. Weight 50–200 g	Double Super	RM20,000 and above /kg
Wood chunk	Smaller blocks of mixed sizes. 80% resin content. Resin is grayish black	Super	RM11,000 and above / kg
Chips	Mixed sizes, resin content 45–90%, colour light brown, brown to blackish, less aromatic	Grade A1 to A7	RM8,000–RM1,500 /kg
Chips	Mixed sizes, light weight, resin content 40% or entire, light brown, black	Grade A8 to A10	RM1,200–RM900 /kg
Small chunk, chips and flakes	Mixed sizes, light weight, less aromatic, resin content 35%, light brown colour, black, brown	Grade B1 to B5	RM800–RM250 /kg
Small chunk, chip	Mixed sizes, light weight, less aromatic, resin content 25%, light brown colour	Grade C1 to C10	RM200–RM80 /kg
Chips, flakes	Mixed sizes, entirely covered with resin, whitish, light brown	Grade D	N/A
Small chunk, chips, flakes	Mixed sizes, light weight, less aromatic, resin content 25%–50%, light brown colour	Grade wood to be processed to powder or dust for oil distillation	RM5–RM50 /kg
Agarwood oil			RM350–RM2200 /tola (1 tola = 12 ml or 11.6 g of oil)

Source: Antonopoulou et al. (2010); Malaysian Timber Industry Board (2010); Mazlan and Dahlan (2010); Mohamed and Lee (2016); Sustainable Asset Management (2018).

TABLE 7.3
Grading Based on Resin Content, Colour and Odour of Agarwood

Percentage of Resin Content	Grade	Price (RM /kg)
Entirely covered with resin or aesthetic natural shape, all colours (black to light brown)	Super (A)	RM30,000/kg
30% and above Shiny black, black and aromatic	A	RM20,000/kg
20%–29.99% Brown to black brown and aromatic	B	RM10,000/kg
9%–19.99% Whitish to whitish yellow	C	RM10,000/kg
<9%	D	N/A

Source: Forestry Department of Peninsular Malaysia (2015); Mohamed and Lee (2016); Nor Azah et al. (2013)

TABLE 7.4
Agarwood Classification based on End Products

Category	End Use	Grade
Aroma	Wood chips and blocks containing fragrant resin for direct burning	Super A and B
Block	Wood blocks of various shapes and sizes, containing moderate to high density of fragrant resin, for use in making end products such as sculptures, beads and bracelets	Tiger stripes and colour of the sculpture
Classic	Wood blocks with fragrant resin of unique natural shapes for sale as an aesthetic product	Classic
Dust	Dust and debris, by-product of washing and oil extraction, but has remaining fragrant resin	Black, grey, yellow, dust, incense powder, debris
Extractable wood	Wood blocks and pieces of various sizes with low fragrant resin, suitable for oil distillation	C
Fragrance	Wood pieces covered uniformly with resin on one side. Low to moderate fragrance	A1, A and B

Source: MTIB (2014); Mohamed and Lee (2016)

powder and dust form. However, in Table 7.2, Table 7.3 and Table 7.4, the term *agarwood* is also used for other grades, which only refer to resinous impregnated wood.

7.5.4 CHEMICAL CONTENT OF AGARWOOD

Agarwood is the resin-impregnated wood of *Aquilaria* trees. It is understood that the resin is produced and deposited in the wood when the tree reacts against pathological infection. It is also revealed in many studies that the endophytes in the tree also

TABLE 7.5
Chemical Content of Agarwood from *Aquilaria* spp.

Derivatives of Sesquiterpenes	Derivatives of Chromones
1. Agarol	1. 2-12 – (4-methoxyphenyl)-ethyllchromone
2. β-dihydro agarofuran	2. 6-methoxy-2-[2-(4 -methoxypheny1)-ethyl] chromone
3. β-agarofuran	3. 2-(2 (4methoxyphenyl)ethyl)chromones 4. 2-(2phenylethyl) chromone
4. α-agarofuran	4. 2-(2phenylethyl) chromone
5. nor-keto agar furan	5. 7,8-dimethoxy-2-[2-(3acetoxyphenyl)ethyl] chromone
6. 4-hydroxydihydroagarofuran	6. 6-Hydtoxy-2-(2-phenylethyl)chromone
7.3,4-dihydroxydihydroagarofuran	7. 6, 7-dimethoxy-2-(2-phenylethyl) chromone
8. Aquillochin	8. 6-methoxy-2-[2-(3 methoxyphenyl)ethyl]chromone
9. Agarospirol	9. (5S,6R,7R,8S)-2-(2-phenylethyl)-5',6,8' tetrahydroxy5,6,7,8-tetrahydrochromone
10. jinkoh-eremol	10. 5, 6, 7, 8, tetrahydroxy2-[2-(4-(Methoxy-phenyl) ethyl]5-6-7-8-tetrahydro chromone
11. epi-γ-eudesmol	11. 5,8-dihydroxy-2-(2-phenylethyl)chromone
12. Valerianol	12. 6,7-dihydroxy-2-[2-(4-methoxyphenyl)ethyl] chromone
13. benzyl acetone	13. (5S,6S,7R)-2-[2-(2-acetoxyphenyl) ethacetoxy-5,6,7,8,8-pentahydrochromone
14. anisyl acetone	14. (5S,6R,7R,8S)-2-(2-phenylethyl)-5,6,7-trihydroxy5,6,7,8-tetrahydro-8-[2-(2-phenylethyl)-7-methoxychromonyl-6-oxy]chromone
15. Kusunol	15. (5S,6R,7R,8S)-2-(2-phenyl-ethyl)-5,6,7-trihydroxy-5,6,7,8-tetrahydro-8-[2-(2-phenyl-ethyl) chromonyl-6-oxy]chromone
16. Dihydrokaranone	16. (5S,6S,7S,8R)-2-(2-phenylethyl)-5,6,7-trihydroxy-5,6,7,8 tetrahydro-5-[2-(2 phenylethyl) chromonyl-6-oxy]chromone +
17. oxo-agarospirol	17. 5, 6,7 trihydroxy-8-methoxy-2-(2-phenylethyl) chromone
18. (−)-selina-3,11-dien-9-one	18. 5, 6,7, 8, tetrahydroxy-2-[2-(2-hydroxyphenyl) ethyl]5,6,7,8-tetrahydrochromone
19. (+)-selina-3,11-dien-9-ol	19. 5hydroxy-6-methoxy-2-(2-phenylethyl)-4H-chromen-4-one
20. (−)-guaia-l(10),11-diene-15-ol	20. 5-Hydroxy-6-methoxy-2-[2-(4-methoxyphenyl) ethyl]-4H-1-benzopyran-4-one
21. (−)-guaia-l(10),l l-diene-15carboxylic acid	21. 6-methoxy-2-[2-(4-methoxyphenyl) ethyl]-4H-chromen-4-one
22. methylguaia-l(10),11-diene-15-carboxyiate	22. 6-methoxy-2-[2-(3-methoxypheny-l) ethyl]-4H-chromen-4-one
23. (+)-guaia-l(10),l l-dien-9-one	23. 1-hydroxy-1,5-diphenylpentan-3-one

(Continued)

TABLE 7.5 (CONTINUED)
Chemical Content of Agarwood from *Aquilaria* spp.

Derivatives of Sesquiterpenes	Derivatives of Chromones
24. (−)-l,10-epoxyguai-ll-ene +	24. 5, 6,7, 8, tetrahydroxy-2-[2-(2-hydroxyphenyl) ethyl]5,6,7,8-tetrahydrochromone
25. (−)-guaia-l(10),11-dien-15,2-olide -	25. 5.6:7,8-diepoxy-2-(2- phenylethyl)-5,6,7,8 tetrahydrochromone, 5.6:7,8-diepoxy-2-[2-(4-methoxyphenyl)ethyl] 5,6,7,8-tetrahydrochromone
26. α-guaiane	26. 5.6:7,8-diepoxy-2-[2-(3-hydr-oxy-4 methoxyphenyl)ethyl]-5,6,7,8-tetrahydrochromone +
27. guaia-l(10),11-dien	27 6-methoxy-2-[2-(4 '-methoxypheny1)-ethyl] chromone
28. 1,5-epoxy-nor-ketoguaiene	28. 5-hydroxy-6-methoxy-2-(2-phenylethyl)chromone
29. dehydrojinkoh-eremol	29. 6-hydroxy-2-(2-hy-droxy-2 phenylethyl)chromone
30. (−)-rotundone	30. 8-chloro-2-(2-phenylethyl)-5,6,7-trihydroxy-5,6,7,8-tetrahydrochromone + 31 6,7-dihydroxy-2-(2-phenylethyl)-5,6,7,8 tetrahydro chromone
31. baimux-3,11-dien-14-al	31. 6,7-dihydroxy-2-(2-phenylethyl)-5,6,7,8 tetrahydro chromone
32. 9,11-eremophiladien-8-one	32. 5,6,7,8-tetra-hydroxy-2-(3-hydroxy-4 methoxyphenyl)-5,6,7,8-tetrahydro-4H-chromen-4-one +
33. selina-3,11-dien-14-ol	33 ,(5S*,6R*,7S*)-5,6,7-trihydroxy-2-(3-hydroxy-4methoxyphenethyl) 5,6,7,8tetrahydro-4H-chromen-4 one +
34. selina-4,11-dien-14-al	34. (5S*,6R*,7R*)-5,6,7-trihydroxy-2-(3-hydroxy-4-methoxyphenethyl)-5,6,7,8-tetrahydro-4H-chromen-4-one
35. Sinenofuranol	35. 6,7-dihydroxy-2-[2-(4-methoxyphenyl)ethyl] chromone,6-Hydroxy-7-methoxy-2-[2-(3-hydroxy-4 methoxyphenyl)ethyl]chromone
36. Karanone	36. 6,7-Dimethoxy-2-[2-(3-hydroxy-4-methoxyphenyl) ethyl]chromone
37. selina-4,11-dien-14-oicacid	37. 7-Hydroxy-6-methoxy-2-[2-(3-hydroxy-4-methoxy-phenyl)ethyl]chromone
38. selina-3,11-dien-14-oicacid	38. 6,7-Dimethoxy-2-[2-(4-hydroxy-3-methoxyphenyl)ethyl]chromone
39. (−)-guaia-l(10),11-dien-15-oicacid	39. 6-Hydroxy-7-methoxy-2-[2-(4hydroxyphenyl) ethyl]chromone
40. 2-hydroxyguaia-l(10),11-dien-15-oicacid	40. 6,8-Dihydroxy-2-[2-(3-hydroxy-4-methoxyphenyl) ethyl]chromone
41. 9-hydroxyselina-4,11-dien-14-oicacid	41. 6-Hydroxy-2-[2-(40-hydroxy-3 methoxyphenyl) ethenyl]chromone
42, E-nerolidol acetate	42. 6,7-Dimethoxy-2-[2-(4-hydroxyphenyl)ethyl] chromone

(Continued)

TABLE 7.5 (CONTINUED)
Chemical Content of Agarwood from *Aquilaria* spp.

Derivatives of Sesquiterpenes

43. (−)- methylselina-3,11-dien-14-oate

44. (+)- methylselina-4,11-dien-14-oate

45. Gmelofuran

46. methyl9-hydroxyselina-4,11-dien-14-oate

47. (−)-l0-epi-γ-eudesmol

48. (−)-guaia-l(10),11-dien-15,2-olide

40. γ-eudesmol

50. Viridiflorol

51. caryophyllene oxide

52. γ-gurjunene

53. Valencene

54. allo aromadendrene epoxide

55. Spathulenol

56. tricyclo[5.2.2.0(1,6)]Undecan-3-ol

57. 2-methylene-6,8,8-trimethyl-

58. Patchoulene

59. eremophila-1(10),11-dien

60. Isolongifolene

61. neoisolongifolene,8,9-dehydro

62. isolongifolen-5-one

63. (.+−.)-cadinene

64. Longiverbenone

65. α-cedrene oxide

66. Dehydrofukinone

67. Dehydroabietane

68. (4R,5R,7R)-1(10)-spirovetiven-11 -ol-2-one

69. (2R,4As)-2-(4a-rnethyl-l,2,3,4,4a,5, 6,7-octahydro-2-naphthyl)-propan-2-ol

70. (S)-4a-rnethyl-2-(1 − methylethyl)-3.4.4a,5,6,7- hexahydronaphthalene

71. (S)-4a-methyl-2-(1–rnethylethylidene) 1,2,3,4,4a,5,6,7-octa-hydronaphthalene +

72. (S)-4a-rnethyl-2-(1 –methylethyl)- 3.4.4a,5,6,7-hexahydronaphthalene

73. (1R,6S,9R)-6,10,10-trirnethyl-11- oxatricyclo[7.2.1.0]dodecane

74. 1R,2R,6S,9R)-6.10,10-trimethyl-l1– oxatricyclo[7.2.1 .0] ydroxyl-2-ol

Derivatives of Chromones

43. 2-[2-Hydroxy-2-(4-hydroxyphenyl)ethyl] chromone

44. 2-[2-Hydroxy-2-(4-methoxyphenyl)ethyl] chromone

(Continued)

TABLE 7.5 (CONTINUED)
Chemical Content of Agarwood from *Aquilaria* spp.

Derivatives of Sesquiterpenes   **Derivatives of Chromones**

75. Rel-(2R8S,8aR)-2-(1,2,3,5,6,7,8,
8a-Octahydro-8,8u-dimethyl-2-nuphthyl)-
prop-2-en-1-ol
76. Rel-(3R,7R,9R,10S)-9,10-Dimethyl-6-
methylene-4-oxatricyclo[7,4,0,0]tridec-1–en
77. Re1-(2R,8S,8aR)-2-(1,2,6,7,8,
8a-Hexahydro8,8a-dimethyl-2-naphthyl)-
propan-2-ol,
78. Rel-(5R,10R)-2-isopropylidene-10-methyl-
spiro[4.5]dec-6-ene-6-carbaldehyde +
79. Rel(5R,7R,10R)-2-lsopropylidene-10-
methyl-6 methylene-Spiro[4.5]decan- 7-ol
80. Rel-(lR,2R)-9-Isopropyl-2-methyl-8-
oxatricyclo[7,2.1.0]dodec-5-ene
81. Rel-(lR,2R)-(9Isopropyl-2-methyl-8-
oxatricyclo[7.2.1. 0]dodeca-4,6-dien,
2-(1,2,3,4,5,6,7.8,8a-Octahydro-8,8a
dimethyl-2-naphthy1)-propanal
82. 3,4dihydroxydihydroagarofuran
83. 9,11-eremophiladien-8-one
84. Epoxybulnesene
85. Elemol
86. caryophyllene oxide
87. Guaiol
88. α-eudesmol
89. ß-maaliene
90. Aromadendrene
91. γ-gurjunene
92. α-muurolen
93. γ-guaiene
94. epi-α-cadinol
95. epi-α- bisabolol
96. α- bisabolol
97. selina-3,11-dien-9-al
98. Eudesmol
99. jinkohol ii
100. β-eudesmol
101. Cyperotundone
102. α-humulene
103. β-elemene
104. Viridiflorol
105. α-gurjunene

(Continued)

TABLE 7.5 (CONTINUED)
Chemical Content of Agarwood from *Aquilaria* spp.

Derivatives of Sesquiterpenes	Derivatives of Chromones

Derivatives of Sesquiterpenes

106. β-gurjunene
107. α-selinene
108. δ-cadinene
109. α-cedrene
110. α-copaene
111. α-funebrene
112. ϒ-muurolene
113. ar-curcumene
114. cis-β-guaiene
115. ϒ-cadinene
116. selina-3,7(11)-diene
117. β-vetivenene
118. Spathulenol
119. Aristolene
120. ß-selinene
121. Copaene
122. trans-α-bergamotene
123. α-caryophyllene
124. Alloaromadendrene
125. 1,2-Epoxide-humulene
126. γ –gurjunenepoxide
127. isoaromadendrene epoxide
128. cis-nerolidol
129. ledene oxide-(II)
130. trans-longipinocarveol
131. jinkohol
132. Nootkatane
133. 2-[(2b,4ab,8b,8ab)-Decahydro-4a-
    hydroxy-8, 8a-dimethylnaphthalen-2-yl]
    prop-2-enal
134. (4ab,7b,8ab)-Octahydro-7-[1-
    (hydroxymethyl)ethenyl]-1, 8a-
    dimethylnaphthalen 4a(2H)-ol,(4ab,7b
135. (8ab)-3,4,4a,5,6,7,8,8a-Octahydro-7-[1-
    (hydroxymethyl)ethenyl]-4a-methylnaphthalene-
    1-carb-oxaldehyde +
136. (1aβ,2β,3β,4aβ,5β,8aβ)-Octahydro-4a,5-
    dimethyl-3-(1-methylethenyl)-3H-
    naphth[1,8a-b]oxiren-2-ol
137. selina-4,11-diene-12,15-dial,
    Eudesm-4-ene-11,15-diol
138. β –acorenone

(Continued)

TABLE 7.5 (CONTINUED)
Chemical Content of Agarwood from *Aquilaria* spp.

Derivatives of Sesquiterpenes	Derivatives of Chromones

Derivatives of Sesquiterpenes

139. baimux-3,11-dien-9-one
140. Cyclocolorenone
141. α-(Z)-santalol acetate
142. α-bisabolol acetate
143. β-E-santalol acetate
144. E-α-bergamotene
145. baimuxifuranic acid
146. γ-selinene
147. baimuxifuranic acid
148. Baimuxinal
149. p-methoxybenzyl acetone +
150. baimuxinic acid
151. 1R,6S,9R)-6,10,10-trirnethyI-11-
     oxatricyclo[7.2. 1.0]dodecane
152. (S)-4a-rnethyl-2-(1methylethy1)-3.4.4a,
     5,6,7-hexahydronaphthalene
153. (1R,2R,6S,9R)-6.10,10-trimethyl-11-
     oxatricyclo[7.2.1 .0] hydroxyl-2-ol
154. epi-ligulyl oxide
155. (1R,2R,6S,9R)-6.10,10-trimethyl-11-
     oxatricyclo[7.2.1 .0] hydroxyl-2-ol
156. (S)-4a-methyl-2-(1–rnethylethylidene)
     1,2,3.4,4a,5,6,7-octa hydronaphthalene
157. 4hydroxydihydroagarofuran
158. α-santalol
159. Hinesol
160. Neopetasane
170. Baimuxinol
171. Aristolenepoxide
172. Isobaimuxinol
173. 4-hydroxyl-baimuxinol
174. 7β-H-9(10)-ene-11,12-epoxy-
     8oxoeremophilane
175. 7α-H-9(10)-ene-11,12-epoxy-
     8-oxoeremophilane
176. Cubenol
177. aromadendrene oxide
178. eudesm-7(11)-en-4a-ol
179. α-copaen-11-o
180. eremophila-7(11), 9-dien-8-one

Source: Ahmed and Kulkarni (2017).

contribute to the production of resin. The combination of the wood cells and endophyte reaction result in the formation of many types of chemical in the agarwood to fight against various fungi and bacteria invading the *Aquilaria* trees. According to many studies, there are several groups of chemical contents that can be obtained from the agarwood oil after the extraction process; these contents are responsible for the colour and aromatic odour of the agarwood oil. Agarwood oil extracted from *A. malaccensis*, *A. crassna*, *A. sinensis* and *A. subintegra* have similar major chemical compounds but differ in their composition. Agarwood oil is composed of sesquiterpenes, monoterpenes, chromones, sterols and fatty acid methyl ester. Aromatic compounds are composed of aldehyde, phenol, ether and ketone groups (Jong et al., 2014). Sesquiterpenes and chromones show various types of derivatives (Table 7.5). These chemical constituents can be divided into several main groups: Agarofurans, Agarospiranes, Guaianes, Eudesmanes, Eremophilanes and Prezizanes (Ahmed and Kulkarni, 2017). Sesquiterpenes in plants are responsible for defensive agents or pheromones. The volatile compounds of the sesquiterpenes of agarwood contribute to its aromatic scent. The chromone (ring system 1-benzopyrane-4-one) is the basic unit of several flavonoids such as flavones, flavonols and isoflavones. The chromones are responsible for the wide variation of colour (Tawfik et al., 2014). The chromones' biological and physiological activities can be used for anti-mycobacterial, anti-fungal, anti-convulsant, anti-microbial, anti-allergic, muscular relaxation, therapeutic, anti-asthma, anti-cancer and anti-inflammation purposes (Dubey et al., 2014). Recently, more than 300 chemical compounds have been isolated from agarwood (Wang et al., 2018). Table 7.5 shows examples of the various chemical compounds isolated from agarwood.

Jong et al. (2014) extracted agarwood from infected and healthy *A. malaccensis* wood and found various chemical compounds. The major compounds were chromone derivative, aromatic compounds, sesquiterpenes, monoterpenes, sterols and fatty acid methyl ester. Aromatic compounds were composed of aldehyde, phenol, ether and ketone groups. In the infected and healthy agarwood extract, the major compounds found were 2-(2-phenylethyl) chromone derivative, 4-phenyl-2-butanone, (1S,4S,7R)-1,4-dimethyl-7-(prop-1-en-2-yl)-1,2,3,4,5,6,7,8-octahydroazulene [guaiene], 1,1,4,7-tetramethyl-2,3,4,5,6,7,7a,7b-octahydro-1aH-cyclopropa[h]azulen-4a-ol [palustrol], 4-(4-methoxyphenyl) butan-2-one [anisylacetone], agarospirol, alloaromadendre oxide (2), α-elemol, γ-eudesmol, and guaiol. Other examples of the chemical compounds found are Propenoic acid, Furanone, Benzaldehyde, Guaiacol, Benzylacetone, Benzenepropanoic acid, Syringol, Benzenepropanoic acid, Vanillin, Aristolene, Anisylacetone, Alloaromadendrene oxide, Caryophyllene oxide, α-Elemol, Guaiene, Palustrol, Humulene, Agarospirol, Guaiol, Benzopheone, γ-Eudesmol, Butanone, Benzenepropanoic acid, Hexadecanoic acid, Octadecenoic acid, Stigmastanol and Stigmasterol.

7.5.5 Uses of Agarwood

Different forms of agarwood have been used continousy from ancient times to the present, and these uses are becoming more important. It is now used mainly for perfumery, incense, medicine, religious, ritual, ceremonial, cultural, health-promoting and psychological purposes (Table 7.6). In religious ceremonies and rituals, agarwood

TABLE 7.6

Types and Uses of Agarwood

Countries	Types of Agarwood for Trading and Usage	Uses	References
Malaysia	Whole plant (seedlings), logs, sawn wood, chunks, chips, fragments, shavings, splinters, powder, dust, aesthetic natural shape and oil.	• Main uses are for aromatic perfumery, fragrance, pharmaceutical uses, medicine, aromatherapy, religious purposes, burnt offerings, idols, beads, rosaries, ornamental items, decorative carvings (selection of agarwood pieces for particular usage depends on the ethnicity of the user and grade of agarwood). • Logs for decorative carving • Aesthetic natural shape for decoration • Sawn wood for beads and carving • Shaped pieces or chunks for spiritual practice purposes • Chips, shavings, splinters for incense burning during prayers by Malays • Smoke as medicine • Ground as libation for gravesides • *Gaharu merupa* shaped pieces resembling such things as birds or humans for spiritual purposes • Agarwood mixed with coconut oil as a liniment • Boiled concoction to treat rheumatism and other body pain • Use by Chinese, Orang Asli and native shamans of Sabah for ritual and spiritual purposes. Chinese use gaharu joss stick as incense. • Penan people of Sarawak use agarwood to treat stomach aches, fevers and as insect repellent • Grated agarwood used for cosmetic purposes, medicine during sickness and childbirth • Essential oil in cosmetic as face mask for wrinkle treatment. • *Kayu gaharu lempong* to treat jaundice and body pain • Muslims burn agarwood splinters or chips for its fragrance during religious occasion and *Ramadan* prayers • Malay tribes fumigate paddy fields with agarwood smoke to appease spirits • Insect repellent against lice, fleas and mosquitoes	Burkhill (1935); Burkill (1966), Bland (1886); Chua (2008); Chakrabarty et al. (1994); Dentan (2001); Goh (2006); Gimlette and Thomson (1939); Gimlette (1915); Hansen (1998); Heuveling van Beek, and Phillips (1999); Lim and Anak (2010); Schafer (1963); Skeats (1900); Wilkinson (1955)

(Continued)

TABLE 7.6 (CONTINUED)
Types and Uses of Agarwood

Countries	Types of Agarwood for Trading and Usage	Uses	References
China	Whole plant, logs, wood blocks, wood chips, oil, powder	Main uses: Perfumery, medicines, works of art, carving, beads • Medicine to promote flow of qi • To treat pain, gastric, vomiting, asthma, diarrhoea, dysentery, gout, rheumatism, cough, fever and various skin diseases • Used as stimulant, antiasthma, carminative, sedative, analgesic, tonic, aphrodisiac and astrigent • Incense stick and coil for fragrance • Sculpture and carving of religious objects • Natural shape of agarwood use in traditional Fengshui to govern the flow of energy • Agarwood incense burned by Buddhists on religious occasion	Yin et al. (2016); Barden et al. (2000); Yaacob (1999); Hashim et al. (2016)
Taiwan	Agarwood logs, timber, wood chips, wood with various sizes, oil, aesthetica natural shapes, powder	• Incense • Sculpture and carving of statues for religious purposes • Beads for wristlets or rosaries (religious objects) • Incense and raw chips burned for religious practice by Buddhists and Taoists (the fragrance is for self-purification) • Agarwood incense stick burned for prayers during traditional festivals and ceremonies to bring safety and good luck • Agarwood as aromatic ingredient of Chu-yeh Ching amd Vo Ka Py wine • Agarwood collections and aroma appreciation as hobby (collections of high end-products (powder, wood chips, natural shape, wristlets, robeads))	Barden et al. (2000); Traffic (2005)

(Continued)

TABLE 7.6 (CONTINUED)
Types and Uses of Agarwood

Countries	Types of Agarwood for Trading and Usage	Uses	References
India and Bangladesh	Agarwood oil	Perfumery, incense, incense during funerals of priests and princes • Agarwood oil is distilled to produce minyak attar to lace prayer clothes • Agarwood incense burned by Hindus for religious ceremonies • Agarwood oil in incense sticks • Used for funeral pyres and preparation of bodies for burial • Agarwood medicine prepared as pills, decoctions and plasters in combination with other ingredients to use as stimulant, carminative, aphrodisiac, antirheumatic, antimalarial, analgesic, deobstruent, tonic, and diuretic • Agarwood prescribed for dropsy, heart palpitations, tonic during pregnancy and after childbirth, to treat disease of female genital organs • Medicine in Ayurvedic practice • Used as medicine for treatment of pleurisy, nervous disorders, digestive, bronchial complaints, smallpox, rheumatism, spasms in the digestive and respiratory systems, fevers, epilepsy, abdominal pain, asthma, cancer, colic, diarrhoea, dysentery, vomiting, anorexia, mouth and teeth disease, facial paralysis, shivering, sprains, bone fractures, nausea, regurgitation, weakness in the elderly, shortness of breath, chills, general pains, kidney problems, cirrhosis of the liver, lung, stomach tumours, leprosy, inflammation, arthritis and gout • Tonic and stimulant to treat mental disorder and malnutrition • Essential oil as agarwood essence, soap and shampoo	Chakrabarty et al. (1994); López-Sampson and Page (2018); Smith and Stuart (2003); Yaacob (1999)

(Continued)

TABLE 7.6 (CONTINUED)
Types and Uses of Agarwood

Countries	Types of Agarwood for Trading and Usage	Uses	References
Japan	Wood chips, wood pieces, small pieces (*kowari*), *jin-koh matsu* (powder), *kizami* (cuts), *kakuwari* (square pieces), incense sticks, cones or pressed-powder shapes, agarwood oil, *Mei-koh* (small pieces), *Sho-koh* (chipped mixed fragrance), *Naru-koh* (blended incense ball), *sen-koh* (blended incense sticks), *Nioi-buruko* (sachets)	Main uses: Incense in cultural and religious, and in medicinal practices • Agarwood with burned firewood to produce smoky perfume during cooking • Wood chips used as jinkoh incense ('supreme fragrance') in Buddhism or religious practice • To perfume clothes through burning incense • Use in the culture of enjoying the scent of burning, called *soratakimono* • In ko-doh ceremony, the burning of agarwood to appreciate the fragrance and in several ceremonies in Japanese culture • Burning agarwood for its fragrance as 'suitable feminine pleasure' • Agarwood as ingredient in Rokushingan use as anaesthetic, zu-sei to treat fatigue • Incense containing agarwood as aromatic calmative in the workplace to increase productivity and to produce sense of mental 'quietness' • Agarwood incense for stomachache and sedatives • Agarwood incense to anoint the dead	Compton and Ishihara (2005); Okugawa et al. (1993)

(Continued)

TABLE 7.6 (CONTINUED)
Types and Uses of Agarwood

Countries	Types of Agarwood for Trading and Usage	Uses	References
Yemen	Agarwood oil, essential oil, sticks or splinters, beads	• Expensive perfume base, pure agarwood oil perfume, perfumery for prayer clothes, oil to rub on prayer wooden beads, Attar oil • Fragrance, soap and shampoo, fragrance • Burned for fragrance during prayer in mosques and parish halls • (Oud fragrance is effective for meditation, enlightenment, bringing deep tranquility and relaxation) • Burned for fragrance at home (to calm the body, remove destructive and negative energies, provide awareness, reduce fear, invoke a feeling of vigour and harmony, and enhance mental functionality) • Oud eases neurotic and obsessive behavior and helps create harmony and balance in the home • Prayer beads • Oud to promote positive energy and good luck • Oud perfumery mixed with other essential oils to embalm corpse for Islamic funerary practice	Jung (2011)
Indonesia	Agarwood	Treatment of joint pain	Grosvenor et al. (1995); Hashim et al. (2016)

(Continued)

TABLE 7.6 (CONTINUED)
Types and Uses of Agarwood

Countries	Types of Agarwood for Trading and Usage	Uses	References
Korea	Agarwood	Used to treat cough, acroparalysis, croup, asthma; and as a tonic, stomachic agent, sedative and expectorant.	Takagi et al. (1982); Yuk et al. (1981); Hashim et al. (2016)
United Arab Emirates	• Agarwood oil / Dihn al oudh ♣ –Arabian perfumes (oil based) ♣ –Arabian perfumes • French-style perfumes • Agarwood chips / oudh , bakhoor -Scented chips • Arabian perfumes in sprays ♣ • Arabian perfumes, • French-style perfumes • Agarwood chips/oudh , bakhoor • Scented chips ♣ -Sprays • Agarwood chips/oudh (high qualities are usually used for special occasions, weddings, etc.) ♣ –Bakhoor ♣ – Scented chips	• Body fragrance (pure oil or incense is applied on the hair, behind ears, on neck and in nostrils as fragrance for personal devotion) • Clothes fragrance for prayers and special occasion • House fragrance • To receive honoured guests as customary practice • Burning agarwood chips as fragrance considered as part of the country's heritage and modern national identity	Antonopoulou et al. (2010)

is used by Buddhist, Jewish, Christian, Muslim and Hindu communities (Barden et al., 2000; Chua, 2008). Its uses and benefits are becoming more important nowadays since studies have shown the many potential uses of agarwood as medicine for anti-cancer, anti-tumour, anti-diabetic, anti-allergic, anti-inflammatory, antinociceptive, antipyretic, analgesic, anti-ischemic/cardioprotective, anti-microbial and antioxidant purposes (Hashim et al., 2016).

Agarwood is traded and used in various forms, includng whole plants (seedlings), logs, chips, flakes, chunks, timber, pure oil, essential oil, powder and blends with other fragrance herbs. The selection of agarwood piece and types by users depends on grade and ethnicity (Jung, 2011).

7.6 CONCLUSION

A. malaccensis is an important species for its scented resinous wood. It has a unique anatomical structure that consists of xylem and interxylary phloems. The xylem is composed of fibres, vessel elements and axial and ray parenchyma cells. The interxylary phloems exist consistently in the stem and are diffusely distributed in the stem. The interxylary phloems consist of phloem fibres, sieve elements, companion cells and axial and ray parenchyma cells. Both types of cells can be differentiated by the thin walls of the interxylary phloem cells. It is suggested that the abundant axial and ray parenchyma cells, together with interxylary phloems, are responsible for the formation of resin in *A. malaccensis* when the tree is attack by fungi or bacteria, leading to the development of agarwood. The formation of resin is the tree's response against any fungi or bacteria attack. The resin that is secreted and accumulated can be found as resin droplets in the xylem parenchyma cells and interxylary phloems. The continuous accumulation of resin will turn the whitish wood to a brown or blackish brown colour, and the wood odour will change to an aromatic odour. The colour and odour changes are due to the chemical compounds of Sesquiterpenes and Chromones derivatives. The resin extracted from agarwood has been used as medicine to treat diseases and illnesses, mainly on the basis of the activities of chromones derivatives. The agarwood resin is made up of various complex chemical compounds responsible for the various quality of the agarwood's odour and colour. The aromatic odour is very important for perfume, fragrance and incense used in religious, ritual, ceremonial, therapeutic, and cultural practices by communities. The odour and colour of the agarwood is usually used to determine its quality and grade. Agarwood oil and chips are used for fragrance and as medicine to treat many diseases and illnesses. Agarwood products become more important with increasing prices caused by high market demand for its use for perfumery, fragrance and medicinal purposes.

ACKNOWLEDGEMENTS

Special thanks and acknowledgement to Mr. Azli Sulid (Senior Lab Assistant of Forestry Complex, Faculty of Science and Natural Resources, Universiti Malaysia Sabah) for helping to prepare wood sections of *Aquilaria malaccensis* for microscopic observation.

REFERENCES

Abd Majid, J., Hazandy, A.H., Paridah, M.T., Nor Azah, M.A., Mailina, J., Saidatul Husni, S. and Sahrim, L. 2018. *Determination of Agarwood Volatile Compounds from Selected Aquilaria Species Plantation Extracted by Headspace-Solid Phase Microextraction (HS-SPME) Method*. IOP Conference Series: Materials Science and Engineering. IOP Publishing. Vol. 368, p. 012023.

Adams, S.J., Manohara, T.N., Krishnamurthy, K.V. and Kumar, T.S. 2014. Histochemical Studies on Fungal-Induced Agarwood, *Indian Journal of Plant Science* 5(1): 102–110.

Ahmed, D.T. and Kulkarni, A.D. 2017. Sesquiterpenes and Chromones of Agarwood: A Review, *Malaysian Journal of Chemistry* 19(1): 33–58.

Akter, S., Islam, M.T., Zulkefeli, M. and Khan, S.J. 2013. Agarwood Production: A Multidisciplinary Field to Be Explored in Bangladesh, *International Journal of Pharmaceutical and Life Sciences* 2(1): 22–32.

Anon. 2008. *Report of the Fourth National Forest Inventory, Peninsular Malaysia*. Forestry Department Peninsular Malaysia, Kuala Lumpur, p. 97.

Antonopoulou, M., Compton, J., Perry, L.S. and Al-Mubarak, R. 2010. *The Trade and Use of Agarwood (Oudh) in the United Arab Emirates*. TRAFFIC South East Asia, Malaysia. http://www.trafficj.org/publication/10_Trade_Use_Agarwood.pdf. Accessed on 8/10/2018.

Barden, A., Anak, N.A., Mulliken, T. and Song, M. 2000. Heart of the Matter: Agarwood Use and Trade and CITES Implementation for *Aquilaria malaccensis*.

Blanchette, R. and van Beek, H.H. 2005. Cultivated Agarwood. US Patent 6,848,211.

Bland, R.N. 1886. Notes on kayu gharu, *Journal of the Royal Asiatic Society, Straits Branch* 18: 359–361 (Singapore. Cited in Lim, T.W. and Anak, N.A. 2010. Wood for the trees: A review of the agarwood (gaharu) trade in Malaysia. TRAFFIC Southeast Asia. Petaling Jaya, Selangor, Malaysia).

Burfield, T. and Kirkham, K. 2005. The Cropwatch Files – The Agarwood Files. http://www.cropwatch.org/agarwood.htm (Cited in Jung, D. 2011. The Value of Agarwood. Reflections Upon Its Use and History in South Yemen. In: HeiDOK, http://www.ub.uni-heidelberg.de/archiv/ (May 30, 2001)).

Burkhill, I.H. 1935. *Dictionary of Economic Products of Malay Peninsula*. Crown Agents for the Colonies, London (Cited in Lim, T.W. and Anak, N.A. 2010. Wood for the trees: A review of the agarwood (gaharu) trade in Malaysia. TRAFFIC Southeast Asia. Petaling Jaya, Selangor, Malaysia).

Burkill, I.H. 1966. *A Dictionary of Economic Products of the Malay Peninsula, I. Government of Malaysia and Singapore*. The Ministry of Agricultural and Cooperatives, Kuala Lumpur (Cited in Barden, A., Anak, N.A. and Mulliken, T. and Song, M. 2000. Heart of the matter: Agarwood use and trade and CITES implementation for *Aquilaria malaccensis*. file:///C:/Users/acer/Downloads/traffic_pub_forestry7%20(3).pdf. Accessed on 8/10/2018).

Chakrabarty, K., Kumar, A. and Menon, V. 1994. Trade in Agarwood WWF-TRAFFIC, India.

Chua, L.S.L. 2008. Agarwood (*Aquilaria malaccensis*) in Malaysia. International Expert Worksop on CITES Non-Dentriment Findings (NDF) WORKSHOP CASE STUDIES. November 17th–22nd, 2008. In: Trees, CASE STUDY 3, *Aquilaria malaccensis*. Cancun, Mexico: WG 1.

Chung, R.C.K. and Purwaningsih. 1999. *Aquilaria malaccensis* (Thymelaeceae). In: L.P.A. Oyen and X.D. Nguyen (eds.) *Plant Resources of South East Asia No. 19: Essential-Oil Plants*. Backhuy Publisher, Leiden, pp. 64–67.

CITES. 2005. Notification to the Parties. No. 2005/0025. https://www.cites.org.

Compton, J. and Ishihara, A. 2005. *The Use and Trade of Agarwood in Japan*. PC15 Inf. 6. TRAFFIC Southeast Asia and TRAFFIC East Asia-Japan, CITES.

Dentan, R.K. 2001. Semai-Malay Ethnobotany: Hindu Influences on the Trade in Sacred Plants, Ho Hiang. In: R. Rashid and W.J. Karim (eds.) *Minority Cultures of Peninsular Malaysia: Survivals of Indigenous Heritage*. Academy of Social Sciences (AKASS), Penang (Cited in Lim, T.W. and Anak, N.A. 2010. Wood for the trees: A review of the agarwood (gaharu) trade in Malaysia. TRAFFIC Southeast Asia. Petaling Jaya, Selangor, Malaysia).

Donovan, D.G. and Puri, R.K. 2004. Learning from Traditional Knowledge of Non-Timber Forest Products: Penan Benalui and the Autecology of *Aquilaria* in Indonesia Borneo, *Ecology and Society* 9(3): 3.

Dubey, R.K., Dixit, P. and Arya, S. 2014. Naturally Occurring Aurones and Chromones – A Potential Organic Therapeutic Agents Improvising Nutritional Security, *International Journal of Innovative Research in Science, Engineering and Technology* 3(1): 8141–8148.

Figueiredo, A.C., Barroso, J.G., Pedro, L.G. and Scheffer, J.J.C. 2008. Factors Affecting Secondary Metabolite Production in Plants: Volatile Components and Essential Oils, *Flavour and Fragrance Journal* 23(4): 213–226.

Gibson, I.A.S. 1977. The Role of Fungi in the Origin of Oleoresin Deposits of Agaru in the Wood of *Aquilaria agallocha* Roxb., *Bano Biggyan Patrika* 6: 16–26.

Gimlette, J.D. 1915. *Malay Poisons and Charm Cures*. 1981 Reprint. Oxford University Press, Kuala Lumpur (Cited in Lim, T.W. and Anak, N.A. 2010. Wood for the trees: A review of the agarwood (gaharu) trade in Malaysia. TRAFFIC Southeast Asia. Petaling Jaya, Selangor, Malaysia).

Gimlette, J.D. and Thomson, H.W. 1939. *A Dictionary of Malay Medicine*. Oxford University Press, Kuala Lumpur (Cited in Lim, T.W. and Anak, N.A. 2010. Wood for the trees: A review of the agarwood (gaharu) trade in Malaysia. TRAFFIC Southeast Asia. Petaling Jaya, Selangor, Malaysia).

Goh, R. 2006. *Illegal Gatherers Making Beeline for Sandalwood*. The New Straits Times, Malaysia, 3 April (Cited in Lim, T.W. and Anak, N.A. 2010. Wood for the trees: A review of the agarwood (gaharu) trade in Malaysia. TRAFFIC Southeast Asia. Petaling Jaya, Selangor, Malaysia).

Grosvenor, P.W., Gothard, P.K., McWdham, N.C., Suprlono, A. and Gray, D.O. 1995. Medicinal Plants from Riau Province, Sumatra, Indonesia. Part 1: Uses, *Journal of Ethnopharmacology* 45(2): 75–95.

Hansen, E. 1998. The Nomads of Gunung Mulu, *Natural History* 107(3) (Cited in Lim, T.W. and Anak, N.A. 2010. Wood for the trees: A review of the agarwood (gaharu) trade in Malaysia. TRAFFIC Southeast Asia. Petaling Jaya, Selangor, Malaysia).

Heuveling van Beek, H. and Phillips, D. 1999. Agarwood: Trade and CITES Implementation in Southeast Asia. Unpublished report prepared for TRAFFIC Southeast Asia, Malaysia (Cited in Lim, T.W. and Anak, N.A. 2010. Wood for the trees: A review of the agarwood (gaharu) trade in Malaysia. TRAFFIC Southeast Asia. Petaling Jaya, Selangor, Malaysia).

Hou, D. 1960. Thymeliaceae. In: C.G.G.J. Van Steenis (ed.) *Flora Malesiana Series I*. Wolter-Noordholf Publishing, Groningen, the Netherlands. Vol. 6, pp. 1–15. https://archive.org/details/floramalesiana81stee/. Accessed on 21/11/2018.

Jabatan Perhutanan Semenanjung Malaysia. 2015. *Manual Penggredan Gaharu*. Alamedia Sdn. Bhd.

Jong, P.L., Tsan, P. and Mohamed, R. 2016. Gas Chromatography-Mass Spectrometry Analysis of Agarwood Extracts from Mature and Juvenile *Aquilaria malaccensis*, *International Journal of Agriculture and Biology (Pakistan)* 16: 644–648.

Jung, D. 2011. The Value of Agarwood. Refelections Upon Its Use and History in South Yemen. In: *HeiDOK*. http://www.ub.uni-heidelberg.de/archiv/. Accessed on 30/05/2001.

Kaiser, R. 2006. *Meaningful Scents Around the World. Olfactory, Chemical, Biological, and Cultural Considerations.* Helvetica Chimica Acta, Zürich; Wiley-VCH, Weinheim (Cited in Jung, D. 2011. The Value of Agarwood. Reflections Upon Its Use and History in South Yemen. In: HeiDOK, http://www.ub.uni-heidelberg.de/archiv/).

Karlinasari, L., Danu, M.I. and Nandika, D. 2017. Drilling Resistance Method to Evaluated Density and Hardness Properties of Resionous Wood of Agarwood (*Aquilaria malaccensis*), *Wood Research* 62(5): 683–690.

Karlinasari, L., Putri, N., Turjaman, M., Wahyudi, I. and Nandika, D. 2016. Moisture Content Effect on Sound Wave Velocity and Acoustic Tomograms in Agarwood Trees (*Aquilaria malaccensis* Lamk.), *Turkish Journal of Agriculture and Forestry* 40: 696–704.

Lee, S.Y. and Mohamed, R. 2016. *Chapter 1: The Origin and Domestication of Aquilaria, An Important Agarwood-Producing Genus. Agarwood, Tropical Forestry.* Springer Science+Business Media, Singapore. doi:10.1007/978-981-10-0833-7_1.

Lim, T.W. and Anak, N.A. 2010. *Wood for the Trees: A Review of the Agarwood (Gaharu) Trade in Malaysia.* TRAFFIC Southeast Asia, Petaling Jaya, Selangor, Malaysia.

Liu, Y.Y., Wei, J.H., Gao, Z.H., Zhang, Z. and Lyu, J.C. 2017. A Review of Quality Assessment and Grading for Agarwood, *Chinese Herbal Medicines* 9(1): 22–30. doi:10.1016/S1674-6384(17)60072-8.

Lok, E.H. and Yahya, A.Z. 1996. The Growth Performance of Plantation Grown *Aquilaria malaccensis* in Peninsular Malaysia, *Journal of Tropical Forest Science* 8(4): 573–575.

López-Sampson, A. and Page, T. 2018. History of Use and Trade of Agarwood, *Economic Botany* 72(1): 107–129.

Malaysia Timber Industry Board (MTIB). 2014. *Panduan pengelasan produk gaharu Malaysia.* 1st ed. Lembaga Perindustrian Kayu Malaysia, Kuala Lumpur.

Malaysian Timber Industry Board (MTIB). 2010. *Merungkai khazanah rimba. Industri Gaharu Malaysia: Cabaran dan Prospek.* MTIB, Selangor, Malaysia.

Mazlan, M. and Dahlan, T. 2010. *Penggredan dan Pemprosesan Gaharu.* Seminar Kebangsaan dan Pameran Gaharu, Selangor (Cited in Mohamed, R. and Lee, S.Y. 2016. Chapter 10 Keeping Up Appearances: Agarwood Grades and Quality. Cited from Agarwood: Science Behind the Fragrance. DOI 10.1007/978-981-10-0833-7_10. Publisher: 1614-9785. Springer Science and Business Media Singapore).

Mohamed, R., Jong, P.L. and Abd Kudus, K. 2014. Fungal Inoculation Induces Agarwood in Young *Aquilaria malaccensis* Trees in the Nursery, *Journal of Forestry Research* 25(1): 201–204.

Mohamed, R., Jong, P.L. and Zali, M.S. 2010. Fungal Diversity in Wounded Stem of *Aquilaria malaccensis, Fungal Diversity* 43(1): 67–74.

Mohamed, R. and Lee, S.Y. 2016. Chapter 10 Keeping Up Appearances: Agarwood Grades and Quality (Cited from Agarwood: Science Behind the Fragrance. DOI 10.1007/978-981-10-0833-7_10. Publisher: 1614-9785. Springer Science and Business Media Singapore).

Mohamed, R., Wong, M.T. and Halis, R. 2013. Microscopic Observation of Gaharu Wood from *Aquilaria malaccensis, Pertanika Journal of Tropical Agricultural Science* 36(1): 43–50.

Naef, R. 2010. The Volatile and Semi-Volatile Constituents of Agarwood, the Infected Heartwood of *Aquilaria* species: A Review, *Flavour and Fragrance Journal* 26(2): 73–87. doi:10.1002/ffj.2034.

Nath, S.C. and Saikia, N. 2002. Indigenous Knowledge on Utility and Utilitarian Aspects of *Aquilaria malaccensis* Lamk. in Northern India, *Indian Journal of Traditional Knowledge* 1(1): 47–58.

National Parks Board, Singapore. 2013. *Aquilaria malaccensis* Lamk. https://florafaunaweb. nparks.gov.sg/. Accessed on 31/10/2018.

Ng, L.T., Chang, Y.S. and Kadir, A.A. 1997. A Review on Agar (Gaharu) Producing *Aquilaria* species, *Journal of Tropical Forest Products* 2(2): 272–285.

Nobuchi, T. and Hamami, M.S. 2008. *The Formation of Wood in Tropical Trees: A Challenge from the Perspective of Functional Wood Anatomy*. Universiti Putra Malaysia, Serdang, Selangor.

Nor Azah, M.A., Chang, Y.S., Mailina, J., Abu Said, A., Saidatatul Husni, S., Nor Hasnida, H. and Nik Yasmin, N.Y. 2008. Comparison of Chemical Profiles of Selected Gaharu Oils from Peninsular Malaysia, *The Malaysian Journal of Analytical Sciences* 12: 338–340.

Novriyanti, E. and Santosa, E. 2011. The Role of Phenolics in Agarwood Formation of Aquilaria *crassna* Pierre ex Lecomte and *Aquilaria microcarpa* Baill Trees, *Journal of Forestry Research* 8(2): 101–113.

Ogata, K., Fujii, T., Abe, H. and Baas, P. 2008. *Identification of the Timbers of Southeast Asia and the Western Pacific*. Kaiseisha Press, Shiga, Japan.

Oldfield, S., Lusty, C. and MacKinven, A. 1998. *The World List of Threatened Trees*. World Conservation Press, Cambridge, UK (Cited in Barden, A., Anak, N.A., Mulliken, T. and Song, M. 2000. Heart of the matter: Agarwood use and trade and CITES implementation for *Aquilaria malaccensis*. file:///C:/Users/acer/Downloads/traffic_pub_forestry7%20(3).pdf. Accessed on 8/10/2018.

Paoli, G.D., Peart, D.R., Leighton, M. and Samsoedin, I. 2001. An Ecological and Economic Assessment of the Nontimber Forest Product Gaharu Wood in Gunung Palung National Park, West Kalimantan, Indonesia, *Conservation Biology* 15(6): 1721–1732.

Peng, Chong S., Rahim, K.A. and Awang, M.R. 2014. Histology Study of *Aquilaria malaccensis* and the Agarwood Resin Formation under Light Microscope, *Journal of Agrobiotechnology* 5: 77–83.

Pojanagaroon, S. and Kaewrak, C. 2005. Mechanical Methods to Stimulate Aloeswood Formation in *Aquilaria crassna* Pierre H. Lec (Kritsana) Trees. ISHS Acta Horticulturae 676: III WOCMAP Congress on Medicinal and Aromatic Plants - Volume 2: Conservation, Cultivation and Sustainable Use of Medicinal and Aromatic Plants. doi:10.17660/ActaHortic.2005.676.20.

Premalatha, K. and Karla, A. 2013. Molecular Phylogenetic Identification of Endophytic Fungi Isolated from Resinous and Healthy Wood of *Aquilaria malaccensis*, a Red Listed and Highly Exploited Medicinal Tree, *Fungal Ecology* 6(3): 205–2011.

Rao, K.R. and Dayal, R. 1994. The Secondary Xylem of *Aquilaria agallocha* (Thymelaeaceae) and the Formation of 'Agar', *IAWA Bulletin n.s* 13(2): 163–172.

Richter, H.G. and Dallwitz, M.J. 2000 Onwards. Commercial Timbers: Descriptions, Illustrations, Identification, and Information Retrieval. delta-intkey.com.

Sathyanathan, V., Saraswathy, A. and Jayaraman, P. 2012. Pharmacognostical Studies on the Wood *Aquilaria malaccensis* Lam., *International Journal of Pharmacy and Industrial Research* 02(04): 416–423.

Schafer, E. 1963. *The Golden Peaches of Samarkand: A Study of T'ang Exotics*. University of California Press, Berkeley and Los Angeles (Cited in Lim, T.W. and Anak, N.A. 2010. Wood for the trees: A review of the agarwood (gaharu) trade in Malaysia. TRAFFIC Southeast Asia. Petaling Jaya, Selangor, Malaysia).

Shoeb, M., Begum, S. and Nahar, N. 2010. Study of an Endophytic Fungus from *Aquilaria malaccensis* Lamk., *Journal of the Bangladesh Pharmacological Society* 5(1): 21–24.

Siburian, R.H.S., Siregar, U.J., Siregar, I.Z., Santoso, E. and Wahyudi, I. 2013. Identification of Anatomical Characteristics of *Aquilaria microcarpa* in Its Interaction with *Fusarium solani*, *Biotropia* 20(2): 104–111.

Skeat, W.W. 1900. *Malay Magic: Being an Introduction to the Folklore and Popular Religion of the Malay Peninsula*. The Malaysian Branch of the Royal Asiatic Society (MBRAS), Reprint No. 24 (2005), Kuala Lumpur (Cited in Lim, T.W. and Anak, N.A. 2010. Wood for the trees: A review of the agarwood (gaharu) trade in Malaysia. TRAFFIC Southeast Asia. Petaling Jaya, Selangor, Malaysia).

Slik, J.W.F. 2009 Onwards. Plants Resources of Southeast Asia (PROSEA). http://www.asian-plant.net/Thymelaeaceae/Aquilaria_malaccensis.htm. Accessed on 25/4/2018.

Smith, F.P. and Stuart, G.A. 2003. *Chinese Medicinal Herbs: A Modern Edition of a Classic Sixteenth-Century Manual*. Dover Publications, Mineola, NY (Cited in López-Sampson, A. and Page, T. 2018. History of Use and Trade of Agarwood, Economic Botany 72(1): 107–129).

Suhartono, T. and Newton, A.C. 2001. Reproductive Ecology of *Aquilaria* spp. in Indonesia, *Biological Conservation* 152(1–3): 59–71.

Sustainable Asset Management. 2018. Prices for Oud Oil. https://www.agarwoodprices.com/prices/oud-oil. Accessed on 29/11/2018.

Tabata, Y., Widjaja, E., Mulyaningsih, T., Parman, I., Wiriadinata, H., Mandang, Y.I. and Itoh, T. 2003. Structural Survey and Artificial Induction of Aloeswood, *Wood Research* 90: 11–12. Kyoto University.

Takagi, K., Kimura, M., Harada, M. and Otsuka, Y. 1982. *Pharmacology of Medicinal Herba in East Asia*. Nanzando, Tokyo.

Tawfik, H.A., Ewies, E.F. and El-Hamouly, W.S. 2014. Synthesis of Chromones and Their Applications During the Last Ten Years, *IJRPC (International Journal of Research in Pharmacy and Chemistry)* 4(4): 1046–1085.

TRAFFIC. 2005. The Trade and Use of Agarwood in Taiwan, Province of China. PC 15 Inf. 7. CITES.

Wang, S., Yu, Z., Wang, C., Wu, C., Guo, P. and Wei, J. 2018. Chemical Constituents and Pharmacological Activity of Agarwood and Aquilaria Plants, *Molecules* 23(2): 342.

Wulffraat, S. 2006. *The Ecology of a Tropical Rainforest in Kayan Mentarang National Parks in the Heart of Borneo*. Published by WWF-Indonesia. The Ministry of Forestry of the Government of Indonesia and The World Wide Fund for Nature Indonesia.

Yaacob, S. 1999. Agarwood: Trade and CITES Implementation in Malaysia. Unpublished report prepared for TRAFFIC Southeast Asia, Malaysia (Cited in Barden, A., Anak, N.A. and Mulliken, T. and Song, M. 2000. Heart of the matter: Agarwood use and trade and CITES implementation for *Aquilaria malaccensis*.

Yamada, I. 1995. Aloeswood Forest and the Maritime World, *Southeast Asian Study* 33(3): 181–196.

Yang, J.L., Dong, W.H., Kong, F.D., Liao, G., Wang, J., Li, W., Mei, W.L. and Dai, H.F. 2016. Characterization and Analysis of 2-(2-Phenylethyl)-Chromone Derivatives from Agarwood (*Aquilaria crassna*) by Artificial Holing for Different Times, *Molecules* 21(7): 911. doi:10.3390/molecules21070911.

Yin, Y., Jiao, L., Dong, M., Jiang, X. and Zhang, S. 2016. *Wood Resource, Identification, and Utilization of Agarwood in China*. Agarwood, Tropical Forestry. Springer Science+Business Media, Singapore, Chapter 2.

Yuk, C.S., Lee, S.J., Yu, S.J., Kim, T.H., Hahn, Y., Lee, K., Moon, S.Y., Hahn, Y.H., M.W. and Lee, K.S. 1981. *Herbal Medicine of Korea*. Gyechukmunwhasa, Seoul.

8 Cellulose-Based Products and Derivatives

Melissa Sharmah Gilbert

CONTENTS

8.1 INTRODUCTION

Cellulose is an organic compound consisting of a linear chain that may comprise hundreds to thousands of glucose monosaccharide units. Cellulose can be located in the cell walls of most plants, except for some marine algae, due to its role as one of the cells' main structural materials (Preston, 1968). Thus, it is the most abundant organic materials on Earth and is found in plants as one of the three lignocellulose "buddies" that intertwine with each other, the other two being hemicellulose and lignin. Cellulose and hemicellulose provide the sturdy building material, while lignin glues it all together to form fibers.

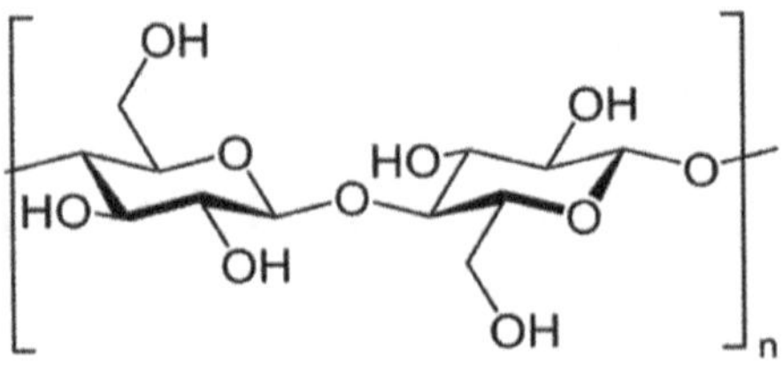

FIGURE 8.1   Cellulose structure. (From Neurotiker 2007.)

Cellulose was first discovered by a French chemist named Anselme Payen while identifying the various types of wood. The similarities between starch and cellulose was what intrigued Payen; both similarly break down into monomeric glucose, but are assembled differently. He later named the new substance *cellulose*, from the "cell" of the cell wall in which he located it. The chemical formula was determined $(C_6H_{10}O_5)n$ with a β-1,4-glycosidic linkage such as in Figure 8.1 (Lavanya et al., 2011).

Plants are not the only organisms that produce cellulose, because interestingly, some bacteria produce a naturally occurring cellulosic bacterial film that encapsulates the biotic agents during infection. As enticing as it may seem, only a minor portion of it is produced by the pathogen, rather than the naturally occurring cellulose on plants (Thongsomboon et al., 2018). In plants, cellulose is one of the main materials for the formation and stabilization of the cell structure, with the assistance of hemicellulose and lignin. The sources of lignocellulose in plants include a range of wood, bast, leaves, fruits, seeds, and many others (Cordeiro, 2016). The composition also differs between different types of wood, as shown in Table 8.1.

8.1.2   Extraction of Cellulose

Since cellulose is a part of the lignocellulose "buddies", which are linked via covalent bond, the separation of the three components has proven strenuous and highly specific. All three components have different characteristics; for example, while cellulose consists of only one type of monosaccharide (i.e. glucose) in a long linear chain, lignin consists of heterogeneous cross-linking phenolic that differs from one

TABLE 8.1
Typical Compositions of Hardwood and Softwood

Compound	Hardwood (%)	Softwood (%)
Cellulose	45 ± 2 %	42 ± 2 %
Hemicellulose	30 ± 5 %	27 ± 2 %
Lignin	20 ± 4 %	28 ± 3 %
Extractives	5 ± 3 %	3 ± 2 %

Source: Fengel and Wegener (1989)

species to another. Because of all these differing compositions, the extraction needs to be highly distinctive and accurate, with methods including mechanical, chemical, and natural.

8.1.2.1 Mechanical

Mechanical treatment is a process that usually includes the beating of wood into pulp. This process is most recognized as mechanical pulping, which is the fundamental process of accessing cellulose by hammering the wood. The production of mechanical pulp involves grinding the wood against a water-lubricated rotating stone. The heat generated by grinding softens the lignin binding the fibers, and the mechanized forces separate the fibers. It is one of the most basic but poor ways to extract cellulose through physically breaking the materials to free the fibers from each other as well as weakens the lignin bond physically. The process includes a series of mechanical pulverizations, which include cutting, chipping and flaking (Keefe et al., 1998).

The product was initially referred to as *ground wood* rather than *pulp* before a more advanced method of thermo-mechanical pulping (TMP) was introduced. The separation process is not as easy with TMP because the eventual chemical composition remains more or less the same. However, the addition of heat in TMP allows a better yield with improved covalent bond breakage and depolymerization of the heterogeneous components. Subsequent modifications of this pulping method include chemi-thermo-mechanical pulping, (CTMP) in which the wood chips are impregnated with chemicals such as sodium sulphite prior to grinding (McDonald et al., 2004).

8.1.2.2 Chemical

Chemical treatment is more efficient than physical treatment in obtaining cellulose from wood. It is commonly used in the pulp and paper industry to produce pulps for papers and other products. It can be divided into two categories: the usage of either acidic or alkaline reagents to obtain cellulose. Similar to the pulp and paper industry, the aim of pulping using chemicals such as acid in acid sulfite pulping and alkaline in Kraft pulping yields better cellulose composition and lowers the lignin content substantially (Islam et al., 2014).

In Kraft pulping, an alkaline reagent or two is used to carry out the treatment. Solutions such as sodium sulfide and sodium hydroxide are used as reagents in both Kraft pulping and alkaline hydrolysis to produce good quality pulp or cellulose. Kraft pulping itself is fundamentally a type of alkaline hydrolysis, where it enables the liberation of the cellulose by disrupting the structure of the lignin using an alkaline solution, consequently weakening its hold on the imprisoned cellulose and hemicellulose. However, Kraft pulping includes high temperature and pressure to improve the yield per process as well as speed it up. Unlike Kraft pulping, alkaline hydrolysis is actually regarded as a pretreatment process to obtain pure cellulose as it locally targets the lignin and disturbs its structure, and is usually followed by acidic hydrolysis, eventually separating the structural linkages between the lignin and the holocellulose (Duprey, 1968).

Acid sulfite pulping uses a similar process to Kraft pulping but differs in the cooking reagents used. As the name suggests, sulfite pulping uses acids such as sulfuric acid in place of the alkaline reagent in Kraft pulping to dissolve the lignin in the wood. Heat and pressure are both needed to cook the wood chips in the digester, which is followed by washing and bleaching. Acid uses its corrosive properties to eliminate most of the lignin binding the cellulose, while also depolymerizing the hemicellulose as well as some of the cellulose. In normal cellulose extraction, acid hydrolysis will be the main process used due to its ability to break linkages (Islam et al., 2014).

8.1.2.3 Natural

Wood consists of lignin binding the holocellulose to build up the structure. The holocellulose itself is also known as *polysaccharides*, which literally means multiple sugars and is a known food source for microorganisms. In the timber industry, attacks from fungi and insects are a well-known problem during the storage phase, where the sugars in the wood are targeted for sustenance. One of the most common kinds of damage comes from the brown and white rot fungi that attack improperly dried wood. In cellulose extraction, wood-decaying fungi actually benefit the outcome as they function as a natural converter, targeting either the lignin or holocellulose, while isolating the other.

Wood-decaying fungi from the basidiomycete are one of the divisions in the subkingdom of dikarya within the fungi kingdom. They are divided into two categories—brown and white rot fungi—which are classified based on the color of the wood following decay. Brown rot fungus, as the name suggests, leaves the wood brown due to its preference for attacking cellulose and hemicellulose, which is lighter in color, and leaves the lignin, which is darker in color, alone.

The popular theory of the white rot fungus (Figure 8.2) mechanism is the opposite of the brown rot, as it favors lignin more, hence the white residue following attack. However, white rot can actually simultaneously degrade all three components at once or can selectively target lignin over holocellulose. Most of the time, white

FIGURE 8.2 White rot fungi. (From Monster 2015.)

rot selectively targets the lignin, followed by the hemicellulose, eventually gulping down the cellulose through hydrolysis (Hastrup et al., 2012).

Another natural method of obtaining cellulose is through the manipulation of enzymes. Similarly with decaying fungi such as brown and white rot fungi, the mechanism of deterioration or isolation relies on the presence of enzymes in the microorganisms that enables them to digest the wood components. Identical to most cellulose digestion in insects and fungi, enzymes like cellulases, cellobiohydrolases, and glucosidase are important in digestion assistance (Martin, 1983). However, the depolymerization of polysaccharides by the aforementioned enzymes is usually inhibited by the presence of lignin, which challenges the degradation through its complex structure (Abdel-Hamid et al., 2013).

As for cellulose extraction, the lignin gains access first to the hemicellulose, which is easier to dissolve, and finally to the targeted cellulose. The enzymes involved in lignin biodegradation consist of two major groups: heme peroxidases and laccases. The peroxidases can be further declassified into different types of enzyme, which include the lignin peroxidase, dye-decolorizing peroxidase, and manganese peroxidase. The mechanism of the peroxidase enzyme relies on the presence of its strong oxidants, which, with a high reduction capacity, oxidize the phenolic and non-phenolic structures of the lignin. Meanwhile, the laccases use the catalysis of electron oxidation to reduce the molecular structure (Abdel-Hamid et al., 2013).

As for the hemicellulose, it contains a variety of branched heterogeneous polysaccharide that contains a number of pentoses and hexoses sugars such as xylan, mannans, and galactans. For the removal of hemicellulose naturally, a number of enzymes are needed due to the numbers of sugar varieties in the content, unlike acid hydrolysis, which simply breaks off the branched polymer. The enzymes accountable for hemicellulose degradation are named according to the substrate they catalyst, such as xylanases for xylan, mannanases for mannans, and galactanase for galactans (Moore et al., 2012).

8.1.3 IMPORTANCE

Although abundant in number and has multiple potentials, cellulose as part of lignocellulose has always been taken for granted due to its abundance. In the form of biomass such as logging residue, oil palm waste, and sugarcane bagasse, these lignocellulose materials are perpetually undervalued and are considered a nuisance due to the costs entailed in their disposal (Abdullah and Sulaiman, 2013). Thereupon another problem emerged, most notably in the oil palm industry, where a "shortcut" was introduced: the open burning of discarded biomass. This involves stacking and burning old palm trunks, pruned fronds, and empty fruit bunches, polluting the immediate and neighboring areas with immeasurable amounts of smoke (Mohd Noor, 2003).

Complications such as this would not have happened if the value of lignocellulose resources had been publicized. As we all know, cellulose is part of the lignocellulose trio that makes up the structural components of the cell wall in plants. Therefore the most primary usages of cellulose are usually in the form of wood, which has been known to be useful as a source of energy in the form of heat if set ablaze, as well as being popularly used for construction purposes. Apart from the basic, unchanged form of cellulose, two other forms of cellulose usage exist.

The second type of cellulose usage is in the purified state, where the lignin and the hemicellulose from the lignocellulose are partially or completely removed. This type of form is used to make a number of vital products such as paper and fabric. Meanwhile for the final form, purified cellulose is modified into a semi-synthetic material through chemical conversion. The new material will have improved properties compared to its predecessor, as some of the unfavorable characteristics of pure cellulose are enhanced; for example, it becomes thermoplastic and soluble in solvents (Shokir and Adibkia, 2013).

8.2 LIGNOCELLULOSE PRODUCTS

Lignocellulose is a material in which cellulose, hemicellulose, and lignin are present. The content of cellulose in this form has evaded modification or human intervention and is readily available in nature. The procurement of the heterogeneous components is via plants, specifically in the form of wood/biomass or residue. This is the most abundant raw material on Earth and, if given the opportunity, it has a lot of potential such as the existing utilization for energy and structure.

8.2.1 Wood Fuel

The terms *fuel* or *wood fuel* mostly refer to the potential of woody biomass to be developed into fuel for energy use. Some engines and turbines still use wood to generate electricity from steam, but most wood fuel is used by the household sector. This kind of fuel is considered the most traditional type of fuel due to the simple mechanism of burning it to obtain heat (Anttilla et al., 2009). Biomass energy accounts for 15% of global energy consumption, but in certain least developed countries (LDC) such as Uganda, majority of the country still relies on it as their main source of energy (Brenda et al., 2017).

These fuels require no or moderate machinery to produce as the mechanism of energy production involves the burning of fuel to produce fire and heat. Production of these wood fuels ranges from basic to intermediate, where the easiest fuel to produce, such as firewood, just needs to be dried and resized, and the hardest requires a compacter and molder to shape it. Examples of wood fuel from lignocellulose sources include firewood, charcoal, briquettes, and pellets, such as in Figure 8.3 (Brenda et al., 2017).

8.2.2 Construction

Lignocellulose in the form of wood is one of the most preferred building materials in the world. It is a renewable material originating in numerous species of woody plants that has various properties based on species and various other factors. There are a number of factors that give wood the upper hand when it comes to construction material selection. One of them is the fact that it is a renewable resource that can be planted and replenished after every felling (Dost and Botsai, 1990).

FIGURE 8.3   Wood pellet.

   The environmental effect of using renewable resources, albeit that trees take a considerable amount of time to renew, is lower compared to the non-renewable concrete and steel. Another advantage is the anisotropic nature of wood, which gives the material three unequal dimensional changes and strength along the longitudinal, tangential, and radial cardinal directions, which is very important for perception of strength. The selection of primary timber products available includes sawn timber and plywood (Dost and Botsai, 1990).

8.3   PURE CELLULOSE PRODUCTS

Like most refined products, pure cellulose is the material obtained after undergoing a long series of intricate purification processes. The extraction of cellulose includes numerous complex processes due to the intertwined structure of the lignocellulose, and extraction has proven to be costly. Methods of extraction include mechanical, chemical, and natural processes, with a chemical-based pulping method being used by many for its purity. The yield after extraction consists of a pure or partially pure raw material, which is then used to make other products such as textiles for clothing, pulp for paper and cardboards, and consumables, which are the edible form of cellulose.

8.3.1   Pulp and Paper

The form of pure cellulose used for pulp or paper may contain other lignocellulose components, as the raw material is obtained through pulping. Pulping is a type of extraction that involves the breaking of bonds within the wood structure via chemical or mechanical methods. Chemical pulps are divided into either alkaline or acidic pulp, produced by Kraft or sulfite pulping, respectively. The chemicals are used to break the bond connecting the lignin and hemicellulose to the cellulose, chemically dissolving or eliminating the residues in the process. Unlike pure cellulose extraction, some residual lignin and hemicellulose are still present after the pulping process in differing quantities (Keefe et al., 1998).

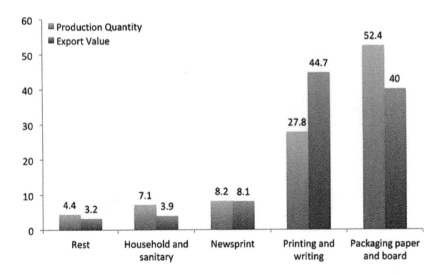

FIGURE 8.4 Production quantity and export value shares (%) of global paper-based products in 2010. (From Hetemäki et al. 2013.)

Meanwhile, mechanical pulp is instead produced initially through grinding the wood against a stone or metal plate, separating the fiber physically. Due to the blunt physical force of shearing and separating, mechanical pulp is weaker than chemical pulp and does not eliminate the unwanted components. However, the more modern mechanical pulp incorporates heat and pressure in the process to obtain better pulp quality. The highest pulp yield is produced by Kraft pulping, followed by mechanical and then sulfite pulping.

Paper or pulp products are commonly categorized into five main groups, including printing and writing paper, packaging paper and board, newsprint, and household and sanitary products, as well as other paper products. Contrary to popular belief, paper is not the main product of the pulp and paper industry. Packaging paper and board has the highest consumption such in Figure 8.4, surpassing printing and writing paper. The major paper products for packaging include corrugated boxes, boxboard, and paper bags. Meanwhile, other products include paper for writing, paper for newsprint, tissue paper, toilet paper, and some medical paper products.

8.3.2 FIBER AND TEXTILE

Cellulose is one of the most highly durable and flexible materials that can be found naturally and is one of the more prominent raw materials used in textile production. Textiles are pliable materials that are usually made into clothes. They are made by mixing fiber components through a blending process, twisted this into balls of yarn, and finally interlaced the yarn into fabrics. There are various blended fibers, which are made out of either natural or synthetic fibers, or both. Fiber blends include polyester, nylon, spandex, and cotton. The different types of textile serve different

FIGURE 8.5   Cotton.

purposes according to their properties, such as high absorption for cotton and quick drying for polyester.

Cellulose-based cotton (Figure 8.5) is one of the materials most in demand for fabric. Cotton itself contains about 85% cellulose in its structure, and is the highest yielding natural resources of cellulose per volume. Apart from being highly absorbent, cotton also has other highly unique characteristics that make it suitable for producing cloth, such as being hypoallergenic, having high heat resistance and being durable. Cotton is also a favorite among naturalists, as it is known as a vegan or natural fiber due that yields an almost pure natural cellulose component. Apart from fabric, cottons are also used in home furnishing items such as bed sheets, mattresses, curtains, and tablecloths (Kihlman, 2012).

8.3.3   Consumables

A consumable is a type of cellulose that is processed into forms that are safe for consumption, hence the name. As humans lack the enzyme to digest cellulose in the same way we digest starch, the fiber and hygroscopicity of wood helps in aiding the digestive tract. The most common type of consumable cellulose is microcrystalline cellulose (MCC). The cellulose in MCC is made out of refined wood pulp and is used in the food industry. To produce MCC, the structure of cellulose is first divided into crystalline or amorphous types. The sturdier crystalline type, which is whiter, odorless, and chemical resistant, is isolated and then undergoes trituration to produce MCC. MCC is commonly used as emulsifiers, stabilizers, anti-caking agents, extenders, and bulking agents in foods (Lavanya et al., 2011). Cellulose ethers also offer a wide array of consumables, mostly for drug encapsulation and buffer release.

8.3.4   Bio-Ethanol

Bio-ethanol in general is the most used liquid type of biofuel. Over 40 billion liters are produced per year worldwide in countries such as Brazil and the United States.

FIGURE 8.6   Flexfuel Chevy Tahoe. (From Schwen 2008.)

The demand for bio-ethanol has been increasing tremendously due to the manufacturing of hydrated ethanol-fueled vehicles called *flexible-fuel* or *flex-fuel* vehicles (Figure 8.6). In its purest state, cellulose is made by linking glucoses with a varying degree of polymerization from hundreds to thousands. Through the breaking of these linkages, countless amounts of glucose are liberated in its simplest energy-producing form for consumption (Anttila et al., 2009).

Apart from producing adenosine triphosphate (ATP), which is energy in animals, glucose can also be converted into biofuels, specifically bio-ethanol. The isolated cellulose obtained from plants is used to produce glucose. The cellulose can be obtained from any type of lignocellulose materials, such as wood, oil palm residues such as empty fruit bunches and fronds, or silviculture waste such as branches. The common methods for bio-ethanol production include cellulolysis, gasification, and consolidated bio-processing (CBP). The cellulolysis process involves the extraction of cellulose from its lignocellulosic material through hydrolysis, followed by the liberation of glucose from its cellulose form through enzyme hydrolysis (Thongsomboon et al., 2018).

The gasification process is conducted without fermentation, involving combusting the biomass into syngas and heat. With enough oxygen, the syngas can be converted into compounds such as methanol and ethanol. Lastly, CBP is actually an integrated single process that involves cellulose production, substrate hydrolysis, and fermentation through cellulolytic microorganisms. This process has proven more economical than other processes due to its lower energy input and higher conversion output.

8.4   CELLULOSE DERIVATIVES: ESTERS

Cellulose is a very versatile compound that contains multiple hydroxyl groups in its bases. The hydroxyl group plays an important role in the modification or derivation of cellulose, as with its substitution, cellulose undergoes a whole new alteration in its chemical structure as well as its properties. The production of cellulose derivatives

FIGURE 8.7   Cigarette filter.

involves a reaction process between the cellulose and its reagent, substituting, partially or fully, the three free hydroxyls in its base. The derivatives are divided into two, based on the categories of their reagent. Cellulose esters are the product of cellulose reaction to acids; meanwhile, cellulose ethers react with halogenoalkanes, epoxides, and halogenated carboxylic acids.

Cellulose esters can further be divided into two categories based on the use of inorganic or organic acids, hence the creation of inorganic and organic esters, respectively. Cellulose esters have a number of upgraded attributes from cellulose, such as the ability to dissolve in most organic solvents and good film formation. Most organic esters are primarily used in the pharmaceutical industry, where they are commonly used as coatings for controlled-release drugs. The various types of cellulose esters include cellulose acetate, cellulose acetate butyrate, cellulose acetate phthalate, and cellulose acetate propionate (Granström, 2009).

8.4.1   Cellulose Acetate

Cellulose acetate is an organic cellulose ester. It is one of the most common cellulose derivatives to be found as it has numerous uses, such as for cigarette filters (Figure 8.7), film base for photography, in playing cards, and as eyeglass frames. It is produced through the reaction of cellulose with acetic acid and acetic anhydride as its main reagent, with sulfuric acid as its catalyst. The process then undergoes hydrolysis to remove sulfates and a number of acetate groups to produce a more uniform degree of substitution. The naming of cellulose as either *diacetate* or *triacetate* refers to the number of hydroxyls substituted in every glucose molecule, with diacetate generally referring to the substitution of two hydroxyls, and triacetate to three.

Cellulose diacetate is the more preferred form of acetate ester due to its stability and potential development, which is quite similar to triacetate. The properties of cellulose after modification to cellulose acetate include a myriad of enhancements such as good absorption, solubility in most organic solvents, good liquid transport, resistance to mold and mildew, and being hypoallergenic and biodegradable. Most of its attributes further support its usage in fabric production. Usually, after it is formed, cellulose acetate is ready to be dissolved in an organic solvent for further processing into products such as acetate fibers and films. Plasticizers, pressure, and heat can also be used to easily bond cellulose acetate. Apart from the aforementioned products, cellulose acetate is also widely used for diapers, buttons, clothing, and upholstery (Bogati, 2011).

8.4.2 NITROCELLULOSE

Nitrocellulose is another type of cellulose ester that is produced through the reaction between cellulose and nitric acid or another strong nitrating agent. Nitrocellulose was the once the most widely known cellulose ester as it was commonly used as gun-cotton, with excellent low-order explosive trait. The mechanisms of modification use to produce nitrocellulose are similar to other cellulose esters, in which the hydroxyl group in the cellulose is substituted with nitrate derived from the nitrating agent. Due to the instability of nitrocellulose, which is why it is used as gunpowder, complex washing and stabilizing are carried out for safety. Sometimes damping agents or plasticizers are also added before marketing (Bogati, 2011).

Nitrocellulose is an inorganic cellulose ester because an inorganic nitric acid is used for modification. Nitrocellulose was named after the *nitro* in nitrate but is also known as *cellulose nitrate*. In liquid form, sometimes mixed with alcohol and ether, nitrocellulose is referred to collodion. The nitrogen content influences the properties of the nitrocellulose, ultimately affecting the potential product development. Nitrocellulose with a nitrogen content below 12.3% is usually more stable and easily dissolves in organic solvent, hence its use in lacquer, coating, and ink production. When the nitrogen content is more than 12.6%, the materials are less stable and are usually used for explosives such as guncotton (Ángeles Fernández de la Ossa et al., 2012).

8.4.3 CELLULOSE SULFATE

Cellulose sulfate is another inorganic cellulose ester that is produced through hydroxyl substitution. The production involves the partial or total substitution of hydroxyl groups positioned in C_2, C_3, and C_6 with sulfate. Although almost always present in the production of other esters such as cellulose acetate due to the usage of sulfuric acid as a catalyst, cellulose sulfate is usually removed during the hydrolysis process. While apparently unpopular, this ester has been a favorite in the medical field, especially the water-soluble type, which is widely used as a drug coating. Its biodegradability and biocompatibility allow it to be developed into assorted other products such as microbicides, anticoagulant agents, and thickeners (Chen et al., 2013).

8.5 CELLULOSE DERIVATIVES: ETHERS

Similar to their counterparts, cellulose ethers react with etherizing agents to produce various ethers. There are several commercially important cellulose ethers and esters in the world, as shown in Table 8.2. The common reagent categories

TABLE 8.2
Commercially Important Cellulose Ethers and Esters

Product	Global Production (t/a)	Functional Group	Degree of Substitution	Solubility	Application
Cellulose Acetate	900 000	-OAc	0.6–0.9 1.2–1.8 2.2–2.7 2.8–3.0	Water 2-methoxy ethanol Acetone Chloroform	Coatings and membranes
Cellulose Nitrate	200 000	$-NO_2$	1.8–2.0 2.0–2.3 2.3–2.8	Ethanol Methanol, acetone Acetone	Membranes and explosives
Cellulose Xanthate	32 000 000	-C(S)SNa	0.5–0.6	NaOH/water	Textiles
Carboxymethyl Cellulose	300 000	$-CH_2COONa$	0.5–2.9	Water	Coatings, paints, adhesives, and pharmaceuticals
Methyl Cellulose	150 000	$-CH_3$	0.4–0.6 1.3–2.6 2.5–3.0	4% aq. NaOH Cold water Organic solvents	Film, textile, food and tobacco industry
Ethyl Cellulose	4 000	$-CH_2CH_3$	0.5–0.7 0.8–1.7 2.3–2.6	4% aq. NaOH Cold water Organic solvents	Pharmaceutical industry
Hydroxyethyl Cellulose	50 000	$-CH_2CH_2OH$	0.1–0.5 0.6–1.5	4% aq. NaOH Cold water	Paints, coatings, films, and cosmetics

Source: (Hamad et al. (2016))

in cellulose ether production include halogenoakanes and epoxides. When the conditions are met, the three-hydroxyl group for each monosaccharide in cellulose is substituted with the prominent functional group in the reagents, such as methyl, to consequently produce methyl cellulose (Ángeles Fernández de la Ossa et al., 2012).

Cellulose ethers have an ameliorated trait after modification if compared to cellulose such as solubility in solutions such as water and organic solvent and high water retention. In turn, these traits allow cellulose ethers to be developed into numerous products such as paper coating, thickeners, and gelling agents. The various ethers modified from cellulose include methyl cellulose, carboxymethyl cellulose, and hydroxypropyl cellulose (Granström, 2009).

8.5.1 Methyl/Ethyl Cellulose

Methyl cellulose is one of the more well-known commercial cellulose ethers and is used in many industrial applications. It is produced from the reaction between cellulose and methyl chloride, whereby the hydroxyl group in Carbon 2, 3, and 6 is substituted with the methoxide (-CH_3) groups. The degree of substitution of the methyl cellulose determines its solubility, either in water or organic solvents. It is white in color and possesses a number of new attributes compared to its predecessor, such as being hydrophilic and able to be formed into films. Methyl cellulose is widely used to treat constipation, hemorrhoids, and diverticulosis due to its high fiber content, as well as in shampoos, soaps, and toothpastes as a thickener and emulsifier (Hamad et al., 2016).

8.5.2 Hydroxyethyl/Propyl Cellulose

Hydroxyethyl and hydroxypropyl cellulose are very similar cellulose ethers that are produced through the reaction of pure cellulose with ethylene and propylene oxide, respectively. The production of hydroxyethyl cellulose involves the hydroxyethylation (-CH_2CH_2OH) of the hydroxyl group in the repeating glucose unit in cellulose. It is hydrophilic and dissolves in both cold and warm water, making it suitable for use in cosmetics and household products as a gelling and thickening agent. Being non-ionic also benefits the utilization of hydroxyethyl cellulose as it is less reactive and safe for use with contact of the skin, such as for product for personal care (Bogati, 2011).

If compared to its counter half, hydroxypropyl cellulose is a more popular cellulose ether. It is produced through the substitution of the hydroxyl group in the cellulose with hydroxypropyl (-$CH_2CH(OH)CH_3$) from the propylene oxide. It is soluble in both water and organic solvents, and is both hydrophilic and hydrophobic. Hydroxypropyl cellulose is commonly used to make artificial tears to treat medical conditions associated with the inability to produce sufficient tears, such as recurrent corneal erosions and keratoconjunctivitic sicca. In the food industry, hydroxypropyl cellulose is an additives that is used as thickener and emulsion stabilizer, as well as a coating for drugs (Figure 8.8) (Banker et al., 2008).

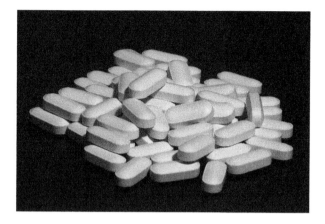

FIGURE 8.8   Calcium tablet coated with hydroxypropyl cellulose. (From Ragesoss 2008.)

8.5.3 CARBOXYMETHYL CELLULOSE

Carboxymethyl cellulose is a carboxyalkyl cellulose ether produced from the reaction between cellulose and chloroacetic acid to substitute the hydroxyl group to carboxymethyl ($-CH_2-COOH$) group. From the reaction, a mixture of 60% carboxymethyl cellulose and 40% salts are produced, which consist of sodium chloride and sodium glycolate. Pure carboxymethyl cellulose is obtained through further purification to remove the salts, but a semi-purified product is also available for use in document restoration. With the polarity of the carboxyl group, it is soluble in water and miscible organic solvents.

The properties of the carboxymethyl cellulose depend on the degree of substitution, where it refers to the number of hydroxyl groups substituted in the monosaccharide units in cellulose by the reagent. Due to its high viscosity and nontoxicity, carboxymethyl cellulose is widely used in the food industry as a viscosity modifier and thickener (E466). However, is one of the cellulose ethers that have wide applications, not just in the food industry (Koh, 2013).

8.6 SUMMARY

From a marketing perspective, the lignocellulosic biomass residues or "wastes" from plantations are simply a nuisance. The processes of silviculture or harvesting are necessary in order to maintain a healthy crop or as a means to auction off the product to gain market. However, the method of disposal has proven to be problematic; the easiest ways of disposing of the biomass, such as burning, cause air pollution, and the most environmental friendly ways, such as nutrient recycling, take longer and are a workplace hazard.

Unbeknownst to most and ignored by others is the potential for cellulose to be developed into numerous products. This chapter has shown cellulose's potential for use in the production of food additives such as emulsifier and the generation of energy primarily in the form of briquettes and the more complex bio-ethanol. Since cellulose is the most abundant biopolymer renewable resource in the world, there will be no drawbacks in terms of sourcing raw materials either. With this much potential, cellulose could literally be the future of renewable product resources.

REFERENCES

Abdel-Hamid, A. M., Solbiati, J. O., and Cann, I. K. O. 2013. Insights into Lignin Degradation and Its Potential Industrial Applications. *Advances in Applied Microbiology*, 82: 1–28.

Abdullah, N., and Sulaiman, F. 2013. The Oil Palm Wastes in Malaysia. *Biomass Now, Sustainable Growth and Use*. Intech Open.

Ángeles Fernández de la Ossa, M., Torre, M., and García-Ruiz, C. 2012. Nitrocellulose in Propellants: Characteristics and Thermal Properties. *Advances in Materials Science Research*, 7: 201–220.

Anonymous. No Yr. *Types of Pulping Processes*. Confederation of European Paper Industries.

Anttila, P., Karjalainen, T., and Asikainen, A. 2009. Global Potential of Modern Fuelwood. Working Papers of the Finnish Forest Research Institute 118.

Banker, G., Peck, G., Ja, S., Pirakitikulr, P., and Taylor, D. 1981. Evaluation of Hydroxypropyl Cellulose and Hydroxypropyl Methyl Cellulose as Aqueous Based Film Coatings. *Journal of Drug Development and Industrial Pharmacy*, 7(6): 693–716.

Bogati, D. R. 2011. *Cellulose Based Biochemicals*. Saimaa University of Applied Sciences: Imatra, Finland.

Brenda, M. G., Innocent, E. E., Daniel, O., and Abdu, Y. A. 2017. Performance of Biomass Briquettes as an Alternative Energy Source Compared to Wood Charcoal in Uganda. *International Journal of Scientific Engineering and Science*, 1(6): 55–60.

Chen, G., Zhang, B., Zhao, J., and Chen, H. 2013. Improves Process for the Production of Cellulose Sulfate Using Sulfuric Acid/Ethanol Solution. *Carbohydrate Polymers*, 95(1): 332–337.

Cordeiro, R. C. 2016. *Plasma Treatment of Natural Fibers to Improve Fiber-Matrix Compatibility*. Universidade Federal do Rio de Janeiro.

Dost, W. A., and Botsai, E. E. 1990. *Wood: Detailing for Performance*. GRDA Publications: Mill Valley, CA.

Duprey, R. L. 1968. *AP-42: Compilation of Air Emissions Factors: Chemical Wood Pulping*. United States Environmental Protection Agency (EPA).

Fengel, D., and Wegener, G. 1989. *Wood: Chemistry, Ultrastructure, Reactions*. Walter de Gruyter: Berlin; New York.

Granström, M. 2009. *Cellulose Derivatives: Synthesis, Properties and Applications*. University of Helsinki: Finland.

Hamad, A. M. A., Ates, S., and Durmazi, E. 2016. Evaluation of the Possibilities for Cellulose Derivatives in Food Products. *Kastamonu University Journal of Forestry Faculty*, 16(2): 383–400.

Hastrup, A. C., Howell, C., Larsen, F. H., Sathitsuksanoh, N., Goodell, B., and Jellison, J. 2012. Differences in Crystalline Cellulose Modification Due to Degradation by Brown and White Rot Fungi. *Fungal Biology*, 116(10): 1052–1063.

Hetemäki, L., Hänninen, R., and Moiseyev, A. 2013. Markets and Market Forces for Pulp and Paper Products. *The Global Forest Sector: Changes, Practices, and Prospects*, 99–128.

Islam, M. T., Patrucco, A., Montarsolo, A., Zoccola, M., and Alam, M. M. 2014. Preparation of Nanocellulose: A Review. *AATC Journal of Research*, 1(5): 17–23.

Keefe, A., Astrakianakis, G., and Anderson, J. 1998. Chapter 72: Pulp and Paper Industry. *Encyclopaedia of Occupational Health and Safety* (4th Ed.).

Kihlman, M. 2012. *Dissolution of Cellulose for Textile Fiber Applications.* Karlstad University.

Koh, M. H. 2013. *Preparation and Characterization of Carboxymethyl Cellulose from Sugarcane Bagasse.* Universiti Tunku Abdul Rahman.

Lavanya, D., Kulkarni, P. K., Dixit, M., Raavi, P. K., and Krishna, L. N. V. 2011. Sources of Cellulose and Their Applications—A Review. *International Journal of Drug Formulation and Research*, 2(6): 19–38.

Martin, M. M. 1983. Minireview: Cellulose Digestion in Insects. *Comparative Biochemistry and Physiology*, 75A(3): 313–324. (Pergamon Press Ltd).

McDonald, D., Miles, K., and Amiri, R. 2004. *The Nature of the Mechanical Pulping Process.* Pulp and Paper Canada: Ontario.

Mohd Noor, M. 2003. Zero Burning Techniques in Oil Palm Cultivation: An Economic Perspective. *Oil Palm Industry Economic Journal*, 3(1).

Monster, H. 2015. Fomes Fomentarius. Retrieved from: https://www.panoramio.com/photo/125474387 (CC BY 3.0 license)

Moore, D., Robson, G. D., and Trinci, A. P. J. 2012. 21st Century Guidebook to Fungi (2nd Ed.). *Quarterly Review of Biology*, 87(4).

Neurotiker. 2007. Cellulose Sessel. Retrieved from: https://commons.wikimedia.org/wiki/File:Cellulose_Sessel.svg (Public domain)

Preston, R. D. 1968. Plants Without Cellulose. *Scientific American*, 218(6): 102–111.

Ragesoss 2008. 500 mg Calcium Supplements with Vitamin D. Retrieved from: https://commons.wikimedia.org/wiki/File:500_mg_calcium_supplements_with_vitamin_D.jpg. (CC BY-SA 4.0 license)

Schwen, D. 2008. Flexfuel Chevy Tahoe. Retrieved from: https://commons.wikimedia.org/wiki/File:Flexfuel_Chevy_Tahoe.jpg (CC BY-SA 4.0 license)

Shokri, J., and Adibkia, K. 2013. Application of Cellulose and Cellulose Derivatives in Pharmaceutical Industries. *Cellulose, Theo van de Ven and Louis Godbout.* Intech Open.

Thongsomboon, W., Serra, D. O., Possling, A., Hadjineophytou, C., Hengge, R., and Cegelski, L. 2018. Phosphoethanolamin Cellulose: A Naturally Produces Chemically Modified Cellulose. *Science*, 359(6373): 334–338.

9 Organosolv Pulping and Handsheet Properties of Acacia Hybrid

Eunice Wan Ni Chong and Kang Chiang Liew

CONTENTS

9.1 INTRODUCTION

The pulp and paper industry is one of the most competitive sectors worldwide. Paper and paperboard consumption has been growing intensely despite the current era of electronic gadgets. This can seem contradictory to sense or logic. However, statistics have confirmed the growth increments in paper and paperboard consumption, where in Asia, it increased from a total of 109.4 million metric tons in 2000 to 178.1 million metric tons in 2010 and is expected to increase to 250.7 million metric tons by 2020 and 323.7 million metric tons by 2030 (Hetemäki et al., 2013). The average amount of consumption per person can be as high as 40 kg per person per year (Bajpai, 2013). Subsequently, this has led to extensive research and development in pulp and paper, ranging from mitigating effluence, saving energy, finding potential raw materials, and increasing pulp yield to enhancing the physical and mechanical properties of paper.

In today's pulp and paper industry, chemical pulping, namely Kraft pulping, is the most common method employed thus far. Statistics have revealed that the world's chemical pulps are widely produced using the Kraft pulping process (Sridach, 2010). This promising method is valued for its higher yield and stronger pulp compared to that produced using semi-chemical and mechanical treatments. Unfortunately, Kraft pulping has generated concern about its environmental effects due to emissions of sulphur dioxide, suspended solids and waste water pollution (Chong et al., 2013). Hence, it has become necessary to pursue a more environmentally friendly pulping process.

The organosolv pulping process is recognized as a more environmentally friendly method compared to Kraft pulping. The process uses organic solvents such as ethanol, methanol, acetic acid, and formic acid as delignifying agents. Ethanol, an easily obtained organic solvent, has been used in organosolv pulping research for more than 100 years. The common practice of using ethanol as organic solvent in organosolv pulping is due to several factors: (i) the ease with which it penetrates through the structure of wood, which leads to uniform delignification (Muurinen, 2000); (ii) its ability to increase pulp yield whilst reducing Kappa number and screening rejects (Akgul and Tozluoglu, 2010); (iii) its ability to reduce the use of energy during the pretreatment process, which enhances the efficiency of wood hydrolysis (Aravamuthan et al., 2002); and (iv) its ability to protect cellulose during delignification (Kleinert, 1974). Thus, ethanol is used as the organic solvent in most organosolv pulping processes.

In the recent years, a shortage of raw material from forestland has become a concern. In order to compensate for the declining supply from natural forests, research and industry have focused on the use of fast-growing plantation timber species instead of forestland to produce pulp and paper (Maraseni et al., 2017). For example, plantation species such as *Acacia mangium* have been used as the raw material in the pulp and paper industry in Sabah, Malaysia. However, Acacia hybrid was identified as another alternative to *Acacia mangium* owing to its higher hollocellulose content compared to *Acacia mangium* and *Acacia auriculiformis* (Chong and Liew, 2014). Moreover, it is one of the fast-growing timbers that has overtaken many species in commercial pulpwood production due to its attractive morphology characteristics, physical properties and mechanical properties (Jahan et al., 2007). This chapter aims to highlight the potential of Acacia hybrid as a fiber source for organosolv pulping to address concerns related to pulping methods and the shortage of raw material from forest land.

9.2 THE FOUNDATION OF PULP AND PAPER MAKING

9.2.1 STRUCTURE AND DEFINITION OF PULP AND PAPER

Pulp and paper are differed from one another in terms of structure. Pulp is an important material resulting from a complex manufacturing process that involves liberating fibers from lignocellulosic plant material by chemical, mechanical, or a combination of chemical and mechanical processes. It basically appears as a fibrous mass and it is named after the pulping process or pulping method whereby the bonds in the wood are ruptured (Lehto and Alén, 2015). In its air-dried form, pulp consists of 10% water and 90% oven-dried pulp. However, this depends on the actual condition of a shipment. On the other hand, paper is defined as thin, flexible web composed of individual cellulosic fibers deposited randomly on top of each other in a water suspension and dried to form inter-fiber hydrogen bonds that hold the fibers together (Walker, 2006).

9.2.2 TYPES OF PULPING PROCESS

There are various types of pulping process, but the most promising type is still chemical pulping. Chemical pulping is an effective substitute mechanical pulping thus far due to its better contribution of pulp and handsheet properties. The process focuses on the use of chemical reactants and heat energy to disintegrate the bonds of lignin in the raw material. The Kraft pulping process, also known as sulphate pulping, is the most popular method used in the pulp and paper industry nowadays. The process involves the action of chemicals and heat on the raw material (i.e. wood chips) in a stainless-steel digester vessel for an extended period of time at elevated pressure; the pulp produced at the end of pulping will be refined in a disintegrator to strengthen the paper properties. This section will briefly discuss the three primary means of chemical pulping, often applied in the research fields or industries: (a) the sulphite process, (b) the sulphate process, and (c) the alternative method, organosolv pulping.

a) **The sulphite process**. The sulphite process was developed in the United States in 1867. Sulphurous acid and calcium bisulphate are the main solvent used in this pulping process to delignify wood fibers. The pulp produced by the sulphite process is usually soft, flexible, moderately strong, and easier to bleach and refine. This process is used to help improve mechanical pulp, which is used for making newspaper, writing paper, and cellophane. However, the recovery of chemicals in the waste liquors for this process is more difficult compared to Kraft pulping and organosolv pulping (Chong, 2015).

b) **The sulphate process**. The sulphate process has been employed since 1884. This process was derived from the soda process used in the 19th century. It was begun by adding sodium sulphate to the soda process; eventually, a stronger pulp was formed. Soon after that, chemist discovered that sodium sulphite was the active ingredient and it was therefore named the Kraft process rather than the sulphate process, as the word *Kraft* means *strength* in German. Kraft pulp is not just stronger but also has a better heat and chemical recovery system that can reduce processing costs. Moreover, it is effective in digesting work well on high-speed presses. However, the process emits sulphuric organic compounds which can cause environmental pollution (Chong, 2015).

c) **Organosolv pulping**. Organosolv pulping appears to mitigate the problems faced in chemical pulping since the process is more environmentally friendly. Organosolv pulping is a process involving using organic solvent in a cooking operation. The most common solvents used are methanol, ethanol, acetic acid, and formic acid while other more rarely used solvents have included phenols, amines, glycols, nitrobenzene, dioxane, dimethylsulphoxide, sulfolene, and carbon dioxide (Akgul and Kirci, 2009).

9.3   ACACIA HYBRID, THE SOURCE OF PULP AND PAPER MAKING

Wood fiber has been an important source of lignocellulose material in pulp and paper production since the late 1800s. In fact, wood fiber is amongst the most widely used source of virgin pulp production mainly because it offers as high as 90% pulp yield (Chong, 2015). Depletion of forest wood and the slow harvest cycle has led to increasing dependency on the use of plantation timber species as the source of pulp and paper at present (Paiman et al., 2018). It is common for monoculture plants to acquire higher intrinsic value compared to natural regeneration timber due to their uniformity of fiber and desirable fiber characteristics. Besides, the advance of the plantation system nowadays means that it is able to yield fast-growing plants to meet the market demands. There are various of timber plantation species planted in Malaysia such as *Hevea brasiliensis, Tectona grandis, Paraserianthes falcataria, Octomeles sumatrana, Azadirachta excels, Neolamarckia cadamba and Khaya ivorensis,* and *Eucalyptus spp.* Nevertheless, *Acacia mangium* is still considered the main plantation timber species to cater for future demands for wood (Jusoh et al., 2013). Previous studies revealed that Acacia hybrid, *Acacia auriculiformis, Acacia dealbata,* and *Acacia melanoxylon* are among potential timber plantation species for pulp and paper making (Chong, 2015; Santos et al., 2010; Jahan et al., 2007). The selection of wood species for pulp and paper making is important, as the types of wood chosen could possibly affect the pulp yield and the properties of the paper.

9.3.1   BACKGROUND OF ACACIA HYBRID

Acacia spp. is native to Australia, Africa, Madagascar, throughout the Asia-Pacific region, and in the Americas. It is famous for its high adaptability in tropical and temperate environments, which can be found in many Asian lands. They belong to the pea flowering tree family of Fabaceae and subfamily of Mimosoideae (Bueren, 2004). The three main *Acacia* spp. woods which can be found in Malaysia are *Acacia mangium, Acacia auriculiformis,* and Acacia hybrid. *Acacia* woods are famous for their use for firebreak, reforestation and rehabilitation purposes to conserve the natural forest. The raw material of *Acacia* woods has been in high demand for pulp and paper production. However, there is a limitation to *Acacia mangium* as it is prone to heart rot diseases due to the presence of fungus or bacteria that causes the heartwood to decay (Barry et al., 2005; Bakri et al., 2018).

Back in 1971, a hybridization of *Acacia mangium* and *Acacia auriculiformis* was spotted near to Ulu Kukut Plantation in north-western Sabah, Malaysia. It was believed that the seedlings collected from the naturally generated hybrid were planted in other localities, which gave rise to Acacia hybrid. Studies have confirmed that hybridization was able to reduce the limitations encountered by the parent plants (Paiman et al., 2018;

FIGURE 9.1 Harvesting Acacia hybrid tree in SAFODA.

Bakri et al., 2018). To date, it has been a significant timber plantation species attributed to the fact that it has better wood properties than its parent plants, as it inherits its straightness and self-pruning ability from *Acacia mangium* and its better stem circulatory system from *Acacia auriculiformis* (Paiman et al., 2018).

Acacia hybrid is the hybridization or a cross between the male plant of *Acacia auriculiformis* and the female plant of *Acacia mangium*. It can be described as a medium-size tree. It can grows up to 8–10 m tall, and its diameter at breast height (DBH) ranges from 7.5 to 9.0 cm within 2 years. It grows well in the temperature range of 12–35°C. The traits of the flower color, leaf shape, size, pod, bark and wood density are the result of the combination of *Acacia mangium* and *Acacia auriculiformis*. Acacia hybrid can be harvested as it grows up to 5 years (Bueren, 2004). Figure 9.1 shows harvesting taking place in Sabah Forestry Development Authority Forest Reserve (SAFODA).

9.3.2 Physical Properties of Acacia Hybrid

a) **General properties**: Acacia hybrid can be easily recognized by the difference between the heartwood and sapwood. The heartwood is dark brown whilst the sapwood is light brown in color. It was reported that the heartwood is preferable in the production of construction beams, parquet flooring and furniture, while sapwood is famous for applications such as pulp and paper, medium-density fiberboard and oriented-strand board (Bueren, 2004). The grain is fine and intertwined with a slightly smooth texture. Fine lines of parenchyma which mimic the presence of growth rings can be observed as well (Rokeya et al., 2010) owing to the good characteristics, especially in growth rate and wood properties (i.e. wood density, moisture content and fiber length) (Rokeya et al., 2010).

b) **Density**: Table 9.1 shows that *Acacia auriculiformis* was found to have the highest wood density, followed by Acacia hybrid and *Acacia mangium*, which is in agreement with the results obtained by Jusoh et al. (2013). Wood density is usually influenced by various factors including silviculture, fertilization, site stocking, genetics, and growth rate, which affect the morphology and chemical compositions of a plant (Rokeya et al., 2010). Although higher wood density generally leads to higher pulp yield, Santos and the team reported that there was no correlation between the wood density and pulp yield but rather associated with the Kappa number (degree of delignification) with the solvents and the composition of the wood (Santos et al., 2012). Recently, studies have shown that it is possible to predict the properties of the end product, paper, by pre-determining wood density using machine learning techniques such as classification and regression trees (CART) and multi-layer perceptron (MLP) via algorithms (Lglesias et al., 2017; Anjos et al., 2015).

c) **Moisture content:** Wood is usually hydrophilic in nature due to the presence of hydroxyl groups. This is the factor that makes wood susceptible to moisture absorption and swelling (Abdul Khalil et al., 2010). The removal of moisture content is also needed for satisfactory performance in wood. Wood with a higher moisture content dries slower compared to wood with a lower moisture content, and its drying process consumes more heat energy, time and cost (Yamamoto et al., 2003). Therefore, it is important to determine wood and pulp moisture content before pulping to avoid wastage of raw material. Acacia hybrid showed a higher moisture content in heartwood with a percentage of 253% compared to sapwood at 154% (Yamamoto et al., 2003). Rokeya et al. (2010) reported that the moisture content of green Acacia hybrid was 98% and of air-dried Acacia hybrid was 12%, while Chong et al. (2013) reported that the moisture content of oven-dried Acacia

TABLE 9.1
Wood Density of Acacia Hybrid and Its Parents

Species	Density (kg/m³)	Reference
Acacia hybrid	490	Yahya et al. (2010)
Acacia mangium	460	
Acacia auriculiformis	520	

hybrid was 10.28%. This shows that moisture content varies under different conditions caused by external factors such as relative humidity and internal factors such as the water absorbed through pulping and pulp washing (Forughi et al., 2016).

9.3.3   FIBER MORPHOLOGY OF ACACIA HYBRID

A fiber morphology test was done to measure fiber length, fiber diameter, fiber lumen diameter, fiber wall thickness, Runkel ratio, slenderness ratio, coefficient of rigidity, and flexibility coefficient, as stated in Table 9.2. This test can be done through a maceration process and observed through microscope image analyzer to assess the properties related to pulp and paper production (Chong et al., 2013). For instance, fiber length can be attribute to the strength of pulp and paper properties. This can be explained by the fact that longer fibers usually consist of more fiber joints than shorter fibers, which promotes a stronger fiber network compared to shorter fiber. Consequently, fiber-to-fiber bonding in the pulp is strengthened. As reported by Yahya et al. (2010), Acacia hybrid showed longer fiber length compared to *Acacia mangium* and *Acacia auriculiformis*. Similar cases can be found in Jusoh et al. (2013).

Besides fiber length, the Runkel ratio is one of the indicators to determine the suitability of raw material for pulp and paper production. A Runkel ratio value of below 1 is often preferable as it shows acceptable tensile strength. Fiber with a lower Runkel ratio value is better in terms of strength as it is more flexible and less rigid. As shown in Table 9.2, the Runkel ratio for Acacia hybrid is lower compared to that for *Acacia mangium* and *Acacia auriculiformis*. This indication showed that pulp and paper produced by Acacia hybrid could possibly exhibit better strength compared to *Acacia mangium* and *Acacia auriculiformis*.

TABLE 9.2

Fiber Dimensions of Acacia Hybrid, *Acacia mangium* and *Acacia auriculiformis*

Fiber Dimensions	Species					
	Acacia Hybrid		*Acacia mangium*		*Acacia auriculiformis*	
Fiber length (mm)	0.96	1.06	0.93	0.98	0.89	0.88
Fiber diameter (μm)	17.16	18.76	17.43	19.39	17.80	16.74
Fiber lumen diameter (μm)	12.75	13.74	13.35	14.29	12.80	11.13
Fiber wall thickness (μm)	2.21	2.51	2.03	2.55	2.50	2.81
Runkel ratio	0.35	0.37	0.31	0.37	0.39	0.55
Slenderness ratio	57.00	57.40	54.00	51.29	50.00	52.65
References	Jusoh et al. (2013)	Yahya et al. (2010)	Jusoh et al. (2013)	Yahya et al. (2010)	Jusoh et al., (2013)	Yahya et al. (2010)

On the other hand, the slenderness ratio is also correlated to bursting and tensile strength (Ona et al., 2001). A slenderness ratio of more than 33 is often considered sufficient for pulp and paper making. A higher value of slenderness ratio often leads to more fiber flexibility (Jusoh et al., 2013). The higher the flexibility of the fiber, the more-inter fiber bonds can be formed, resulting in stronger pulp and paper (Yahya et al., 2010). In this case, Acacia hybrid showed the highest slenderness ratio as compared to all other species (Table 9.2). In short, the longer and better flexibility of Acacia hybrid fibers make it a good alternative candidate for raw material for pulp and paper making to the two other species of *Acacia*.

9.3.4 Chemical Composition of Wood

Proximate chemical compositions including lignin, cellulose, hemicelluloses, hollocellulose, extractives, and ash can be determined and extracted according to TAPPI standards as shown in Table 9.3. The test is done to evaluate the chemical content of the raw material and to assess the properties of pulp and paper production at the same time.

Lignin is an important component in plants that helps to control fluid flows, protect polysaccharides from pathogen attack, and give support and strength to the cell well. Lignin also acts as an adhesive in the wood to hold cellulose and hemicellulose fibrils together. Lignin can be found concentrated in the middle lamella. It is a complex amorphous polymer attached with polysaccharides, consisting of phenyl propane-based monomeric units that are linked together by several types of ether linkages and also various kinds of carbon–carbon bonds. Wood with a high lignin content is usually not suitable for pulping due to the strong carbon–carbon bond in the lignin. It requires more energy and chemicals to break down stronger bonds in lignin during pulping. The process of lignin removal, often known as *delignification*, can be done through pulping and/or bleaching. Yahya et al. (2010) also indicated that a higher lignin content would reduce pulp yield and strength. The lower content of lignin in Acacia hybrid compared to its parent plants showed that Acacia hybrid has great potential for pulping (Table 9.4).

Holocellulose represents the total content of carbohydrate materials, being composed of 40%–45% cellulose and 15%–25% hemicellulose, making up 65%–70% of a wood's dry weight (Rowell, 2013). Therefore, a high hollocellulose content

TABLE 9.3
Proximate Chemical Compositions Analyses

Chemical composition	Testing method
Lignin	TAPPI T 222 OS-74
Cellulose	TAPPI T 203 om-93
Hemicelluloses	Hollocellulose – cellulose
Hollocellulose	TAPPI T 9m-54
Extractives	TAPPI T 204 cm-97
Ash	TAPPI T 211 om-93

is desirable to produce a high pulp yield. A higher pulp yield can be expected for Acacia hybrid as Acacia hybrid shows a higher content of hollocellulose compared to *Acacia mangium* and *Acacia auriculiformis* (Yahya et al., 2010).

The hemicellulose content can be obtained by deducting the cellulose content from the hollocellulose content. Hemicellulose is a white solid material, which is considered to be amorphous. It consists of polysaccharide polymers including pentoses (D-xylose, D-arabinose), hexoses (D-mannose, D-glucose, D-galactose) and sugar acids. It acts as a supporting material to the plant, with a degree of polymerization (the number of glucose units in a cellulose molecule) ranging from 200 to 300 units in native wood. It can be classified into glucomannans and xylans. It is easily degraded and dissolved into an oligomeric form via pulping, mainly because of the lack of crystallinity (Ban and Van Heiningen, 2011). Although the removal of hemicelluloses from the wood usually occurs during the pulping stage, increasing the pore size of lignin and speeding up the delignification rate, studies have verified that hemicelluloses play an important role in improving pulp yield, pulp properties, and pulp strength (Ban and Van Heiningen; Yahya et al., 2010). The pulping method can influence levels of hemicellulose retention. Organosolv pulping is one pulping method that could assist hemicellulose retention during pulping (Palamae et al., 2014).

Cellulose is made up of D-glucopyranose units connected together by ß-(1 → 4)-glucosidic bonds. Cellulose has a more irregular crystalline structure than hemicellulose. There is no change in the structure even if it is heated to as high as 200°C. The average degree of polymerization for cellulose is approximately 9000–10,000 and sometimes can be as high as 15,000. Table 9.4 shows that the α-cellulose content of the hybrid determined by Yahya et al. (2010) was similar to *Acacia mangium* and the holocellulose content was significantly higher than its parents as shown in Table 9.4.

Wood contains only about 5% w/w of extractives, which are the waxes, fatty acids, resins, and terpenes of a tree. Extractives play an important role in contributing to a wood's color, odor, and taste. They also help a plant to resist decay. The content of

---

TABLE 9.4
Chemical Composition of Acacia Hybrid, *Acacia mangium,* and *Acacia auriculiformis*

Chemical composition	Acacia hybrid		*Acacia mangium*	*Acacia auriculiformis*
Extractives (%)	4.92 ± 0.12	2.9	5.38	5.96
Holocellulose (%)	83.00 ± 0.09	82.88	80.43	71.33
Cellulose (%)	40.74 ± 0.21	45.45	45.71	40.57
Hemicellulose (%)	42.26 ± 0.08	37.43	34.72	30.76
Lignin (%)	30.21 ± 0.33	30.91	31.30	34.10
Ash (%)	1.22 ± 0.20	ND	ND	ND
References	Chong et al. (2013)	Yahya et al. (2010)	Yahya et al. (2010)	Yahya et al. (2010)

ND = Not Determined

extractives varies widely from species to species and from heartwood to sapwood. In Kraft pulping, some resins and extractives are eradicated and about 50% of them remain. On the other hand, ash content in a wood can vary from as low as 0.5% to as high as 13%. The ash content of a wood is usually obtained after combustion at $575 \pm 25°C$ according to TAPPI standard as stated in Table 9.3. The ash content of wood is made up of inorganic content containing silica and some minerals, mainly calcium, magnesium, and potassium. However, a high content of extractives and ash may cause some difficulties in wood processing and pulping as they could cause deposits on the equipment and in the end product which may reduce the aesthetic value of the end product. Besides, large amounts of extractives will also produce an unpleasant odor which makes the product unsuitable for food packaging purposes. According to the study done by Chong et al. (2013), the ash content of Acacia hybrid was as low as 1.22%. A low extractives content is preferred in pulp and paper making (Kube and Raymond, 2002).

9.4   WHY ORGANOSOLV PULPING?

A number of organosolv methods have been suggested to date. Kleinert, Alcell (Ethanol-water), MD organocell (ethanol-soda), organocell (methanol-soda-anthraquinone), ASAM (Alkali-sulphite-anthraquinone-methanol), and ASAE (alkali-sulphite-anthraquinone-ethanol) are among the prominent organosolv pulping processes (Biermann, 1996; Sridach, 2010). All these methods differ from one another in the types of solvent used for pulping. Organosolv solvents have included Glycerol, formic acid, acetic acid, phenol, acetone, methanol and ethanol, but low-molecular-weight aliphatic alcohols such as methanol and etha-nol are the most prevalent organic solvent used thus far (Saberikkhah et al., 2011). Organosolv pulping offers several main advantages over conventional Kraft pulping:

a) Organosolv pulping uses organic solvents, either low boiling solvents such as methanol, ethanol, and acetone or high boiling solvents such as ethylene-glycol and ethanolamine, in delignification. Such organic solvents and other by-products including sulphur-free lignin, hemicelluloses, furfural, sugars and volatile components are easily recovered (Sridach, 2010; Li et al., 2012).

b) It was also reported that organosolv pulping was able to reduce the COD and BOD levels in bleach effluents, which is an advantage in environmental terms over conventional pulping as it improves bleachability, reduces the emission of pollutants, reduce bad odors, and lowers toxicity (Akgul and Kirci, 2009).

c) The entire process results in substantial capital cost savings (Xu et al., 2007; Li et al., 2012).

9.5   PROCESSING

The method of pulp and handsheet production at laboratory scale can be divided into three main stages: the preparation of the raw material, the pulping process, and handsheet making and testing. This section describes the basic yet crucial procedures to follow at laboratory scale.

9.5.1 PREPARATION OF RAW MATERIAL FOR PULPING

Raw materials can be obtained through a series of processes. The first step is to harvest wood logs from a plantation, followed by debarking the wood log. The purpose of debarking is to remove the bark, which contains mostly short and small fibers and a high proportion of impurities, such as silica and calcium, that will affect the chemical recovery process and the end product (i.e. handsheet). Wood logs can be debarked using a large rotating ring steel drum, debarker, cambio debarker, cambial shear barker, or hydraulic barker or even manually by using a debarking tool (frequently used for research purpose) (Biermann, 1996).

After debarking, the stems can be cut into wood strips by using a band saw, and this is followed by chipping to obtain wood chips with a size of approximately 1 cm width × 2 cm length × 0.5 cm thick, as shown in Figure 9.2. It should always be borne in mind that a uniform size of chip is essential in chemical pulping. This is because oversized chips will remain undercooked while small chips will cause low yield and weak pulp due to their short fibers and the high amount of chemicals consumed during cooking. Besides, this may also clog the liquor circulation system, which is not preferable during pulping (Biermann, 1996).

In order to obtain uniform chip size, a screening process has to be done to sort the chips according to their thickness and length with the aim of promoting uniform pulping. Wood chips are then air-dried to prepare them for the next stage of pulping. The moisture content of the wood chips as well as the wood density will be determined prior to pulping to ensure the right amount of raw material needed for pulping. Figure 9.2 shows the process of raw material preparation for Acacia hybrid. This process can be followed for other timber plantation species as well.

9.5.2 PULPING PROCESS

There are various types of pulping methods, as discussed earlier. The common chemical pulping method requires the use of a digester, as shown here. The raw materials together with the solvents and/or catalyst are poured into the vessel of the digester. The pressure and time needed for pulping are pre-set before the process. Pulping conditions such as temperature, pressure, solvents, and catalyst applied usually depend on the parameters of the study and the raw materials used for the study (Figure 9.3). For example, 1 kg of Acacia hybrid was digested using ethanol as solvent (organosolv pulping), with the pressure range being 1.1–1.2 MPa for 3 hours (Chong et al., 2013).

9.5.3 BEATING PROCESS

The beating process is also known as the *refining* process. Pulp that has not undergone a beating process is usually weak in terms of mechanical strength. Thus, the pulping process is often followed by beating to increase fiber fibrillation, which

Harvested billets

Debarking

Screening

Wood strips obtained after sawing

Screened wood chips

Air-drying wood chips

FIGURE 9.2 Preparation of raw material for pulping. (From Chong, 2015.)

improves the network of individual fiber bonding as well as bonding between fibers (Phiong et al., 2014). The details of the beating process using a valley beater can be referred to as TAPPI T 200 sp-01, while the beating process using a PFI mill (Figure 9.4) can be referred to TAPPI T 248 sp-00 (Chong, 2015; Phiong et al., 2014). Figure 9.5 shows the valley beater.

9.5.4 Process of Handsheet Making and Testing

The total pulp from oven drying will be used to form the handsheet. The handsheet is usually made for testing purposes at laboratory scale (Figure 9.6). The handsheet has to be conditioned prior to testing in the room temperature until

Pulping in a digester

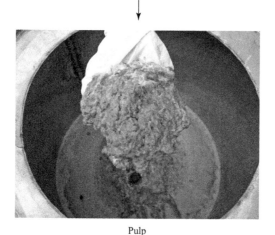

Pulp

FIGURE 9.3   Pulping process. (From Chong, 2015.)

equilibrium moisture content is achieved. The handsheet can be cut to the appropriate size and dimensions based on ISO or TAPPI standards. The properties of handsheet testing that are commonly determined are presented in Table 9.5 (Chong, 2015; Phiong et al., 2014).

9.6   YIELD PRODUCTION OF ACACIA HYBRID ORGANOSOLV PULP

Pulp yield can be affected by many factors, including the composition of the raw materials and the pulping method, where solvent concentration, pulping temperature, pressure and time need to be taken into consideration. Nevertheless, the

FIGURE 9.4   Beat in PFI mill.

FIGURE 9.5   Beat in valley beater. (From Chong, 2015.)

pulping method is one of the most crucial factors that has been studied widely due to its impact on the pulp yield and pulp properties. In the case of Acacia hybrid, two types of different methods (i.e. organosolv and Kraft pulping) with the same conditions in terms of time, pressure and temperature have been compared. Results revealed that pulp yield increased as the ethanol concentration increased from 70% to 90%, and the average screened yield using 90% ethanol was higher compared to Kraft pulping (Table 9.6). This effect could be due to the sorption or redeposition of hemicelluloses and lignin onto the fiber surface during organosolv pulping (Liew and Chong, 2016).

It is believed that ethanol has gained much interest as a solvent for pulping as it is not merely easy to find, abundant, and cost-effective but also able to produce valuable

FIGURE 9.6   Handsheet making.

TABLE 9.5

Dimension of Test Strips' Specimen According to the Standards

Properties	Standards
Tensile strength	ISO 1924-2:1994/TAPPI T 494 om-01
Folding strength	ISO 1974:1990/ TAPPI T 423 om-98
Bursting strength	ISO 2758:2001/ TAPPI T 403 om-02
Tearing strength	ISO 5626:1993/ TAPPI T 414 om-04

(Chong, 2015; Phiong et al., 2014)

TABLE 9.6

Kraft and Organosolv Pulp and Handsheet Properties of Acacia Hybrid

Ethanol Properties	70% Ethanol (Organosolv Pulping)	80% Ethanol (Organosolv Pulping)	90% Ethanol (Organosolv Pulping)	Kraft Pulping
Average screened yield (%)	39.96	42.72	44.19	43.82
Tensile (N.m/g)	16.90	17.32	17.45	17.28
Folding endurance (times)	1.52	1.58	1.85	1.78
Burst (kPa.m²/g)	3.22	3.83	3.90	3.53
Tear (mN.m²/g)	4.08	3.93	3.56	3.60

(Liew and Chong, 2016; Chong, 2015)

byproducts, such as lignin and carbohydrates. Moreover, it helps in the recovery of the solvent by rectification, producing an acceptable quality of pulp and selectivity in pulp yield (Oliet et al., 2002). It is usually used in organosolv pulping for hardwoods due to the ease with which it penetrates into the structure of the wood, which resulted in uniform delignification. It also has the ability to reduce the surface tension of the pulping liquor for the diffusion of other pulping chemicals (Muurinen, 2000).

Previous studies have reported on the effectiveness of using ethanol as the organic solvent for pulping. A study on *Nypa fruticans* pulping done by Akpakpan et al. (2012) showed that the addition of ethanol in soda liquor produced a higher pulp yield and lower lignin residue compared to soda liquor. The author explained that this phenomenon could be due to the lower effect of cellulose degradation in the pulp as well as good selectivity in delignification. Similar cases can be found in experiments with cotton stalk (*Gossypium hirsutum L.*), where the addition of ethanol into soda pulping at 160°C for 90 mins showed an increment in pulp yield from 7.71% to 14% as the ethanol increased from 20% to 50% (Akgul and Tozluoglu, 2010). This is because adding ethanol could effectively improve delignification in terms of the selectivity and physical properties of pulps, as explained earlier.

9.7   HANDSHEET PROPERTIES OF ACACIA HYBRID

The mechanical properties of pulp and handsheets, including tensile, folding, burst, and tear strength, are usually determined to assess the quality and performance of handsheets. It can be observed that as the ethanol concentration increases from 70% to 90%, the tensile, folding, and burst index increases from 16.90 N.m/g to 17.45 N.m/g, from 1.52 to 1.85, and from 3.22 kPa.m²/g to 3.90 kPa.m²/g respectively; the tear index, however, decreased from 4.08 mN.m²/g to 3.60 mN.m²/g (Table 9.6). This can be due to the higher rate of delignification, where a stronger chemical reaction between ethanol and the pulp fibers occurred, causing the fiber to be fibrillated. Fibrillated pulp fibers tend to form stronger bonds between the fibers. Hence, the fiber network forming handsheets is enhanced. However, the reduction of tear strength in this study could be affected by fiber length and fiber bonds. Although increasing the ethanol concentration increased internal and external fibrillations, it also happened to reduce fiber length, which decreased the tear index (Oluwadrare and Osakwe, 2014).

Based on the findings from Yamada et al. (1992), the pulp yield of Acacia hybrid was 55% and the pulp properties, including tensile, tear, burst, and folding indexes, are superior to that produced by *Acacia mangium* and *Acacia auriculiformis* via Kraft pulping. Although a comparison between Acacia hybrid, *Acacia mangium*, and *Acacia auriculiformis* using organosolv pulping has not been made to date, it can be perceived that Acacia hybrid has a lot of potential to produce higher pulp yield and better pulp properties compared to its parents, regardless of the pulping method applied. Yahya et al. (2010) and Chong et al. (2013) have also validated the superiority of Acacia hybrid in terms of fiber length and chemical composition prior to pulping. Nevertheless, a higher pulp yield, superiority in handsheet properties, and a more environmentally friendly approach can be achieved through organosolv pulping than by Kraft and soda pulping due to its efficiency in cellulose and hemicellulose retention as well as its good selectivity properties in delignification (Muurinen, 2000).

9.8 CONCLUSIONS

Acacia hybrid is one of the fast growing timber plantation species that is worth investing in to overcome the shortage of wood supply as the demand for wood production keeps increasing from year to year. It has been a promising raw material in the pulp and paper industry due to its process-ability and ease of engineering. From the few indicators of pulp yield and pulp properties, including fiber morphology and chemical composition, Acacia hybrid has shown many advantages over *Acacia mangium* and *Acacia auriculiformis*. It posseses longer fiber length and a higher content of hollocellulose compared to *Acacia mangium* and *Acacia auriculiformis*, leading to higher pulp yield and better pulp and handsheet properties. Although the composition and the fiber morphology of a raw material play a major role in obtaining sufficient pulp yield and handsheet properties, the pulping method and conditions are equally important to ensure higher pulp yield and better pulp properties. Further work could be carried out to compare the pulp yield and pulp properties of different species of *Acacia* using organosolv pulping.

REFERENCES

Abdul Khalil, H. P. S., Khairul, A., Backare, I. O. and Bhat, Irshad-UI-Haq. 2010. Thermal, Spectroscopic, and Flexural Properties of Anhydride Modified Cultivated *Acacia* spp. *Wood Science and Technology*, 45(3): 315–323.

Akgul, M., and Kirci, H. 2009. An Environmentally Friendly Organosolv (Ethanol-Water) Pulping of Poplar Wood. *Journal of Environmental Biology*, 30(5): 735–740.

Akgul, M., and Tozluoglu, A. 2010. Alkaline-Ethanol Pulping of Cotton Stalks. *Scientific Research and Essays*, 5(10): 1068–1074.

Akpakpan, A.E., Akpabio, U.D., Ogunsile, B.O., and Eduok, U.M. 2012. Influence of Cooking Variables on the Soda and Soda-Ethanol Pulping of Nypa Fruticans Petioles. *Australian Journal of Basic and Applied Sciences*, 5(12): 1202–1208.

Anjos, O., García-Gonzalo, E.,Santos, A.J.A., Simões, R., Martínez-Torres, J., Pereira, H., and García-Nieto, P. 2015. Using Apparent Density of Paper from Hardwood Kraft Pulps to Predict Sheet Properties, Based on Unsupervised Classification and Multivariable Regression Techniques. *BioResources*, 10(3): 5920–5931.

Aravamuthan, R.J., Lechlitner, J., and Lougen, G. 2002. High Yield Pulping of Kenaf for Corrugating Medium. Proceedings of Paper Presented at the TAPPI, Fall Technical Conference.

Bajpai, P. 2013. *Recycling and Deinking of Recovered Paper*. Elsevier, Amsterdam.

Bakri, M.K.B., Jayamani, E., Hamdan, S., Rahman, M.R., and Kakar, A. 2018. Potential of Borneo Acacia Wood in Fully Biodegradable Bio-Composites' Commercial Production and Application. *Polymer Bulletin*, 75: 1–22.

Ban, W., and Van Heiningen, A. 2011. Adsorption of Hemicellulose Extracts from Hardwood onto Cellulosic Fibers. I. Effects of Adsorption and Optimization Factors. *Cellulose Chemistry and Technology*, 45(1): 57.

Barry, K.M., Mihara, R., Davies, N.W., Mitsunaga, T., and Mohammed, C.L. 2005. Polyphenols in Acacia mangium and Acacia auriculiformis Heartwood with Reference to Heart Rot Susceptibility. *Journal of Wood Science*, 51(6): 615–621.

Biermann, C. J. 1996. *Handbook of Pulping and Papermaking*. Second edition. Academic Press, United Kingdom.

Bueren, M.V. 2004. *Acacia Hybrid in Vietnam*. Australian Centre for International Agricultural Research, Australia.

Chong, E.W.N. 2015. *Yield Study of Acacia Hybrid Using Organosolv Pulping and Their Relationship with Handsheet Properties*. Univeristi Malaysia Sabah.

Chong, E.W.N., and Liew, K.C. 2014. Comparative Study Between Organosolv Pulping Using Different Concentrations of Ethanol and Kraft Pulping of Acacia Hybrid. *Agriculture & Forestry/Poljoprivreda i Sumarstvo*, 60(2): 47–57.

Chong, E.W.N., Liew, K.C., and Phiong, S.K. 2013. Preliminary Study on Organosolv Pulping of Acacia Hybrid. *Journal of Forest and Environmental Science*, 29(2): 125–130.

Foroughi, A.F., Green, S.I., and Stoeber, B. 2016. Optical Transparency of Paper as a Function of Moisture Content with Applications to Moisture Measurement. *Review of Scientific Instruments*, 87(2), 023706.

Hetemäki, L., Hänninen, R., and Moiseyev, A. 2013. *Markets and Market Forces for Pulp and Paper Products*. Global Forest Products: Trends, Management, and Sustainability. Taylor and Francis Publishers, Boca Raton, FL.

Iglesias, C., Santos, A.J.A., Martínez, J., Pereira, H., and Anjos, O. 2017. Influence of Heartwood on Wood Density and Pulp Properties Explained by Machine Learning Techniques. *Forests*, 8(1): 20.

Jahan, M.S., Rubaiyat, A., and Sabina, R. 2007. Evaluation of Cooking Processes for Trema Orientalis Pulping. *Journal of Scientific and Industrial Research*, 66(10): 853–559.

Jusoh, I., Zaharin, F.A., and Adam, N.S. 2013. Wood Quality of Acacia Hybrid and Second-Generation Acacia mangium. *BioResources*, 9(1): 150–160.

Kleinert, T.N. 1974. Organosolv Pulping with Aqueous Alcohol. *Tappi*, 57(8): 99–102.

Kube, P. D., and Raymond, C. A. 2002. Prediction of Whole-Tree Basic Density and Pulp Yield Using Wood Core Samples in Eucalyptus Nitens. *Appita Journal*, 55(1): 43–48.

Lehto, J.T., and Alén, R.J. 2015. Chemical Pretreatments of Wood Chips Prior to Alkaline Pulping-A Review of Pretreatment Alternatives, Chemical Aspects of the Resulting Liquors, and Pulping Outcomes. *BioResources*, 10(4): 8604–8656.

Li, M.F., Sun, S.N., Xu, F., and Sun, R.C. 2012. Formic Acid Based Organosolv Pulping of Bamboo (Phyllostachys acuta): Comparative Characterization of the Dissolved Lignins with Milled Wood Lignin. *Chemical Engineering Journal*, 179: 80–89.

Liew, K.C., and Chong, E.W.N. 2016. The Relationship of Pulp Yield with Ethanol Pulping Concentrations on Acacia Hybrid. *Journal of the Indian Academy of Wood Science*, 13(1): 44–47.

Maraseni, T.N., Son, H.L., Cockfield, G., Duy, H.V., and Dai Nghia, T. 2017. The Financial Benefits of Forest Certification: Case Studies of Acacia Growers and a Furniture Company in Central Vietnam. *Land Use Policy*, 69: 56–63.

Muurinen, E. 2000. *Organosolv Pulping - A Review and Distillation Study Related to Peroxyacid Pulping*. Oulu University, Oulu, Finland.

Oliet, M., García, J., Rodríguez, F., and Gilarrranz, M.A. 2002. Solvent Effects in Autocatalyzed Alcohol–Water Pulping Comparative Study Between Ethanol and Methanol as Delignifying Agents. *Chemical Engineering Journal*, 87(2): 157–162.

Oluwadare, D.A., and Osakwe, U.C. 2014. Effects of Applied Organic Materials on Physical Properties of Intensively Cropped Ultisol in North-Eastern Nigeria. *Journal of Recent Advances in Agriculture*, 2(3): 199–207.

Ona, T., Sonoda, T., Ito, K., Shibata, M., Tamai, Y., Kojima, Y., Ohshima, J., Yokota, S., and Yoshizawa, N. 2001. Investigation of Relationship between Cell and Pulp Properties in Eucalyptus by Examinations of Within-Tree Property Variations. *Wood Science and Technology*, 35(3):363–375.

Paiman, B., Lee, S.H., and Zaidon, A. 2018. Machining Properties of Natural Regeneration and Planted Acacia mangium × A. auriculiformis Hybrid. *Journal of Tropical Forest Science*, 30(1): 135–142.

Palamae, S., Palachum, W., Chisti, Y., and Choorit, W. 2014. Retention of Hemicellulose During Delignification of Oil Palm Empty Fruit Bunch (EFB) Fiber with Peracetic Acid and Alkaline Peroxide. *Biomass and Bioenergy*, 66: 240–248.

Phiong, S.K., Liew, K.C., and Chong, E.W.N. 2014. Effect of Refining on Organosolv Acacia Hybrid Pulp Fibers and Handsheet Properties. *Poljoprivreda i Sumarstvo*, 60(3): 187–195.

Rokeya, U.K., Akter Hossain, M., Rowson Ali, M., and Paul, P. 2010. Physical and Mechanical Properties of (Acacia auriculiformis × A. mangium) Hybrid Acacia. *Journal of Bangladesh Academy of Sciences*, 34(2): 181–187.

Saberikhah, E., Mohammadi Rovshandeh, J., and Rezayati-Chanrani, P. 2011. Organosolv Pulping of Wheat Straw by Glycerol. *Cellulose Chemistry and Technology*, 45(1–2): 67–75.

Santos, A., Anjos, O., Amaral, M.E., Gil, N., Pereira, H., and Simões, R. 2012. Influence on Pulping Yield and Pulp Properties of Wood Density of Acacia melanoxylon. *Journal of Wood Science*, 58(6): 479–486.

Santos, A.J., Anjos, O., and Simões, R.M.D.S. 2010. Papermaking Potential of Acacia dealbata and Acacia melanoxylon. *Appita*, 59(1).

Sridach, W. 2010. The Environmentally Benign Pulping Process of Non-Wood Fibers. *Suranaree Journal of Science and Technology*, 17(2): 105–123.

Walker, J.C. 2006. Pulp and Paper Manufacture. In: *Primary Wood Processing*. Springer, Dordrecht.

Xu, Y., Li, K., and Zhang, M. 2007. Lignin Precipitation on the Pulp Fibers in the Ethanol-Based Organosolv Pulping. *Colloids and Surfaces A: Physicochemical and Engineering Aspects*, 301(1–3): 255–263.

Yahya, R., Sugiyama, J., Silsia, D., and Gril, J. 2010. Some Anantomical Features of an Acacia hybrid, A. mangium, and A. auriculiformis Grown in Indonesia. *Journal of Tropical Forest Science*, 22(3): 343–351.

Yamada, N., Khoo, K.C., and Yusoff, M.N.M. 1992. Sulphate Pulping Characteristics of Acacia Hybrid, Acacia mangium and Acacia auriculiformis from Sabah. *Journal of Tropical Forest Science*, 4(3): 206–214.

Yamamoto, K., Sulaiman, O., Kitingan, C., Choon, L.W., and Nhan, N.T. 2003. Moisture Distribution in Stems of Acacia mangium, A. auriculiformis and Hybrid Acacia Trees. *Japan Agricultural Research Quarterly: JARQ*, 37(3): 207–212.

10 Water Absorption Properties of Chemically Modified Sungkai (*Peronema canescens*) Wood Fibers in Medium Density Fiberboard

Kang Chiang Liew, Hui Ching Chong and Su Xin Ng

CONTENTS

10.1 GETTING STARTED

Chemical modification of wood began in the 1950s and interest in it has since grown. The aim of chemical modification is to increase the biological resistance and enhance the properties of wood, such as dimensional stability. This chemical modification process is non-toxic to humans compared with conventional wood preservatives and biocides (Hon, 1996). Basically, chemical modification is the chemistry of cell wall polymers being altered, which can change important properties of wood including durability, hardness, dimensional stability, water repellency, thermal properties and UV-stability (Peydecastaing, 2008). Controlling the moisture content in wood is an effective way to enhance wood durability by protecting it from biological attack, especially fungal attack.

Most fast-growing wood species such as Acacia (*A. mangium*) and Sungkai (*P. canescens*) are prone to deteriorate rapidly under physical and biological influence, especially fungi, due to high moisture sorption. To overcome this problem, the chemical modification of wood appears to be a multi-purpose method that can simultaneously improve wood durability. The chemical modification process involves the formation of covalent bonds with OH groups from cellulose, hemicellulose or lignin. Thus, the chemical nature of wood is changed and its properties can be enhanced. The modification of wood with anhydrides has proved to be efficient in improving dimensional stability.

Chemical modification is widely accepted as one of the effective methods to change wood's properties, including increasing dimension stability. The stability of modified wood brings great interest to environmental changes regarding commercial application of this technology. Chemical modification of wood in order to change its water absorption properties changes wood dimensions as well. Rowell (2005b) reported that this is because cell-wall polymers contain hydroxyl and other oxygen-containing groups that will attract water moisture through hydrogen bonding. This uptake of water moisture swells the cell wall and causes the wood to expand. Wood shrinkage occurs as moisture is lost, and this process is reversible. The low water absorption of wood can decrease the chances of wood defects. Thus, this can decrease the loss of profit in the wood industry because of the wood's good stability.

Chemical modification can change other wood properties aside from water absorption. As well as Acacia, Sungkai was also chosen for this study because it is still a less well known species and more research is needed on it. Currently, there has still been no research carried out regarding the modification of Sungkai wood fiber by using anhydride-based chemicals.

This experiment aimed to investigate the effect of chemically modified wood fibers on the water absorption properties of Acacia and Sungkai Medium Density Fiberboard (MDF). The water absorption properties of wood fibers after chemical modification were investigated through this study to:

1) Determine the Weight Percent Gain (WPG) of wood fibers, moisture sorption properties (weight increase rate and thickness change rate) and pH of MDF for Acacia and Sungkai after reacting with 40%, 60%, and 80% (w/w) of Acetic anhydride (AA), Maleic anhydride (MA) and Succinic anhydride (SA);
2) Evaluate the relationship between the type and concentration of anhydrides and the water absorption properties of Acacia and Sungkai MDF.

10.2 MODIFICATION OF FIBER AND TESTING

10.2.1 WOOD FIBER PREPARATION

A. mangium and *P. canescens* obtained from Sabah Forestry Development Authority (SAFODA) Kinarut Station were used in this experiment. The trees chosen had a diameter at breast height (DBH) in the range of 20–25cm. Using a chainsaw, we cut the logs into billets and debarked them.

Wood fibers were extracted from the whole billet, excluding the bark. First, we reduced the billets to strips and chipped them. The knife flake ring mill was then used to produce the finer wood component of wood fibers by flaking the wood chips. Using a circular screening machine, we obtained <3 mm wood fibers (Figure 10.1) with a moisture content of 4%–8% approximately.

Based on ASTM D 4442-92 (direct moisture content measurement of wood), the wood fibers were stored in individual vapor-tight containers and weighed using a balance consistent with the desired precision. Scoops of fibers were randomly weighed to get the moisture content, and approximately 60g of wood fiber was put into an oven for 24 hours at 103 ±2°C. Then, we put the dried fibers in a desiccator with fresh desiccant until they reached room temperature. The whole weighing process was carried out in closed weighing jars. By using Equation 10.1, we calculated the moisture content:

$$MC \% = (A - B) / B \times 100 \tag{10.1}$$

where
A = original mass, g,
B = oven-dry mass, g.

10.2.2 CHEMICAL MODIFICATION PROCESS

We carried out modifications with AA ($C_4H_6O_3$), MA ($C_4H_2O_3$) and SA ($C_4H_4O_3$) by mixing the anhydride with wood fibers in nine batches for each wood species.

The 60g of oven-dried samples of wood fibers were oven-dried to a constant weight at 103 ± 2°C. The concentration level of anhydride in the solution was

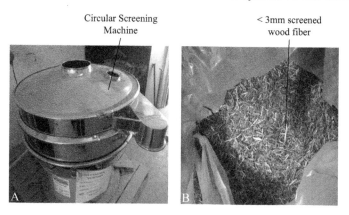

FIGURE 10.1 (a) Circular screening machine, (b) < 3mm screened wood fiber.

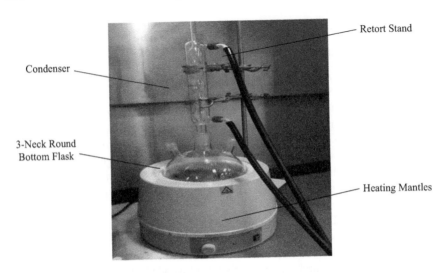

FIGURE 10.2 Set-up for chemical modification process.

established at 40%, 60% and 80% (w/w), respectively (Teaca et al., 2014). Then, we refluxed the wood fibers with anhydride for 3 hours based on Azeh et al. (2013) with a cooking temperature of 120 ± 5°C (Figure 10.2).

The ratio amount of the anhydride added to the fiber was 1:5 based on the dry weight of the wood fibers. After the modification process (Figure 10.3), the residues were washed thoroughly with distilled water and then oven-dried for 3 hours (Figure 10.4). The best way to determine the effectiveness of modification was calculated as WPG and this was based on the difference between the oven-dried weight of the test pieces before and after modification (M_1) and after modification (M_2) according to Equation 10.2.

$$\text{WPG } (\%) = \frac{M_2 - M_1}{M_1} \times 100\% \tag{10.2}$$

where
  $M_1 =$ before modification
  $M_2 =$ after modification

10.2.3 FABRICATION ON SAMPLE MANUFACTURING

The test piece for a water absorption test for wood-based fibers and particle panel material was 25.4 mm (width) × 127 mm (length) × 3 mm (thickness) in dimension, according to ASTM D 1037 – 99, with all four edges smoothly and squarely trimmed. 50g of wood fibers were coated with 72 g of the Urea formaldehyde (UF) (Li et al., 2009). The coated fibers were manually loaded into a 25.4 mm (width) × 127 mm (length) × 3 mm (thickness) aluminum mold (fiber mat), which was then pre-pressed at 0.06 MPa for 10 minutes. Finally, it was pressed into fiberboard using a heat compressor at force 0.5 MPa applied at press temperature of 110°C for 20 minutes (Figure 10.5). The final thickness of the MDF was 3 ± 0.5 mm.

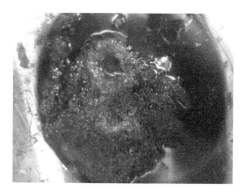

FIGURE 10.3 Wood fibers were cooked with chemicals at $120 \pm 5°C$ for 3 hours.

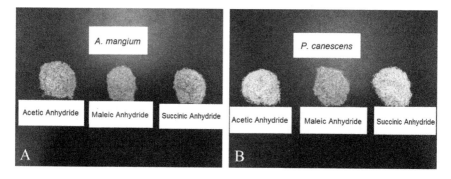

FIGURE 10.4 Dry look of three types of chemically modified wood fibers (a: *A. mangium*; b: *P. canescens*.

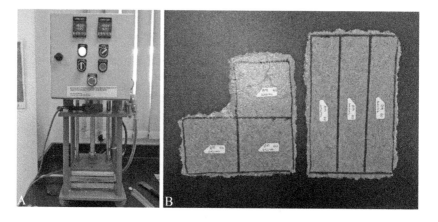

FIGURE 10.5 Heat compressor machine: (a) the dimension required was marked by marker on the MDF; (b) (Left): Test piece for water swelling test and water soaking test, (Right): Test piece for water absorption test).

The targeted density was 0.70 g/cm³. The density for each fiberboard panel was obtained by dividing the fiberboard mass (wet basis) by its volume, and it was given as 0.70 g/cm³. The MDF was placed in an oven at 23°C for 2 days for further analysis. Each MDF panel was cut into three rectangular test pieces of 25.4 mm (width) × 127 mm (length) for a water absorption test.

10.2.4 WATER ABSORPTION TEST

From the different fiber batches, the test pieces underwent a 24-hour water absorption test. Measurement of the water absorption properties of the test pieces was carried out according to ASTM D1037-99 (2000). The 25.4mm × 127mm test pieces (Figure 10.6) were soaked in water for 24 ± 1 hours at room temperature. Then, the test pieces were conditioned to a constant weight and a temperature of 20 ± 3°C for 2 days. After that, the test pieces were left to cool in a desiccator for 15 minutes and weighed to the nearest 0.001g. The test pieces were fully immersed in distilled water for 24 hours. Then, the test pieces were removed from the water one at a time, all the surface water on the test pieces was wiped off with a dry cloth, and they were immediately weighed to the nearest 0.001g (Figure 10.7).

Calculation of the percentage change in weight during immersion was carried out using Equation 10.3. Three replicates were produced for each variable, including the control. A total of 120 replicates with three samples for each concentration value and three non-treated samples for each tree species were produced.

Calculation of percentage changes in weight during immersion was calculated using Equation 10.3.

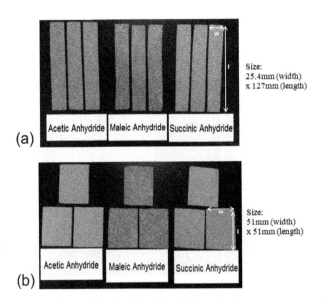

FIGURE 10.6   Test pieces (a) of three chemical types for water absorption, thickness swelling and pH. Test pieces (b) ready for water swelling and soaking test.

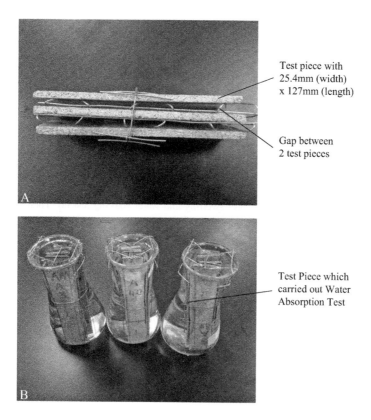

Test piece with
25.4mm (width)
x 127mm (length)

Gap between
2 test pieces

Test Piece which
carried out Water
Absorption Test

FIGURE 10.7 Test pieces were tied with iron wire to keep the gap or distance between another two test pieces (a). Then, the test pieces were fully immersed in distilled water for 24 hours (b).

Weight Increase (WI) Rate (%):

$$\frac{\text{wet weight} - \text{conditioned weight} \times 100\%}{\text{conditioned weight}} \qquad (10.3)$$

For water swelling tests, each fiberboard test pieces, each 51 mm × 51 mm, was placed in a container 10 cm × 10 cm square and 5 cm deep (Figure 10.8). Water was added to the container and the thickness recorded as a function of time. Measurements were taken every 5 minutes for the first hour, every hour for the next 6 hours, and then once a day for 5 days based on Rowell and Keany (1988). All of the water tests were done in duplicate. Water soaking tests were run on test pieces (51 × 51 mm) as previously described by Rowell and Ellis (1978). Each one of three cycles consisted of water soaking for 5 days followed by oven-drying at 103 ± 2°C for 2 days. Thickness of swelling was calculated as a percentage of the original oven-dried thickness. Reconditioning involved oven-drying for 24 hours at 60°C and then 2 weeks of conditioning at 20°C and 65% relative humidity (RH), based on Kojima and Suzuki (2010).

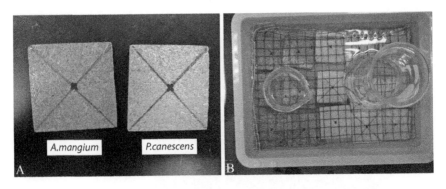

FIGURE 10.8 Test pieces for water swelling and water soaking test (a). Then, test pieces were fully immersed in the distilled water for 5 days (b).

10.2.5 DATA COLLECTION AND ANALYSIS

The water absorption properties (Water Increase Rate, WIR), thickness change (TC) rate: Water Swelling Test and Water Soaking Test and pH of wood fibers were determined and analyzed using SPSS software. A total of 120 replicates were done for the whole experiment (2 species × 3 anhydrides × 3 concentration × 3 replicates × 2 tests + 12 controls). The average mean for each group of water absorption tests was calculated and recorded. The data was analyzed by using two-way ANOVA tests to evaluate the relationship between the anhydride and the concentration of anhydride and the water absorption properties of *A. mangium* and *P. canescens*.

10.3 EXPERIMENTAL RESULTS

10.3.1 WEIGHT PERCENT GAIN (WPG)

The relationship between anhydride types and WPG of treated wood fibers were shown in Figures 10.9 and 10.10. It was found that the WPG values of wood fibers were dependent on wood species, anhydride and anhydride concentration.

It was noticed that the treatment parameter (anhydride: Acetic Anhydride [AA], Maleic Anhydride [MA] and Succinic Anhydride [SA]; anhydride concentration) that performed best in terms of WPG value was 60% of AA. It contributed the highest WPG value for both species: 13.79% for *A. mangium* and 24.91% for *P. canescens*. The results also showed that 60% was the optimum concentration for both AA and MA. However, 80% was the optimum temperature for SA. The wood fibers of *P. canescens* modified with SA showed a significant increase when the anhydride concentration was increased.

Wood fibers of *P. canescens* showed a higher WPG value than *A. mangium*. The WPG value of wood fibres for *P. canescens* was 21.95% which is two times higher than *A. mangium* and shows that *P. canescens* fibers absorb more chemical than *A. manguim* fibers. The air-dried density range of *A. mangium* (650 kg/m³) is slightly higher than *P. canescens* (640 kg/m³). Of these two species, *A. mangium* has a higher density than *P. canescens*. With the same weight, the volume of *P. canescens* wood fibers was higher than that of *A. mangium*. Stamm (1964) had earlier found that higher density hardwood

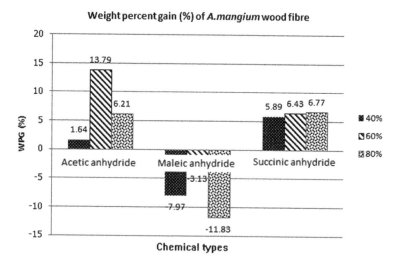

FIGURE 10.9   Weight percent gain (%) of wood fiber for *A. mangium*.

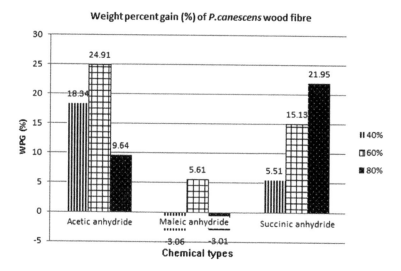

FIGURE 10.10   Weight percent gain (%) of wood fiber for *P. canescens*.

had greater activation energies. Thus, the wood fibers of *P. canescens* can absorb faster than the wood fibers of *A. mangium*.

MA exhibited a negative value of WPG on the wood fibers after modification. The negative WPG value of MA can be explained by the fact that the action of maleic acid at a high temperature can break down wood macromolecules. This process starts with hemicellulose followed by the amorphous part of cellulose and lignin (Bodirlau et al., 2008). The degradation of hemicellulose to xylose by maleic acid starts at 100°C and ends at 150°C (Lu and Mosier, 2008). Hemicellulose is the least thermally stable wood component due to the presence of the acetyl group (Bourgeois et al., 1989). Rowell (1984) also said that hemicellulose and cellulose polymers are degraded by heat well before lignin. However, there was a single positive value of WPG for the 60% MA for *P. canescens*, as shown in Figure 10.10. This is due to the bigger pores and pore arrangement types of wood fibers for *P. canescens*, which are able to absorb more anhydride and reach the level of WPG. Weight gain in fibers from absorbing chemicals is higher than weight loss, which can cause fiber degradation in the presence of high temperature ($120 \pm 5°C$).

A. *mangium* wood is diffuse and porous (Krisnawati et al., 2011), *P. canescens* showed well-defined growth rings, featuring ring-porous porosity characteristics (Azim, 2014). Fibers of diffuse-porous for *A. mangium* has vessels and pores that are uniform in size across the entire growth ring. These vessels are usually small, uniform in size and are very difficult to see with the naked eye. However, *P. canescens* has ring-porous; the earlywood and latewood transition occurs abruptly and is very distinct, a band of large earlywood vessels is clearly visible to the naked eye within each growth ring.

10.3.2   Moisture Sorption Determination

Wood fiber is hygroscopic; the hydroxyl groups in the cell wall polymers are attracted and form hydrogen bonds with the moisture in the atmosphere. As water enters the cell wall, the wood volume increases nearly proportionally to the volume of said water (Stamm, 1964). Then, a swelling reaction occurs. Swelling of the wood fiber continues until the cell reaches the Fiber Saturation Point (FSP) and does not undergo any further swelling. Weight Increase Rate (WIR) and Thickness Change Rate (TCR) (water swelling test and water soaking test) were carried out to measure moisture sorption resulting from interaction with water.

10.3.2.1   Weight Increase Rate (WIR)

The obtained modified MDF test piece was subject to a weight increase test based on the examined weight before and after the water swelling test (24 hours water immersion).

Based on Table 10.1, the analysis of variance for weight increase rate between anhydride types showed a significant difference in WIR at $p \leq 0.05$. The uses of AA to modify the *A. mangium* MDF test piece all showed significantly different values when different anhydride concentrations (40%, 60% and 80%) were used. These show that the concentration of AA has a significant effect on *A. mangium* MDF test pieces, with the lowest water absorption rate (16.28%) of the MDF test piece being reached at 60% (optimum concentration). However, it made no difference to the *P. canescens* MDF piece although the same anhydride (AA) is used. Analyzing the obtained results, it can be said that the treatment of *A. mangium* and *P. canescens* MDF test pieces with all anhydrides and concentrations had significant differences compared to the control (without chemical modification). However, the only result that did not have a significant effect in terms of lowering the weight increase rate was 80% MA. SA showed no difference with AA and MA with the same anhydride concentration, clearly showing in the WIR of 40% of AA, MA and SA for both *A. mangium* and *P. canescens* MDF test pieces.

TABLE 10.1
Weight Increase Rate of MDF Test Piece for Water Absorption Test of
A. mangium* and *P. canescens

Different Concentration of Chemical	*A. mangium*	*P. canescens*
Control (without anhydride)	$30.24^{a}{}_{w} \pm 1.3668$	$54.21^{a}{}_{w} \pm 6.0632$
Acetic Anhydride		
40%	$22.02^{b}{}_{x} \pm 1.4683$	$27.20^{bc}{}_{x} \pm 2.6493$
60%	$16.28^{c}{}_{x} \pm 0.3176$	$23.18^{cd}{}_{x} \pm 1.4367$
80%	$19.82^{d}{}_{x} \pm 1.1389$	$29.39^{db}{}_{x} \pm 0.4267$
Maleic Anhydride		
40%	$27.1^{bc}{}_{yw} \pm 0.4073$	$37.2^{bc}{}_{y} \pm 1.7246$
60%	$25.22^{c}{}_{y} \pm 1.8196$	$35.18^{cd}{}_{y} \pm 0.8901$
80%	$28.72^{ab}{}_{y} \pm 2.1855$	$40.39^{db}{}_{y} \pm 0.5050$
Succinic Anhydride		
40%	$24.58^{bc}{}_{zxy} \pm 0.4073$	$33.16^{bc}{}_{zxy} \pm 0.9933$
60%	$23.55^{cd}{}_{zy} \pm 1.8196$	$32.01^{cd}{}_{zy} \pm 2.0845$
80%	$23.02^{db}{}_{zw} \pm 2.1855$	$29.79^{db}{}_{zx} \pm 3.0260$

Note: Values are mean ± standard deviation.
Value in the same column with different letters (a, b, c, d) within a column (for each anhydride concentration) indicates a significant difference at p≤0.05 for each different weight increase rate of *A. mangium* and *P. canescens* MDF.
Value in the same column with different letters (w, x, y, z) within a column (for each anhydride) indicates a significant difference at p≤0.05 for each different weight increase rate of *A. mangium* and *P. canescens* MDF.

Based on the obtained results, it has been discovered that acetylation of the fibers has a positive effect in reducing water absorption rates such as WIR. In fact, it can be said that the WIR is decreased by acetylation treatment. To explain the possible reason why this happens during the acetylation reaction, Rowell (2006) said that the hydrophobic acetyl groups in AA replaced the hydroxyl groups of the wood fibers. So, the fibers have limited water absorption due to their hydrophobic nature. Also, moisture is presumed to be absorbed by MDF test pieces either as primary or secondary water. *Primary water* is water absorbed to primary sites such as the hydroxyl groups with high binding energy. *Secondary water* is water absorbed to sites with less binding energy, where water molecules are absorbed on top of the primary layer. Since some hydroxyl sites of wood cells are esterified with acetyl groups, there are lesser primary sites where water ca ben absorbed. Also, there may be fewer secondary binding sites since the fibers are more hydrophobic as a result of acetylation.

Figure 10.11 presented the results of the weight increases of chemically-modified MDF test pieces. When weight increase was measured after 24 hours, it was found that there was a progressive increase in weight for all types of anhydride. Chemical modification of an MDF test piece successfully brings down the WIR. The WIR of the non-modified *P. canescens* MDF test piece (control) was so high until it exceeded more than 50% and it was able to decrease to the average 23.18% by using 60% AA.

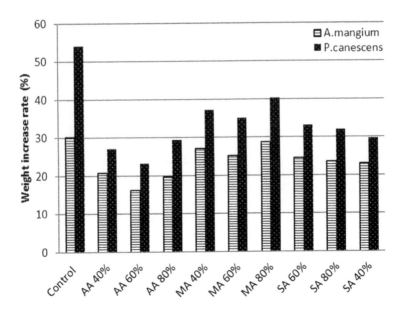

FIGURE 10.11   WIR of test piece for water absorption test of *A. mangium* and *P. canescens*.

For wood, water repellency is a rate phenomenon and dimensional stability is an equilibrium phenomenon (Rowell and Banks, 1985). Chemical modification is one water repellent treatment, and it can prevent absorption or slow down the rate that moisture or liquid are taken up by the wood. Thus, the WIR of both *A. mangium* and *P. canescens* MDF pieces were reduced, and the water absorption rate was also slowed down.

10.3.3   Thickness Change Rate (TCR)

Two tests were carried out to determine the TCR of MDF test pieces: a water swelling test (cyclic water swelling test) and a water soaking test (cyclic water soaking and oven-drying test).

10.3.3.1   Water Swelling Test

The obtained modified MDF was subjected to water swelling test investigation. The rate and extent of thickness swelling of *A. mangium* and *P. canescens* MDF test pieces in water was shown in Figures 10.12 and 10.13.

During the first 60 minutes, MDF test pieces made from 80% MA modified fibers and the control swelled faster than others for *A. mangium* MDF test pieces, and both had almost the same TCR for the first 6 hours. The observed lowest value of the TCR was 60% AA MDF for both species, which were 1.7%–2.31% for *A. mangium* MDF and 3.5%–4.07% for *P. canescens* MDF test pieces by the end of the first 60 minutes. This trend continued through 6 hours of water swelling, but at the end of 5 days, both control pieces of *A. mangium* and *P. canescens* MDF exhibited dramatic swelling compared to the other pieces with modified MDF. It shows that chemically modified

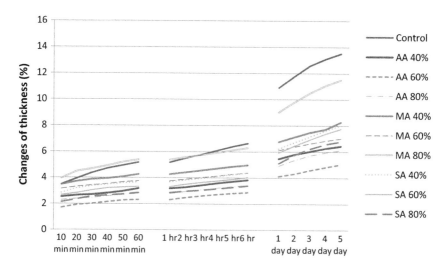

FIGURE 10.12   Thickness change rate (%) of *A. mangium* MDF test pieces for water swelling test.

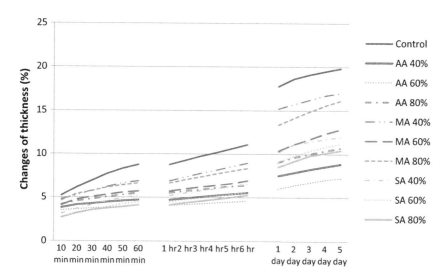

FIGURE 10.13   Thickness change rate (%) of *P. canescens* MDF test pieces (for water swelling test).

MDF test pieces show an obvious decease in TCR for both MDF pieces. The TRC of 80% MA MDF test pieces, which was almost same as control pieces during the first 6 hours, slowed down after 24 hours and in the rate and extend of 9%–11.5%.

The TCR data of the *A. mangium* and *P. canescens* MDF test pieces for various anhydrides (AA, MA and SA) and concentrations (40%, 60% and 80%) was analyzed. The analysis of the thickness change of the MDF test pieces revealed a significant lowering of TCR in the *A. mangium* and *P. canescens* MDF test pieces with various anhydrides and concentrations compared with the control test piece. However, there were still a few MDF test pieces that did not show a significant difference with control: the test piece of 40% MA at first 10 min, the piece of 80% MA at first 1–3 hours for *A. mangium* and the piece of 80% SA at first 40 min for *P. canescens*.

Overall, the TCR data of both non-chemically modified and chemically modified MDF test pieces of *P. canescens* was higher than *A. mangium*. Of these two species, *A. mangium* has higher density than *P. canescens*. The higher density hardwood had greater activation energies, as Stamm had found earlier in 1964. Thus, *P. canescens* swells faster than *A. mangium*. The rate of swelling of wood in water is dependent on several factors: hydrogen bonding ability, molecular size of the reagent, extractives content, temperature, and test piece size (Banks and West, 1989). There is an initial induction period due to the diffusion of water into the cell wall structure, and water then penetrates the cell wall capillaries and moves from lumen to lumen in the direction of the fiber.

The swelling that occurs in wood composites is much greater than in wood itself. This is due to the release of compressive forces as well as normal wood swelling (Rowell, 2005a). The compressive forces are a result of the physical compression of the wood elements during pressing of the board such as cold pressing and hot pressing. A dimensional stability treatment is one that reduces or prevents swelling in wood no matter how much time it is in contact with moisture or liquid water. Chemical modification is also a dimensional stability treatment. Examples of other dimensional stability treatments are penetrating polymers, cross-linking cell wall polymers, bulking the cell wall with polyethylene glycol, or using bonded cell wall chemicals (Rowell and Youngs, 1981).

10.3.3.2 Water Soaking Test (WST)

The thickness change analysis done for the *A. mangium* and *P. canescens* MDF test pieces showed significant differences for the control and the chemically modified MDF test pieces. The range of thickness change for the cyclic water soaking of *A. mangium* was 3–4% for the control MDF and reduced to less than 2% for modified MDF. However, the MDF test pieces of *P. Canescens* decreased from 4%–5% (control) to less than 3% (modified) for the thickness increase range. Anhydrides successfully controlled changes in thickness for the MDF test pieces; however, the trend still increased after oven drying. This means that it was not possible to bring MDF test pieces back to their original thickness even if they were re-dried in the oven for 2 days. Reversible swelling (cyclic water soaking test 2 [C2] – Oven-drying 1 [OD1] and C3 – OD2). which is normal wood cell-wall swelling, was much greater in the control MDF test pieces, being about 9.76%–12.8% for MDF made from *A. mangium* fibers and about 13.97–17.45% for MDF made from *P. canescens* fibers. It reduced to 3.71%–3.98% for 60% acetylated MDF made from *A. mangium* fibers and about 5% for 60% acetylated MDF made from *P. canescens* fibers.

The TCR of 60% acetylated MDF made from *A.mangium* fibers was 5.03% at a WPG of 13.79, and for MDF made from *P. canescens* fibers, it was 7.24% after a 5-day water soaking test. As a result of acetylation, the rate and extent of thickness swelling in the control MDF at the same level of acetylation is greatly reduced. At the end of 5 days of soaking, the control MDF test pieces for *A. mangium* fibers swelled by 13.51% and the control MDF test pieces for *P. canescens* fibers swelled by 19.83%, whereas MDF made from acetylated fibers swelled less than 8.48% percent for *A. mangium* MDF test pieces and reduced by 12.59% for *P. canescens* MDF test pieces. AA was the most efficient anhydride compared to MA and SA, shown by comparing the results of using the same concentration of all anhydrides. The control MDF test pieces exhibited a greater degree of irreversible swelling compared to the MDF test pieces made from acetylated fibers after drying at the end of the test. Rowell (2006) said that the mechanism of dimensional stability or low TCR resulting from acetylation is a result of the bulking of the bonded acetyl groups in the cell wall polymer hydroxyl groups. Only a little swelling can occur when water enters wood because the volume of the cell wall is swollen to near the original green volume of the wood. Acetylated MDF test pieces can absorb water through capillary action and to some extent in the cell wall. Some swelling can occur in "completely acetylated wood" since water molecules are smaller than those of the acetyl group in wood cell wall, but swelling does not exceed the elastic limit of the cell wall.

Thickness changes in the cyclic water soaking/oven-drying test of *A. mangium* and *P. canescens* MDF test pieces with various anhydrides and concentrations are shown in Figures 10.14 and 10.15. There was increasing thickness in every subsequent cycle of both *A. mangium* and *P. canescens* MDF test pieces. Irreversible swelling caused by the release of residual compressive stresses imparted during board pressing during MDF production was greatest in the control MDF test pieces and lowest in the acetic-anhydride-reacted boards.

Both reversible and irreversible swelling took place (Rowell, 2005a). The release of compressive stresses imparted during board pressing and during the first wetting of composites is also known as *irreversible* swelling. It is known as irreversible

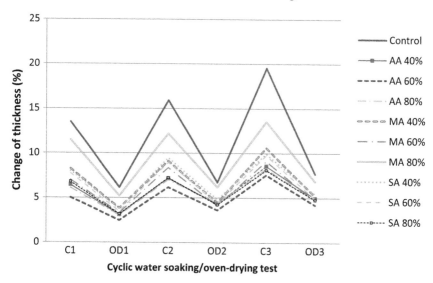

FIGURE 10.14   Cyclic water soaking/oven-drying test of *A. mangium* MDF test pieces.

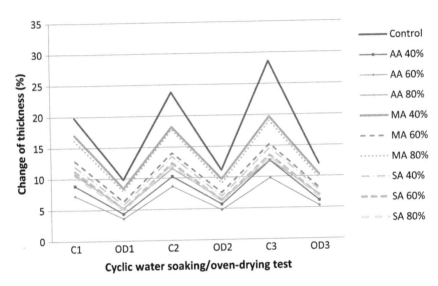

FIGURE 10.15 Cyclic water soaking/oven-drying test of *P. canescens* MDF test pieces.

swelling because it is not reversible upon re-drying. *Reversible* swelling also occurs during wetting and wood shrinks again as a result of re-drying. Irreversible swelling and reversible swelling were greatest in the control MDF pieces and lowest in the MDF test pieces treated with 60% of acetic anhydride. Rowell et al. (1991) also found that acetic-anhydride-reacted boards showed the lowest irreversible swelling and reversible swelling value. All the MDF test pieces showed an increase in permanent thickness swelling. Permanent swelling was probably caused by adhesive failures resulting from test conditions.

Drying (oven-drying) and rewetting (cyclic water soaking) causes an increase in both the rate of swelling and the extent of swelling (Rowell, 2005a). The degradation and extraction of hemicelluloses and extractives as well as some degradation of the cell wall structure during wetting, drying, rewetting and re-drying cycles results in the cell wall being more accessible to water. A significant amount of cell wall polymers can be lost when wood is exposed to high relative humidity in repeated cycles. Although no cell wall polymers are extracted, repeated humidity cycles result in a slight increase in moisture content with each cycle.

10.3.4 pH

The MDF test pieces were immersed in water for 24 hours and the pH values are presented in Table 10.2. According to the analysis of variance for pH between anhydride types, there was a significant different in pH value at $p \leq 0.05$. Acid anhydrides are molecules that form acidic solutions in water, and acid anhydrides are the oxides of non-metals that can react with water. Basically, the three acid anhydrides ("acids without water") used for this study had acid properties and exerted an influence on *A. mangium*, changing the alkali (pH >7) properties of *A. mangium* to become acidic

TABLE 10.2

pH Value of *A. mangium* and *P. canescens* MDF after Immersion in Water for 24 hours

Different Concentration of Chemical	*A. mangium*	*P. canescens*
Control (without anhydride)	$8.14^a{}_w \pm 0.0100$	$7.13^a{}_w \pm 0.03606$
Acetic Anhydride	$6.87^b{}_x \pm 0.0264$	$6.50^{bc}{}_x \pm 0.0458$
40%		
60%	$6.71^c{}_x \pm 0.0360$	$6.44^{cd}{}_x \pm 0.0360$
80%	$6.48^d{}_x \pm 0.0458$	$6.41^d{}_x \pm 0.0173$
Maleic Anhydride	$6.37^{bc}{}_y \pm 1.7490$	$6.17^b{}_y \pm 0.0173$
40%		
60%	$6.28^{cd}{}_y \pm 0.0264$	$6.02^c{}_y \pm 0.0300$
80%	$6.12^{db}{}_y \pm 0.0458$	$5.91^d{}_y \pm 0.0100$
Succinic Anhydride	$6.85^b{}_{zx} \pm 0.0264$	$6.22^{bc}{}_{zy} \pm R\ 0.0360$
40%		
60%	$6.72^c{}_{zx} \pm 0.0264$	$6.19^c{}_z \pm 0.0360$
80%	$5.91^d{}_z \pm 0.0200$	$6.03^d{}_z \pm 0.0264$

Note: Values are mean ± standard deviation.

Value in the same column with different letters (a, b, c, d) within a column (for each anhydride concentration) indicates a significant difference at $p \leq 0.05$ for each different pH value of *A. mangium* and *P. canescens* MDF.

Value in the same column with different letters (w, x, y, z) within a column (for each anhydride) indicates a significant difference at $p \leq 0.05$ for each different pH value of *A. mangium* and *P. canescens* MDF.

(pH <7). Significant differences were caused by using different concentrations of the same anhydride and also the same concentration but different anhydrides. The uses of higher concentrations of anhydride result in greater changes in pH scale. This clearly shows in the effect of MA on the *A. mangium* MDF test pieces, which had a pH of 6.37 for 40% concentration, which then decreased to a pH of 6.28 (60%) and then pH 6.12 (80%). These prove that MA had the highest acidic properties among the three anhydrides.

10.4 CONCLUSIONS

Modification of the wood fibers of Acacia and Sungka by using chemical anhydrides (AA, MA, and SA) has a significant effect on WPG. MDF pieces made from chemically modified fibers were slightly denser and have lower water absorption properties, lower weight increase rate (WIR) and lower thickness change rate (TCR) than MDF made from unmodified fibers.

The effectiveness degree of wood modified percentage was determined using WPG, and there was relationship between WPG and moisture sorption rate. Usually, a higher WPG of wood fibers results in lower moisture sorption uptake. However, this still depends on the type of anhydrides used. The treatment parameter (anhydride type, anhydride concentration) that performed best in terms of WPG value was 60% of AA. It contributed the highest WPG value for both Acacia and Sungkai wood species, with 13.79% for Acacia and 24.91% for Sungkai. The WIR of the non-modified Sungkai MDF test piece increased until it exceeded more than 50% of the weight for the test piece, but it decreased to an average 23.18% using 60% AA.

The rate and extent of swelling in liquid water and water vapor were much greater in unmodified MDF test pieces than in chemically modified MDF test pieces. The observed lowest values of thickness change rate were in the 60% AA MDF test pieces for both species, which was 1.7%–2.31% for Acacia MDF test pieces and 3.5–4.07% for Sungkai MDF test pieces after the first 60 minutes. This trend continued through 6 hours of water swelling, but at the end of 5 days, the swelling of both unmodified Acacia and Sungkai MDF test pieces increased dramatically more than the modified MDF. The range of thickness change for the cyclic water soaking of Acacia was 3%–4% for unmodified MDF test pieces and reduced to less than 2% for modified MDF. However, MDF test pieces of Sungkai (unmodified) decreased from 4%–5% to less than 3% (modified) for the thickness increase range. AA is the best anhydride among the three anhydrides and it has significant effect in terms of slowing down or preventing moisture sorption. MDF modified by AA showed the lowest WIR and thickness change (TC).

REFERENCES

ASTM D1037-99 2000. Standard test methods for evaluating properties of wood-base fiber and particle panel materials. West Conshohocken, PA.

ASTM D4442-92 2003. Standard test methods for direct moisture content measurement of wood and wood-base materials. West Conshohocken, PA.

Azeh, Y., Olatunji, G. A., Mohammed, C., and Mamza, P. A. 2013. Acetylation of wood flour from four wood species grown in Nigeria using vinegar and acetic anhydride. *International Journal of Carbohydrate Chemistry*, 2013(141034): 6.

Azim, A. A. A. 2014. Growth ring formation of selected tropical rainforest trees in peninsular Malaysia. https://doi.org/10.14989/doctor.k18338

Banks, W. B., and West, H. 1989. A chemical kinetics approach to the process of wood swelling. In: Schuerch, C. (ed.), *Proc. Tenth Cellul Conf.*, John Wiley & Sons, New York.

Bodirlau, R., Teaca, C. A., Resmerita, A. M., and Spiridon, I. 2008. Chemical modification of beech wood: Effect on thermal stability. *Bioresource Technology*, 3(3): 789–800.

Bourgeois, J., Bartholin, M. C., and Guyonnet, R. 1989. Thermal treatment of wood: Analysis of the obtained product. *Wood Science and Technology*, 23(4): 303–310.

Hon, D. N. S. 1996. *Chemical Modification of Lignocellulosic Materials*, Marcel Dekker Inc, New York.

Kojima, Y., and Suzuki, S. 2010. Evaluating the durability of wood-based panels using internal bond strength results from accelerated aging treatments. *Journal of Wood Science*, 57(1): 7–13.

Krisnawati, H., Kallio, M. H., and Kanninen, M. 2011. *Acacia Mangium Willd.: Ecology, Silviculture and Productivity*, CIFOR.

Li, X., Li, Y., Zhong, Z., Wang, D., Ratto, J. A., Sheng, K., and Sun, X. S. 2009. Mechanical and water soaking properties of medium density fiberboard with wood fiber and soybean protein adhesive. *Bioresource Technology*, 100(14): 3556–3562.

Lu, Y., and Mosier, N. S. 2008. Kinetic modelling analysis of maleic acid-catalyzed hemicellulose in corn stover. *Biotechnology and Bioengineering*, 101(6): 1170–1181.

Peydecastaing, J. 2008. Chemical modification of wood by mixed anhydrides (Doctoral dissertation), Université de Toulouse.

Rowell, R. M. 1984. *The Chemistry of Solid Wood*, Advance in Chemistry Series No. 207, American Chemical Society, Washington, DC.

Rowell, R. M. 2005a. Moisture properties. In: Rowell, R. M. (ed.), *Handbook of Wood Chemistry and Wood Composites*, CRC Press, Inc, Boca Raton, FL, Ch 4.

Rowell, R. M. 2005b. Chemical modification of wood. In: *Handbook of Wood Chemistry and Wood Composites*. p. 381.

Rowell, R. M. 2006. Acetylation of wood: Journey from analytical technique to commercial reality. *Forest Products Journal*, 56(9): 4–12.

Rowell, R. M., and Banks, W. B. 1985. Water repellency and dimensional stability of wood. USDA Forest Service General Technical Report FPL 50. Forest Products Laboratory, Madison, WI.

Rowell, R. M., and Ellis, W. D. 1978. Determination of the dimensional stability of wood using the water soak method. *Wood and Fiber Science*, 10(2): 104–111.

Rowell, R. M., and Keany, F. M. 1988. Fiberboards made from acetylated bagasse fiber. *Wood and Fiber Science*, 23(1): 15–22.

Rowell, R. M., and Youngs, R. L. 1981. Dimensional stabilization of wood in use. USDA Forest Serv. Res. Note. FPL-0243. Forest Products Laboratory, Madison, WI.

Rowell, R.M., Youngquist, J.A., Rowell, J.S., and Hyatt, J.A. 1991. Dimensional stability of Aspen fiberboard made from acetylated fiber. *Wood and Fiber Science*, 23(4): 558–566.

Stamm, A. J. 1964. *Wood and Cellulose Science*. The Ronald Press Company, New York.

Teaca, C. A., Bodirlau, R., and Spiridon, I. 2014. Maleic anhydride treatment of softwood—effect on wood structure and properties. *Cellulose Chemistry and Technology*, 48(9–10): 863–868.

Properties of *Acacia mangium* Wood Particles and Bioplastic Binder

Yu Feng Tan and Kang Chiang Liew

CONTENTS

11.1 WOOD BINDER INDUSTRY

One of Malaysia's largest timber exports to other countries is particleboard. According to the Malaysia Timber Industry Board (2015), particleboard contributed a total export of 0.54 million m³, which is RM 371.1 million, in 2015. Particleboard is a wood-based panel composite that is manufactured by mixing wood particles with an adhesive as a binder. Moreover, an adhesive is defined as the substance applied to surfaces of materials in order to join them together such that they resist separation. Adhesives have wide applications among different types of materials such as metal, paper, wood, and so on (Kinloch, 1987). Nonetheless, the performance of these adhesively bonded materials depends on the ability of the adhesive to transfer mechanical stresses across the joint interface (Paris and Kamke, 2015). Wood is a natural, porous material where the nature of this interface is highly variable. Therefore, liquid adhesive can flow and penetrate the cellular substrate during bonding.

Wood adhesives are mainly made from a petroleum-based polymer. The main wood adhesives used in the bonding of wood composites include phenol-formaldehyde (PF), urea-formaldehyde (UF), and melamine formaldehyde (MF). These commonly used wood adhesives have highly mechanically and physically stable properties. PF was the earliest wood adhesive used for the mass production of panels (Walker, 2006). As reported by Anon. (2015), UF, PF, and MF resins account for approximately 70% of the global demand for formaldehyde. The increasing demand for wood-based panels on the market subsequently has increased the demand for wood adhesives.

However, the acceptable level of use of these formaldehyde-based resins has gradually reduced over the years. There has been an increasing awareness that these synthetic resins pose a threat to human health and cause pollution to the environment. According to Wang (2006), it is estimated that the world's demand for oil is expected to grow by 50% by 2025. As petroleum production is likely to increase within 20 years and then will experience a gradual fall, a huge reliance on fossil fuel for making adhesive materials would be a problem.

Therefore, much research and development (R&D) has been carried out in the past decades to find alternatives for these petrochemicals that would be renewable and environmentally friendly. Numerous studies have been carried out using agricultural crops such as corn, tapioca, cassava, wheat, and rice to produce bio-based adhesives, known as 'green' adhesives. These biological resources have the advantage of being available in bulk quantities (Saraswat et al., 2014).

11.2   ACACIA WOOD (*ACACIA MANGIUM*)

This tree increases rapidly in height and diameter for the first 4–5 years; the height can reach 5 m while the mean diameter can grow at increments of 5 cm annually (Orwa et al., 2009). According to Krisnawati et al. (2011), Acacia trees can grow to a height of 30 m with a straight bole that is not more than half of the total tree height. This fast-growing species produces 300 clean saw logs per ha of 6 m length in rotations of 15 years, which normally could have reached an average size of 40–45 cm over the bark at breast height (Ahmad, 1992). Acacia wood properties such as valuable wood quality and high tolerances of a wide range of soils and environments are the reason for the pioneer option of plantation wood species in Malaysia under the Compensatory Forest Plantation Scheme in the past decades. The species can even regenerate in sites that were disturbed as reported by Gunn and Midgley (1991); it has been spotted in abundance after forest disturbance, along roads and at the site of slash-and-burn agriculture in Indonesia and Papua New Guinea.

The use of the renowned timber of Acacia has spread since its introduction, and it is widely used in construction, boat building, and the production of pulp and paper, boxes and crates, sawn timber, particleboard, and veneer. Branches or disqualified timber can be chipped, flaked, screened and manufactured into particleboard, fiberboard, and so on. It can also be further processed into attractive doors and windows, furniture, and cabinets, as well as moldings. Table 11.1 describes the anatomy of *A. mangium* wood while Table 11.2 summarizes its physical and mechanical properties (Krisnawati et al., 2011; Gérard et al., 2011; Mohd Hamami et al., 1998).

11.3 ADHESIVES

An adhesive is used as a bonding mechanism to join two substrates together. This bonding can be explained in terms of adhesion and cohesion theory, whereby the adhesive must be able to develop an adhesion to the substrate and at the same time able to acquire bond strength (cohesion) after it sets (Glavas, 2011). According to Glavas (2011), the adhesive should be able to wet the substrates in order to obtain surface adhesion. Furthermore, the wettability of the substrates depends on the adhesive and the surface energy of the substrates, where the lower energy of the adhesive will spontaneously spread over the surface of the substrate and wet it. From the article edited by Anon (1998), the main requirements for a material to perform as an adhesive are as follows:

* The material must "wet" the surfaces – it must be able to spread over the surfaces that are being bonded and displace all the air and other contaminants that are present on the surface.

TABLE 11.1
Summary of Anatomy Description of *A. mangium* Wood

Anatomy	Description
Growth rings	Indistinct or sometimes absent.
Timber grain	Straight.
Texture	Moderately coarse and even.
Fiber length	Short with an average range of 1.0–1.2 mm
Pores and vessels	Diffuse-porous with mostly solitary vessels. Silica crystals present in the vessels.
Rays	Uniseriate and fine.
Perforation plate	Simple.
Intervessel pits	Arranged alternately.
Parenchyma	Low proportion of parenchymatous cell. Scanty paratracheal in thin vasicentric form. Small longitudinal with calcium crystals.
Tyloses	Absent.
Color	Sapwood – narrow and light to pale yellow color. Heartwood – medium to dark brown.

TABLE 11.2

Physical and Mechanical Properties of *A. mangium* Wood

Physical/Mechanical Properties	Value
Density	450–690 kg/m³ at 15% moisture content
	500–600 kg/m³ at dry state
Specific gravity	0.4–0.45 for plantation timber
	0.52–0.6 for natural stands
Tangential shrinkage / radial shrinkage ratio	2/3
Modulus of elasticity (MOE)	Mean value of 10800 MPa
Static bending strength (MOR)	105 MPa
Shear strength	16.4 MPa
Compression parallel to grain	51.8 MPa

- It must also adhere to the surface after the material flows over the whole surface area and stay in position as well as become "tacky."
- It also must develop strength, where the structure will become strong or non-tacky but still adherent to the substrates.
- Lastly, it should remain stable and unaffected by age, environmental conditions, and other factors.

Applications of adhesive can be found in many industries nowadays, especially in such advanced technical domains like the aeronautical and space industry, automobile manufacture, and electronics. It has also been introduced into the field of dentistry and surgery (Pizzi and Mittal, 2003). According to Faherty and Williamson (1995), the advent of the use of adhesives in the wood industry has provided an effective means of connecting a variety of wood components in a way that the strength of the joints can be improved in the structural system. Paris and Kamke (2015) further stated that the ability of the adhesive to transfer mechanical stresses across the joint interface has influenced the performance of all adhesively bonded materials.

The porous nature of wood allows a properly applied adhesive to penetrate the wood surface and hence form a strong bond (Farherty and Williamson, 1995). According to the book edited by Pizzi and Mittal (2003), the penetration of adhesive and its subsequent solidification in the porous wood structures will increase the surface area over which the intermolecular chemical interactions governing wood adhesion operate effectively. Subsequently, solidified adhesives provide mechanical interlocking resistance to hold the substrates together. Moreover, the penetration of the adhesive will transfer joint stresses, past inherent surface irregularities, and machining defects into the undamaged wood cells, which are believed able to enhance bond strength, toughness, and durability (Paris et al., 2014). Three main parameters that are used to determine the properties and performance of the wood-based panels and beams which show the quality of bonding include (Pizzi and Mittal, 2003):

- The wood or wood surface, including the interface between wood surface and the bondline.

- The applied adhesive.
- The process parameters and working conditions.

One of the most important reasons for using adhesive in the construction of wood is to construct composite systems which economize material usage. Adhesives allow different grades and types of wood materials to be assembled in the most efficient manner, also minimizing the effect of defects on strength and stiffness. Another advantage of adhesive bonding is that the building components can be preassembled. Adhesive bonding also increases the rigidity of the wood joint and enables the strength of the materials to be fully utilized. Pizzi and Mittal (2003) also summarized the proper requirements for the production of wood adhesives:

i. Shorter press time, shorter cycle times.
ii. Better hygroscopic behavior of boards (e.g., lower thickness swelling, higher resistance to humidity and water, better outdoor performance).
iii. Cheaper raw materials and alternative products.
iv. Modification of the wood surface.
v. Life cycle assessment, energy and raw materials balance, recycling and reuse.
vi. Reduction of emissions during the production and use of wood-based panels.

11.3.1 Bio-Based Adhesive

Bio-based adhesive is a biodegradable polymer derived from natural polymers such as starch, protein, and cellulose. However, these polymers cannot form adhesives or act as binders alone due to their hydrophilic nature. In other words, these biodegradable adhesives are predominantly water soluble, which limits their applications outdoors or in a moisture-rich environment (Johnson and Yunus, 2009). The adhesive form has limited long-term stability which is affected by water absorption, age-related retrogradation, poor mechanical properties and low processability (Gaspar et al., 2005). Therefore, plasticizers such as glycerol, vinegar, and water are added to enhance the binding properties and to decrease the water uptake problems associated with bio-based adhesives (Liew and Khor, 2015; Saraswat et al., 2014). Many studies have also been carried out by adding different enhancers to these polymer materials, such as nanoparticles, which were added to starch-based wood adhesive to improve bonding strength and water resistance (Gu et al., 2011); and sodium dodecyl sulfate, which was added to improve the performance of the adhesive (Li et al., 2014).

Starch is the major carbohydrate in plant tubes and seed endosperm. It is found in the form of granules, where each granule contains a few million amylopectin molecules accompanied by a larger number of smaller amylose molecules (Gaspar et al., 2005). Amylose and amylopectin in the starch contain several hydroxyl groups that can act as hydrogen bond acceptors and donors. Retrogradation is a formation of a network when a linear polymer that consists of hydrogen bonds and amylose is dispersed in water. This network is able to trap water within itself, forming a gel (Davidson, 1980). Consequently, retrogradation does not occur easily with amylopectin because it is a highly branched polymer. The ratio of amylose to amylopectin in starch will affect the final properties of the starch, such as gelatinization, viscosity, solubility, gel stability, and tackiness (Glavas, 2011).

Gelatinization is closely related to the properties of starch as it can be used to measure the strength of hydrogen bonds between the starch molecules (Davidson, 1980). Starch is present in plants as particles known as granules. These granules are insoluble in water but are capable of dispersal in water. The granules will start to swell when an aqueous dispersion of starch is heated to a certain temperature called *gelatinization* or *pasting temperature*, and this process is irreversible (Davidson, 1980). According to Glavas (2011), gelatinization occurs when the temperature is high enough that it will weaken the hydrogen bonds that hold the starch granules together and allow water molecules to penetrate between the polymer chains. The swelling takes place when granules take up water, and they will collapse when heating is continued. Figure 11.1 shows the variation of viscosity during heating – the gelatinization process.

As gelatinization continues, the viscosity of the dispersion will increase until it reaches *peak viscosity*. The peak viscosity is a measure of the thickening power of starch. It is the highest viscosity that the user may encounter during the preparation of starch paste. Furthermore, according to Anon (2007), the peak viscosity can be measured by using the Brabender Visco-amylograph. This measures the viscosity of starch–water dispersions when stirred and heated at a uniform rate, held at the desired temperature for a specific time, and then left to cool. However, different types of starch have different gelatinization temperatures (Table 11.3) (Anon, 2007).

The reason for using starch in adhesive formulations is that this makes it relatively easy to produce colloidal, an aqueous solution of this high molecular weight of polymer. Subsequently, the solid content of starch should not be too high due to its high molecular weight, or it will cause the solution to be highly viscous, creating

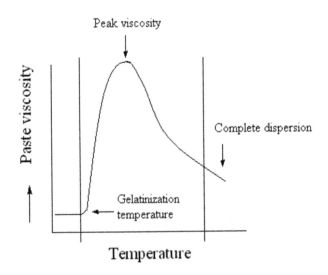

FIGURE 11.1 Variation of viscosity during heating, the gelatinization process. (From Davidson, 1980.)

TABLE 11.3
The Gelatinization Temperature
of Different Native Starches

Starch	Gelatinization Temperature ($^\circ$C)
Tapioca	60–65
Potato	60–65
Wheat	80–85

(Anon, 2007)

difficulties when it is included in the adhesive application. On the other hand, the low amount of solid content may lead to problems like a long drying time, which is not favorable (Onusseit, 1992; Imam et al., 2001). Besides, starch is a polar material, so it will have a high affinity towards other polar materials such as cellulose. Hence, it is specifically useful in the formulation of wood adhesives. The high affinity towards cellulose will minimize the contact angle between substrate and adhesive, allowing the adhesive to wet the substrate and eventually form strong bonds (Imam et al., 1999).

Moreover, the performance of bio-based adhesives is often questioned with their comparison to synthetic polymers especially the poor water resistance characteristics of bio-based adhesives which lead to the short lifespan of wood-based panel. Thus, a specially formulated bio-based adhesive which includes starch, glycerol, vinegar, and water will be incorporated in this study to evaluate the possibility of using a different concentration of starch as a bio-adhesive from the aspects of the physical, mechanical, and dimensional stability of the particleboard.

In this chapter, the effects of different amounts of starch ($^\circ$10%, 15%, and 20%) on the physical, mechanical and dimensional stability of particleboard produced were evaluated and discussed. The effect of different adhesive contents (20%, 25%, and 30%) on the physical, mechanical and dimensional stability properties of the particleboard produced from these adhesives were investigated. Urea formaldehyde (UF) was used as a control to compare the performance between the synthetic adhesive and the bio-based adhesive produced.

11.3.2 UREA FORMALDEHYDE

Urea formaldehyde (UF) is a common adhesive that is widely found in wood industries such as the manufacture of particleboard, plywood, and other wood-based panels. It is based on the manifold of two monomers: the urea and formaldehyde. According to Pizzi and Mittal (2003), the advantages of UF adhesives include their water solubility, which renders them ideal for use in the woodworking industry; their hardness, nonflammability, and good thermal properties; the fact that they are colorless in cured polymers; and their easy adaptability to a variety of good curing conditions. UF adhesives also perform well in terms of strength and durability, which are essential properties in wood-based products.

The major drawbacks of UF adhesives are their moisture and water resistance. Their low resistance to moisture and water causes the urea formaldehyde bond to deteriorate easily due to the hydrolysis of the aminomethylenic bond (Pizzi and Mittal, 2003). According to Dunky (1998), the reaction between urea and formaldehyde is a two-step process involving methylolation and condensation. The reaction involved in the methylolation process of the urea and formaldehyde molecules is reversible because of the low resistance of UF resins to hydrolysis.

Particleboard manufactured by using UF resin as a binder possess a good elasticity, modulus of rupture, internal bond, and other mechanical properties. However, Pizzi and Mittal (2003) reported that the mechanical properties of such particleboard could be affected by the heating rates of the board and resin content of UF. According to the study by Ashori and Nourbakhsh (2008), resin content and press time were the main factors affecting the physical and mechanical properties of particleboard. The mechanical properties of the board increase when the press time and resin content of the adhesive are increased.

11.4 PARTICLEBOARD

Wood-based composites have become an important component in the wood processing chain. Wood-based panels have now become an integral part of the wood product market as these panels are widely used in the construction of furniture, cabinets, and joinery, as well as for their ability to be 'engineered' in order to meet specific performances and requirements. As the utilization of these products has developed significantly, the technology to produce these composites has advanced and automation such as sawing, machining, and finishing has been implemented. This allows better quality components to be produced at low cost. Table 11.4 shows the evolutionary sequence of wood-based panels. Table 11.5 shows the world production of particleboard, oriented strand board (OSB), and medium density fiberboard (MDF) in 2004.

TABLE 11.4
Evolutionary Sequence for Wood-Based Panels

Year	Product	Country
1830	Mechanically sliced veneer	France
1896	Rotary peeled veneer	Estonia
1898	Fiberboard (wet formed)	UK
1906	Plywood	USA
1914	Insulating board	Germany
1925	Hardboard	USA
1941	Particleboard	Germany
1945	Dry process fiberboard	USA
1966	MDF	USA
1969	OSB	Germany

(Clark, 1991)

TABLE 11.5

World Production of Particleboard, OSB, and MDF for 2004

Type of composite	Million m³	%	Average Growth, 1995–2005 Million m³/Year
Particleboard	81.5	54.8	2.4
OSB	26.5	17.8	2.1
MDF	40.7	27.4	3.5
Total	148.6		

(Walker, 2006)

Particleboard, also known as chipboard, utilizes residue materials from other wood-processing operations and has recently become one of the most widely used wood-based panels on the market. Particleboard can be classified as an engineered wood-based product manufactured from various sizes and particles of wood chips and a binding resin. Moreover, the board can be formed in a press using a hot platen mat forming process. Sawdust, planer shavings, edgings, and other wood residues can also be used in the manufacture of particleboard (Stark et al., 2010).

The requirement for particleboard varies from applications to panels, where the surface smoothness and strength is essential (Walker, 2006). Particleboard can be manufactured in different sizes, thicknesses, densities, and grades for a variety of applications. The homogenous properties add value to the core material for furniture manufacturing, while homogeneous raw materials provide the boards with good strength, smooth surfaces, and equal swelling. The most desirable blend features particles with a high degree of slenderness (long, thin particles), no oversized particles, no splinters, and no dust (Stark et al., 2010). Many efforts have been made throughout the years to improve the board's surface smoothness in order to improve the finishing and laminating of the board.

As mentioned, particleboard can be manufactured for various uses, including furniture, cabinets, flooring systems, manufactured houses, and underlay. The mechanical properties (density, tensile strength, MOE and MOR) of particleboard bond with soy protein adhesive, which is also one of the bio-based adhesives, vary according to the board density shown in Table 11.6.

11.5 PERFORMANCE OF STARCH-BASED ADHESIVE AS A WOOD BINDER

Pizzi (2006) reported that wood adhesives are used in three main ways:

- Directly as wood adhesives.
- As modifiers for synthetic adhesives such as phenol and UF, which are more expensive.
- As building blocks for new adhesives.

TABLE 11.6

Mechanical Properties of Wheat Straw Particleboard Bond with Soy-Protein-Based Adhesives

Density (g/cm³)	Tensile Strength (MPa)	MOR (MPa)	MOE (MPa)
0.8–0.9	6.0–7.0	18–21	3000–3300
0.6–0.7	3.5–4.5	7.0–9.0	1700–1900
0.72 (UF resin)	3.9	6.3	1805
0.75 (commercial wood particleboard)	-	13–17	2200–2400

(Wool and Sun, 2011)

According to Wang et al. (2013), although starch is widely used as binders, sizing materials, glue, and pastes, its bonding capacity is still not strong enough to glue wood. Therefore, modified starches that are made from graft polymerization with vinyl acetate and butyl acrylate have been included in the application of wood adhesive (Wu et al., 2009). However, the properties and functions were still not appropriately investigated. As reported by Tanrattanakul and Chumeka (2010), biopolymer-based adhesives are too weak for practical use. Therefore, other additives such as silica nanoparticles were investigated and included in the starch to produce starch-based wood adhesive (Gu et al., 2011). The result shows an improvement in both bonding strength and water resistance.

Moreover, the graft copolymerization of synthetic polymers onto a starch backbone was said to be the best way to improve the bonding properties of starch (Lei et al., 2014). In order to improve the performance of starch-based adhesive, sodium dodecyl sulfate, urea, silane coupling agent, and olefin monomer are added to the oxidized starch to produce graft-copolymerized wood adhesive (Li et al., 2014; Wang et al., 2013; Tanrattanakul and Chumeka, 2010). The outcome has improved bonding strength and water resistance significantly. Zhu et al. (2015) have carried out an investigation to improve the bonding strength and water resistance performance of starch-based wood adhesive through grafting copolymerization with the addition of silane coupling agent.

A study carried out by Yu et al. (2015), meanwhile has shown that the compatibility between starch particles and polyvinyl alcohol (PVA) can be improved by increasing the treatment temperature to give a better bonding quality. According to a study by Liew and En (2016), it was reported that three-ply plywood manufactured by using cassava and sago starch-based adhesives at press temperatures of 100°C and 120°C gave good performance in terms of bending while a press temperature at 140°C yielded better shear performance. In addition, the properties of particleboard made from oil palm starch-based adhesive were claimed to be better than board manufactured from wheat starch by Hashim et al. (2015). Therefore, it can be said that the different types of starch might also give different properties.

Oakley (2010) claimed that different types of plasticizers have certain influences on water absorption of thermoplastic starch (TPS) as it is largely responsible for its material properties, such as the diffusion coefficient. It was found that the commonly used polyol plasticizers continuously decrease the water uptake and diffusion coefficient as the molecular weight increases. These polyols include glycerol, glycol, sorbitol, and xylitol. (Mekonnen et al., 2013). Furthermore, Lopez-Gil et al. (2014) claimed that the mechanical properties of starch-based materials can be improved by reducing the amount of plasticizer, such as glycerol, by up to 20%. Water is also known to be a good plasticizer for starch, but it is preferable not to use it as the only plasticizer with starch as the end product will be brittle due to the volatilization of water itself (Mekonnen et al., 2013). Therefore, glycerol is added to water to use a plasticizer to increase the properties of starch.

11.6   PREPARATION OF ACACIA WOOD PARTICLES

Wood residue or small logs were obtained from the hardwood species of *Acacia mangium*. The wood residue or small logs were then fed into a chipper machine to produce wood chips. These wood chips were then further put into a drum flaker machine to produce particles 1.0–5.0 mm long and 0.5 mm wide. The wood particles were then screened to particles with length of 0.5–2.0 mm in a vibrator screener machine. Then, the particles were oven-dried at 70°C for 24 hours to obtain a moisture content of approximately 2%–4%. The particles were then used to produce a single-layer particleboard.

11.7   ADHESIVE PREPARATION

Tapioca flour was used in the formation of a bio-based adhesive. Besides tapioca starch, vinegar, glycerol, and water were also added to the formulation of starch-based adhesive. Batches of this starch-based adhesive were produced in small amounts to ensure that the adhesive was well cooked. Starches with 10%, 15%, and 20% based on 295 g were incorporated into the mixture. Then, the mixture was stirred using a glass rod and the beaker was then placed on a hot plate to be cooked at 70 °C–80 °C for about 20 minutes. The mixture needs to be constantly stirred during the cooking process to avoid the mixture becoming sticky.

The viscosity of each starch-based adhesive produced using different amounts of starch was measured by using a Brookfield Viscometer. The urea formaldehyde resin obtained from a local glue supplier, which was UF-SU955P, was used as a control. The amount of particles used in the production of the particleboard was calculated according to the standard technical sheet provided by the glue supplier.

11.8   PRODUCTION OF PARTICLEBOARD

Particleboard was formed to achieve the targeted density of board, which was 600 kg/m^3. Particles were mixed with adhesives of different content, either 20%, 25%, or 30% each time, through manual or hand mixing. The product was then placed into a mat-forming frame measuring 300 mm long, 300 mm wide, and 10 mm thick. Subsequently, a conventional hot press was used to press the board at 165°C, 135 MPa, and 5 minutes respectively. The board was then left to cool and conditioned for 72 hours with a relative humidity of 65% at 20 ± 2°C. These methods were repeated for different amounts of starch (10S, 15S, and 20S) and different adhesive content (20%, 25%, and 30%) for each board.

11.9   TYPES OF TESTING FOR PARTICLEBOARD PRODUCED

The particleboards produced in this study were tested according to Japanese Industrial Standard (JIS-A-5908) for water absorption, thickness swelling, density, moisture content, modulus of elasticity (MOE), modulus of rupture (MOR), and internal bond strength.

11.9.1 Physical Properties

i. *Density*

The lengths, widths, and thicknesses of the test pieces were measured according to the JIS a-5908 standard. Then, the mean values of length, width, and thickness were obtained to calculate the volume (V) of the test pieces. After that, the mass (m_1) was measured and the density was calculated using Equation 11.1. The thickness, length, width, and mass were measured to the nearest 0.05 mm, 0.1 mm, 0.1 mm, and 0.1 g respectively. The density was calculated to the nearest 0.01 g/cm³.

$$Density\ (g/cm^3) = \frac{m_1}{V} \qquad (11.1)$$

where
   m_1:   mass (g)
   V: *volume (cm³)*

ii. *Moisture Content*

To obtain the moisture content, the mass (m_1) of a test piece was measured before being put in an oven to dry at 103 ± 2 °C. The mass (m_0) was measured after the constant mass and moisture content was obtained to the tenth's place by using the formula shown in Equation 11.2.

$$Moisture\ content\,(\%) = \frac{m_1 - m_0}{m_0} \times 100\% \qquad (11.2)$$

where,
   m_0:   mass (g) after drying.
   m_1:   mass (g) before drying.

11.9.2 Mechanical Properties

i. *Bending strength test*

The test apparatus was set up as shown in Figure 11.2 according to the JIS-A-5908 standard (JIS, 2003), and a load of approximately 10 mm/min was applied at a mean deformation speed from the surface of the test piece. Maximum load (P) was then measured and recorded. Then, the bending strength of individual test pieces was

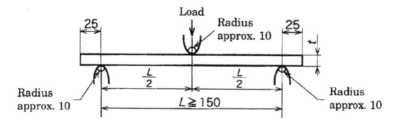

FIGURE 11.2 Test apparatus set up for bending strength (JIS-A-5908).

calculated by using Equation 11.3. MOE and MOR were obtained by using Universal Testing Machine computerized software.

$$\text{Bending Strength } (N/mm^2) = \frac{3PL}{\left(2bt^2\right)} \tag{11.3}$$

where
- P: maximum load (N)
- L: span (mm)
- b: width of test piece (mm)
- t: thickness of test piece (mm)

ii. *Internal bond test*

A test piece was adhered to steel blocks using hot-melt glue. Then, a tension load was applied vertically to the board face and maximum load (P') was measured at the time of failing force (breaking load of perpendicular tensile strength to the board). In this test, the tension loading speed was set to 1 mm/min. The set up for the internal bond test was shown in Figure 11.3. Lastly, the internal bond was calculated according to Equation 11.4.

$$\text{Internal Bond } (N/mm^2) = \frac{P'}{2bL} \tag{11.4}$$

where
- P': maximum load (N) at the time of failing force
- b: width (mm) of sample
- L: length (mm) of sample

11.9.3 DIMENSIONAL STABILITY

i. *Water absorption*

The weight of each test piece was measured before being horizontally immersed in water of 20 ± 1 °C about 3 cm below the water surface for 24 hours. After 24 hours, the test pieces were taken out and excessive water was wiped off before the weight was measured again on the same weighing balance. The weight of the test piece

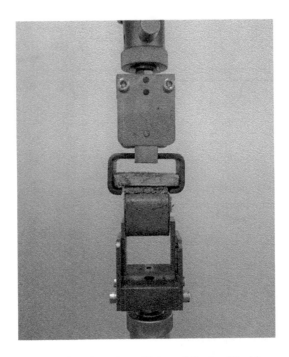

FIGURE 11.3 Internal bond test by using a Universal Testing Machine.

before and after immersion was recorded and the percentage of water absorption was calculated according to Equation 11.5.

$$\text{Water Absorption, WA}\left(\%\right) = \frac{W_1 - W_0}{W_0} \times 100\% \qquad (11.5)$$

where
  W_0: weight of test piece before submersion.
  W_1: weight of test piece after submersion.

ii. *Thickness swelling*

The thickness in the center of the test piece was measured to the nearest 0.05 mm with a micrometer. Then, the test pieces were horizontally immersed in water of $20 \pm 1^\circ\text{C}$ about 3 cm below the water surface for 24 hours. After 24 hours, the test pieces were taken out and excessive water wiped off before the thickness was measured again in the same manner as before. The increase in thickness of the test pieces after immersion in water was calculated using Equation 11.6.

$$\text{Thichness Swelling}\left(\%\right) = \frac{t_2 - t_1}{t_1} \times 100\% \qquad (11.6)$$

where
  t_1: thickness (mm) before immersion in water.
  t_2: thickness (mm) after immersion in water.

11.10 PHYSICAL PROPERTIES OF PARTICLEBOARD PRODUCED

The physical properties of particleboard formed with starch-based adhesive and UF as a control in this study were density and moisture content. The targeted density of the particleboard was 0.60 g/cm³ whereas the average value of density between the starch-based adhesives' test pieces (0.5 cm × 0.5 cm × 0.01 cm) were recorded in the range of 0.56 g/cm³ to 0.59 g/cm³. Meanwhile, the average value of density for UF as a binder ranged from 0.58 g/cm³ to 0.62 g/cm³. Furthermore, the moisture content of particleboard produced with starch-based adhesives and UF as a binder were also recorded. The moisture content of boards made with starch-based adhesives and UF were around 7.83 %–9.56% respectively.

11.10.1 DENSITY

Density did not vary much in terms of average value for particleboard produced with starch-based adhesives and UF as shown in Figure 11.4. Based on the graph displayed, it can be clearly observed that the density of board with a 30% adhesive content gives the highest average value for each type of adhesive from 10S to 20S. The average value for increasing the amount of adhesive content should give increasing average value of density, such as the pattern shown in 15S as well as 20S, when adhesive content increases by 5%.

However, in 10S and UF adhesives, there were slight contrasts in that the average value decreased when the amount of adhesive content was 25% and soared up again at 30%. A possible explanation for this was that the test pieces chosen at random were mostly at the edge of the board produced where materials were less compact, thus giving a lower density. The graph shows that the average value for density of particleboard produced using 30% adhesive content for each type of adhesive was higher. On the other hand, the average density of board with UF as binder was generally

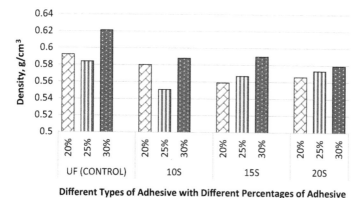

FIGURE 11.4 Average density (g/cm³) for particleboards bound using different amounts of adhesive content (20%, 25%, and 30%) prepared from adhesives with different amounts of starch incorporated (10S, 15S, and 20S).

higher at 20%, 25%, and 30% of adhesive content when compared to starch-based adhesives. It is possible that the level of tackiness of UF was higher than that of starch-based adhesives; more tack indicates less loss of particles at the edges, which results in boards with a higher density (Khosravi, 2011).

11.10.2 MOISTURE CONTENT

The average moisture content for UF and starch-based adhesive did not vary much either. Based on Figure 11.5, the average moisture content for different amounts of starch (10S, 15S, and 20S) in starch-based adhesive only differed by small range at 20% adhesive content. Besides, starch-based adhesive with 25% adhesive content showed a slightly odd pattern whereby it increased at 10S and 15S before it started to experience a drop at 20S. As for 30% adhesive content, the average moisture content decreased in an ascending order of starch content (10S, 15S, and 20S).

As shown in Figure 11.5, the average moisture content of particleboard with UF as a binder was recorded in the range of 8.81%–9.56%. Particleboard with starch-based adhesives as binder, on the other hand, has a lower average moisture content in the range of 7.83%–9.17%. It is possible that the moisture in starch-based adhesives evaporated to the environment throughout the study. As reported by Banker (1966), plasticizers such as sorbitol and glycerol in starch film will affect the water vapor and solute permeability characteristics.

11.11 MECHANICAL PROPERTIES OF PARTICLEBOARD PRODUCED

Mechanical testing was carried out on particleboard with different percentages of starch in the starch-based adhesive and different amount of adhesive content in the board. The mechanical test includes the MOE, MOR, and internal bonding (IB) of the board produced. The average MOE value obtained was in the range of 1697.64

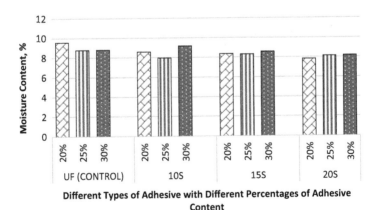

FIGURE 11.5 Average moisture content (%) for particleboards bound using different amounts of adhesive content (20%, 25%, and 30%) prepared from adhesives with different amounts of starch incorporated (10S, 15S, and 20S).

N/mm^2–2282.12 N/mm^2 for particleboard manufactured using UF. However, the average value of starch-based adhesives was recorded in the range of 761.84 N/mm^2–1115.07 N/mm^2 of MOE.

11.11.1 MODULUS OF ELASTICITY

As shown in Figure 11.6, it was obvious that the starch-based adhesives have a lower mean value overall as compared to UF. The highest mean value of MOE among starch-based adhesives was 15S with 30% adhesive content, but the mean value drops when it reaches 20S with 30% adhesive content. Overall, the mean value does not differ much with different percentages of starch.

There were no distinct changes between different amounts of starch (10S, 15S, and 20S) in starch-based adhesives or with different amounts of adhesive content (20%, 25%, and 30%). A huge difference in MOE performance can be seen between starch-based adhesives and urea formaldehyde as a control in particleboard production. Urea formaldehyde performs better due to the presence of polar groups in the urea formaldehyde. These polar groups contribute to the electrostatic absorption between urea formaldehyde and the particles (Tay et al., 2016). The high electrostatic charges that act upon urea formaldehyde on the surface of materials to be bonded have strengthened the interface by holding them together and make it less likely that deformation will occur (Idris et al., 2011).

The elastic property of starch-based adhesives could be related to the plasticizers used in this study. One of the plasticizers used in the formulation of starch-based adhesive is glycerol. According to the study carried out by Lopez-Gil et al. (2014), the mechanical properties of starch-based adhesives are affected by the different amount of plasticizers used. It was reported that the MOE of starch-based adhesive will increase when the amount of glycerol is decreased. However, in this study, the proportion of glycerol remained constant at 18% of the starch-based adhesive

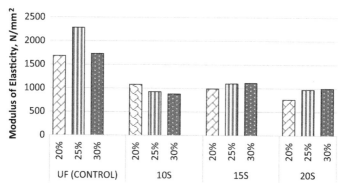

FIGURE 11.6 Average MOE (N/mm^2) for particleboards bound using different amounts of adhesive content (20%, 25%, and 30%) prepared from adhesives with different amounts of starch incorporated (10S, 15S, and 20S).

produced. Hence, it does not reflect much on the MOE among the starch-based adhesives. Saraswat et al. (2014) have explained that the direct interactions with and proximity to the starch chain will reduce the incorporation of plasticizers such as glycerol in the starch network.

11.11.2 MODULUS OF RUPTURE

From Figure 11.7, the overall average value of MOR among the starch-based adhesives was the highest at 15S with 25% adhesive content. Furthermore, the average values at 15S with 20% as well as 30% adhesive content also showed the highest mean among different starch-based adhesives with the same adhesive content. The pattern of the bar graph shows that the average value of MOR among starch-based adhesives with different amounts of starch did not differ much. On the other hand, the comparison of MOR between the bar graph of UF (control) and the starch-based adhesives show a big difference for adhesive contents of 20%, 25%, and 30%.

The contrast shown in Figure 11.7 indicates that UF performs better than starch-based adhesives. UF is a synthetic resin with formaldehyde, which gives better bonding capabilities and higher MOR. Starch-based adhesive, on the other hand, is a pure starch-based adhesive, and its chemical properties cause the starch adhesive solution to have the tendency to gel or retrograde within hours or days. Hence, its shelf life is reduced and it has poor performance in terms of strength and bonding ability (Pizzi and Mittal, 2003). Apart from that, no distinct difference was seen when the usage of adhesive content was increased from 20% to 30% for starch-based adhesives and UF as a binder in particleboard. Therefore, it can be concluded that changes in the amount of adhesive content (20%, 25%, and 30%) will not affect the overall MOR.

The poor performance in MOR for starch-based adhesive could be explained by the chemical composition of starch, which is made up of amylose and amylopectin.

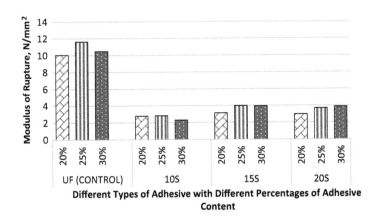

FIGURE 11.7   Average MOR (N/mm²) for particleboards bound using different amounts of adhesive content (20%, 25%, and 30%) prepared from adhesives with different amounts of starch incorporated (10S, 15S, and 20S).

Lourdin et al. (1995) claimed that the mechanical properties of starch films are determined by the ratio of amylose and amylopectin present in the starch. Upon heating of starch granules in water, swelling takes place, resulting in the collapse and rupture of the amylose and amylopectin. Amylopectin that has ruptured shows a low affinity to interaction and thus makes for weak, cohesive and flexible films. Meanwhile, amyloses present in a high solution have a higher tendency to react with the hydrogen bonds and to form stronger and stiffer gels. Furthermore, the same testing has been carried out on MDF and it was claimed that there was no increase in loading at a higher starch content due to the lowest composition of natural latex present in the starch-based adhesive (Akbari et al., 2014). This also gives support to the fact that pure starch-based adhesive has the least ability against the loading strength.

11.11.3 INTERNAL BOND

According to Figure 11.8, within the different amounts of adhesive content used in the manufacture of particleboard, the IB strength of the three consecutive amounts of adhesive content (20%, 25%, and 30%) at 20S of starch-based adhesive has the lowest mean value among all. Furthermore, the mean value of particleboard manufactured using starch-based adhesives experiences increases from 10S to 15S when the amount of adhesive content increases from 20% to 30%. However, particleboard manufactured using UF as binder performs better as compared to starch-based adhesives.

When the amount of starch in starch-based adhesives increases from 10S to 20S, there is an effect on the IB strength of the particleboard. As the amount of starch increases from 10S to 15S, the void between particles is filled, improve the quality of contacts between particles. However, the performance of starch can be seen to decrease when the amount of starch in starch-based adhesive increases from 15S

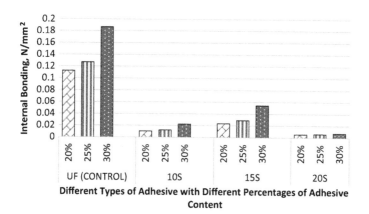

FIGURE 11.8 Average IB (N/mm^2) for particleboards bound using different amounts of adhesive content (20%, 25%, and 30%) prepared from adhesives with different amounts of starch incorporated (10S, 15S, and 20S).

to 20S. This means that starch-based adhesives give better bonding with a 15% starch content, but when it reaches 20%, the bonding between particles might start to weaken, which makes the IB strength drop back to the same level as a 10% starch content. A possible explanation for this trend can be related to the viscosity between starch-based adhesives, which will be discussed later in this section.

The bar graph in Figure 11.8 shows clearly the same MOE and MOR pattern found in this study, in which the mean value of IB increases from 10S to 15S and experiences a drop in 20S for each amount of adhesive content. The viscosity of starch-based adhesives prepared with 10% starch, 15% starch, and 20% starch might have some relationship to this trend. Table 11.7 shows the viscosity and solid contents of the different types of glue used in this study.

According to Siddaramaiah et al. (2004), a higher solid content in an adhesive gives a higher bonding strength. Table 11.7 shows that the solid content of UF is indeed higher than starch-based adhesive, and this could explain the higher strength of the particleboard. The higher viscosity of the adhesive reduced the flowability and sheared the lower of adhesive dispersion in particleboard (Prompunjai and Sridach, 2010). The viscosity of 20S is quite high, so that during the mixing process, it was difficult to blend it with the particles, causing the distribution of the adhesive to be inconsistent. This also affects the wetting ability of the glue in contact with the wood particles. In addition, a 10% starch content in starch-based adhesive gives less IB strength, which increases when a 15% starch content is reached. This statement could be supported by the fact that additional starch is able to fill the void between particles and improve the quality of contact between particles. Therefore, it can be concluded that starch-based adhesive performs better until a 15% starch content is reached; more than this causes low flowability of the adhesive and results in poor penetration as well as low bonding strength.

Moreover, the influence of viscosity can be seen in Figure 11.9, which illustrates the problem faced during internal bond tests of starch-based adhesive particleboard. Delamination occurs at the center of the test piece where it can be clearly seen that the glue did not bind well with the particles, and this is responsible for the poor internal bond strength of particleboard. This causes the failure of some test pieces from 10S, 15S, and 20S as they broke apart in the middle before any load was applied. Particleboard bonded with 10S had the most number of test pieces that experienced delamination.

TABLE 11.7
Viscosity and Solid Content for Different Types of Glue Used in This Study

Types of glue	Viscosity (cps)	Solid Content (%)	Contact Angle (°)
UF (urea formaldehyde)	170	64.1	22
10S (10% starch)	70–80	20.1	43
15S (15% starch)	330–360	23.1	57
20S (20% starch)	570–590	27.0	77

FIGURE 11.9   Delamination that occurs in the middle of test pieces.

11.12   DIMENSIONAL STABILITY OF PARTICLEBOARD PRODUCED

The average dimensional stability for water absorption of particleboard produced with different concentrations of starch was recorded in the range of 125.55%–172.85% for different amounts of adhesive. On the other hand, the swelling of particleboard produced with different concentrations of starch in the starch-based adhesive, with different amount of adhesive added during board production was recorded in the range of 71.18%–84.43%. Both tests differ from the result of particleboard manufactured by using UF in which the water absorption and swelling of the test pieces ranged from 61.38% to 74.02% and 7.58% to 12.06% respectively.

11.12.1   WATER ABSORPTION

As shown on the graph in Figure 11.10, the starch-based adhesives have a higher mean value as compared to UF at different percentages of adhesive added in the particleboard. However, in a comparison of starch-based adhesives only, adhesive with 10% starch (10S) has the highest mean value while 20S has the overall lowest mean value of water absorption.

The water absorptivity of starch-based adhesive could be explained by the hydrophilic nature of starch. According to Oakley (2010), starch molecules consists of two main functional groups, the OH group and C-O-C bond. The OH group is responsible for the high affinity to water that causes many of the problems relating to water absorption, as this functional group is susceptible to substitution reaction. The C-O-C bond present in the starch molecules is also susceptible to chain breakage. Furthermore, Hemmila et al. (2013) reported that the weak hydrogen bonds in starch cause hydrogen groups to break down easily and form another hydrogen bond with water molecules. This has resulted in poor water resistance of the particleboard.

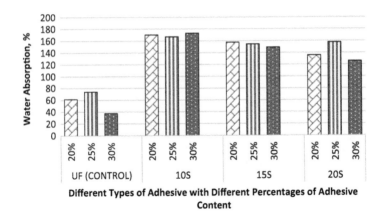

FIGURE 11.10   Average water absorption (%) for particleboards bound using different amounts of adhesive content (20%, 25%, and 30%) prepared from adhesives with different amounts of starch incorporated (10S, 15S, and 20S).

This could explain why different percentages of starch incorporated into starch-based adhesive will affect the end result. By increasing the amount of starch in a starch-based adhesive, more hydrogen bonds were breaking apart, allowing for more water molecule substitution. Figure 11.11 shows the particleboard test piece that was produced using starch-based adhesive before and after immersion in water. It was observed that the test piece swelled after 24 hours of soaking in water. When the test piece was crushed softly, it crumbled easily, as shown in the photo.

Particleboard that uses urea formaldehyde as a binder shows less water absorption where the mean value is lower than a starch-based adhesive. This could be explained by the fact that the urea formaldehyde resin has a stronger and more sustainable cohesive and binding force (Abdul Khalil et al., 2011). Moreover, better interfacial contact in fiber-matrix bonding in UF results in good bonding properties, which then prevent the parenchyma tissues from absorbing water (Abdullah et al., 2012).

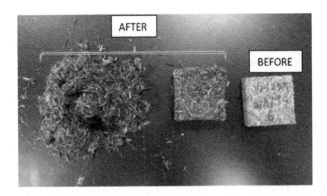

FIGURE 11.11   Condition of starch-based adhesives' test piece before and after 24 hours' water immersion.

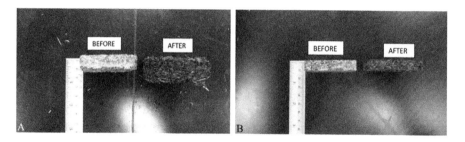

FIGURE 11.12 (a) Thickness swelling of starch-based adhesive particleboard test piece before and after 24 hours' water immersion (b). Thickness swelling of UF particleboard test piece before and after 24 hours' water immersion.

11.12.2 THICKNESS SWELLING

In this study, the changes in the thickness of the test piece between particleboard bound with UF and with a starch-based adhesive can be observed clearly in Figure 11.12. The test piece of starch-based adhesive experienced higher swelling in thickness compared to the test piece of urea formaldehyde after 24 hours of water immersion.

As shown in Figure 11.13, the graph did not fluctuate much for the same amount of adhesive content in each starch-based adhesive. Meanwhile, it can be concluded that among different starch-based adhesives, 10S with 20% adhesive content gave the highest mean value of thickness increase. When compared to urea formaldehyde (control), it was observed that particleboard manufactured using urea formaldehyde does not swell as much as that made using starch-based adhesive, as shown in Figure 11.12.

The better dimensional stability of UF might due to its strong resin properties with stable adhesive forces (Malhotra et al., 2012). Therefore, urea formaldehyde provides better water resistance than starch-based adhesive. The sufficient bonding

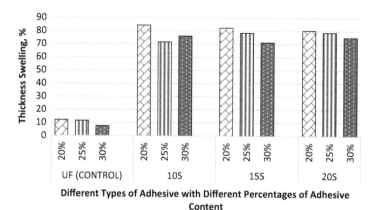

FIGURE 11.13 Average thickness swelling (%) for particleboards bound using different amounts of adhesive content (20%, 25%, and 30%) prepared from adhesives with different amounts of starch incorporated (10S, 15S, and 20S).

strength with a reduction in the internal force generated by water is also one of the reasons urea formaldehyde performs better than starch-based adhesive (Abdul Khalil et al., 2011).

The swelling of the composite material was influenced by several factors such as the quality and distribution of adhesive, moisture content, compatibility, and chemical composition of the furnish (Iswanto et al., 2014). In this case, it was possible that the starch-based adhesive has poor dispersion when mixed with particleboard. Moreover, dissolution of the material due to weak bonding between neighboring polymer also causes the material to swell (Oakley, 2010).

11.13   CONCLUSIONS

In this chapter, the effect of different amounts of starch (10S, 15S, and 20S) on the physical properties, which include density and MC; mechanical properties, which are MOE, MOR, and IB; and the dimensional stability (WA and TS) of particleboard produced were evaluated. From the results obtained, it can be determined that different amounts of starch influence the performance of WA, MC, and IB. Furthermore, the MC test registers an effect when the amount of starch in starch-based adhesives was 10% difference whereas, in the IB test, an effect can only be detected when the amount of starch was 5% difference from 10S to 20S. When different amounts of starch (10S, 15S, and 20S) were incorporated in starch-based adhesive and tested for TS, density, MOE, and MOR, no distinct effect could be detected.

The overall summaries of results show that different amounts of starch with different adhesive contents influence the performance of density and IB only. Significant changes can be observed in density when the amount of starch increases from 10S to 20S at 30% adhesive content. Subsequently, IB test also, show significant influences when the amount of adhesive content increases by 10% for each starch-based adhesive. In a nutshell, different adhesive content gives no significant effect especially in WA, TS, MC, MOE, and MOR test.

In short, starch-based adhesives which have higher values of WA and TS as compared to UF were higher. This indicates that more water was being absorbed, meaning that water resistance was poor. These results are especially unfavorable in the wood industry as poor water resistance of wood will cause more problems in its performance. The presence of starch is what is responsible for the highwater absorptivity and swelling of particleboard.

REFERENCES

Abdul Khalil, H. P. S., Jawaid, M., and Abu Bakar, A. 2011. Woven hybrid composites: Water absorption and thickness swelling behaviors. *BioResources*, 6(2): 1043–1052.

Abdullah, C. K., Jawaid, M., Abdul Khalil, H. P. S., Zaidon, A., and Hadiyane, A. 2012. Oil palm trunk polymer composite: Morphology, water absorption, and thickness swelling behaviors. *BioResources*, 7(3): 2948–2959.

Ahmad, Z. Y. 1992. Growth of *Acacia mangium* during three years following thinning: Preliminary results. *Journal of Tropical Forest Science*, 6: 171–180.

Akbari, S., Gupta, A., Khan, T. A., Jamari, S. S., Ani, N. B. C., and Poddar, P. 2014. Synthesis and characterization of medium density fiber board by using mixture of natural rubber latex and starch as an adhesive. *Journal of the Indian Academy of Wood Science*, 11(2): 109–115.

Anon. 1998. *Polymers and Surface Coating. Adhesive* (2nd edition). New Zealand: Chemical Processes.

Anon. 2007. Industrial starch chemistry. http://www.agrobynature.com/IndustrialStarchC hemistry.pdf. Accessed on 26 March 2016.

Anon. 2015. *Chemical Economics Handbook*. Formaldehyde. London: IHS Inc.

Ashori, A., and Nourbakhsh, A. 2008. Effect of press cycle time and resin contents on physical and mechanical properties of particleboard panels made from the underutilized low-quality raw materials. *Industrial Crops and Products*, 28(2): 225–230.

Banker, G. S. 1996. Film coating theory and practice. *Journal of Pharmaceutical Sciences*, 55(1): 81–89.

Bin Awang, K. 2010. *Formaldehyde Emissions from Wood Based Panels*. Gambang: University Malaysia Pahang.

Clark, P. A. 1991. Panel products: Past, present and future developments. *Journal of the Institute of Wood Science*, 12(4): 233–241.

Davidson, R. L. 1980. *Handbook of Water-Soluble Gums and Resins*. New York, NY: McGraw-Hill, Inc.

Dunky, M. 1998. Urea-formaldehyde (UF) adhesive resins for wood. *International Journal of Adhesion and Adhesives*, 18(2): 98–107.

Faherty, K. F., and Williamson, T. G. 1995. *Wood Engineering and Construction Handbook* (2nd edition). New York, NY: McGraw-Hill, Inc.

Gaspar, M., Benko, Zs., Dogossy, G., Reczey, K., and Czigany, T. 2005. Reducing water absorption in compostable starch-based plastics. *Polymer Degradation and Stability*, 90(3): 563–569.

Gérard, J., Guibal, D., Paradis, S., Vernay, M., Beauchêne, J., Brancheriau, L., and Thibaut, A. 2011. Tropix 7 (Version 7.5.1): Acacia mangium's data sheet. CIRAD. doi:10.18167/74726f706978.

Glavas, L. 2011. *Starch and Protein-Based Wood Adhesives*. Degree project in Polymer Technology. Nacka, Sweden.

Gu, Z., Wang, Z., Hong, Y., Cheng, L., and Li, Z. 2011. Bonding strength and water resistance of starch-based wood adhesive improved by silica nanoparticles. *Carbohydrate Polymers*, 86(1): 72–76.

Hashim, R., Mohd Salleh, K., Wan Nadhari, W. N. A., Abd Karim, N., Jumhuri, N., Ang, L. Z. P., and Hiziroglu, S. 2015. Evaluation of properties of starch-based adhesives and particleboard manufactured from them. *Journal of Adhesion Science and Technology*, 29(4): 319–336.

Hemmila, V., Trischler, J., and Sandberg, D. 2013. Bio-based adhesives for the wood industry – An opportunity for the future? *Pro Ligno*, 9(4): 118–125.

Idris, U. D., Aigbodion, V. S., Atuanya, C. U., and Abdullahi, J. 2011. Eco-friendly (watermelon peels): Alternatives to wood-based particleboard composites. *Tribology in Industry*, 33(4): 173–181.

Imam, S. H., Gordon, S. H., Mao, L., and Chen, L. 2001. Environmentally friendly wood adhesives from a renewable plat polymer: Characteristics and optimization. *Polymer Degradation and Stability*, 73(3): 529–533.

Imam, S. H., Mao, L., Chen, L., and Greene, R. V. 1999. Wood adhesive from crosslinked poly (vinyl alcohol) and partially gelatinized starch: Preparation and properties. *Starch/Stärke*, 51(6): 225–229. doi:10.1002/(SICI)1521-379X(199906)51:6<225::AID-STAR225>3.0.CO;2-F.

Iswanto, A. H., Azhar, I., Supriyanto, Ir., and Susilowati, A. 2014. Effect of resin type, pressing temperature and time on particleboard properties made from sorghum bagasse. *Agriculture, Forestry and Fisheries*, 3(2): 62–66.

Japanese Industrial Standard (JIS). 2003. *JIS A 5908 – Particleboards*. Tokyo: Japanese Standard Association.

Johnson, A. C., and Yunus, N. 2009. *Particleboards from Rice Husk: A Brief Introduction to Renewable Materials of Construction*. Malaysia: The Institute of Engineers.

Khosravi, S. 2011. *Protein-Based Adhesives for Particleboards*. Licentiate Thesis. Sweden: School of Chemical Science and Engineering, Royal Institute of Technology.

Kinloch, A. J. 1987. *Adhesion and Adhesives*. Science and Technology. Springer Science and Business Media.

Lei, H., Du, G., Wu, Z., Xi, X., and Dong, Z. 2014. Cross-linked soy-based wood adhesives for plywood. *International Journal of Adhesion and Adhesives*, 50: 199–203.

Li, Z., Wang, J., Cheng, L., Gu, Z., Hong, Y., and Kowalczyk, A. 2014. Improving the performance of starch-based wood adhesive by using sodium dodecyl sulfate. *Carbohydrate Polymers*, 99: 579–583.

Liew, K. C., and En, G. Y. C. 2016. Mechanical properties of 3 ply plywood made from *Acacia mangium* veneers and green starch-based adhesives. 2nd International Conference on Green Design and Manufacture 2016 (IConGDM 2016), 1–2 May 2016. Phuket, Thailand.

Liew, K. C., and Khor, L. K. 2015. Effect of different ratios of bioplastic to newspaper pulp fibres on the weight loss of bioplastic pot. *Journal of King Saud University-Engineering Sciences*, 27(2): 137–141.

Lopez-Gil, A., Rodriguez-Perez, M. A., and De Saja, J. 2014. Strategies to improve the mechanical properties of starch-based materials: Plasticization and natural fibers reinforcement. *Polímeros*, 24(spe): 36–42. doi:10.4322/polimeros.2014.053.

Lourdin, D., Della Valle, G., and Colonna, P. 1995. Influence of amylose content on starch films and foams. *Carbohydrate Polymers*, 27(4): 261–270.

Malhotra, N., Sheikh, K., and Rani, S. 2012. A review on mechanical characterization of natural fiber reinforced polymer composites. *Journal of Engineering Research and Studies*, 3(1): 75–80.

Mekonnen, T., Mussone, P., Khalil, H., and Bressler, D. 2013. Progress in bio-based plastics and plasticizing modifications. *Journal of Material Chemistry A*, 1(43): 13379–13398. doi:10.1039/C3TA12555F.

Mohd Hamami, S., Ashaari, Z., Abdul Kader, R., and Abdul Latif, M. 1998. Physical and mechanical properties of *Acacia mangium* and *Acacia auriculiformis* from different provenances. *Journal of Tropical Agricultural Science*, 21(2): 73–81.

MTIB. 2015. *MTIB E-Statistics – Export*. Vol. 12. Kuala Lumpur: Malaysia Timber Industry Board.

Oakley, P. 2010. *Reducing the Water Absorption of Thermoplastic Starch Processed by Extrusion*. Degree of Master of Applied Science. Thesis. Canada: University of Toronto.

Onusseit, H. 1992. Starch in industrial adhesives. *Industrial Crops and Products*, 1(2–4): 141–146.

Orwa, C., Mutua, A., Kindt, R., Jamnadass, R., and Simons, A. 2009. *Agroforestree Database: A Tree Reference and Selection Guide Version 4.0*. Kenya: World Agroforestry Centre.

Paris, J. L., and Kamke, F. A. 2015. Quantitative wood-adhesive penetration with X-ray computed tomography. *International Journal of Adhesion and Adhesives*, 61: 71–80.

Pizzi, A., and Mittal, K. L. 2003. *Handbook of Adhesive Technology* (2nd edition, revised, and expanded). New York: Marcel Dekker, Inc.

Prompunjai, A., and Sridach, W. 2010. Preparation and some mechanical properties of composite materials made from sawdust, cassava starch, and natural rubber latex. *International Journal of Chemical, Molecular, Nuclear, Materials and Metallurgical Engineering*, 4(12): 772–776.

Raj, B., and Somashekar, R. 2004. Structure-property-relation in polyvinyl alcohol/starch composites. *Journal of Applied Polymer Science*, 91(1): 630–635.

Saraswat, Y., Patel, M., Sagar, T., and Shil, S. 2014. Bioplastics from starch. *International Journal of Research and Scientific Innovation*, 1(8): 385–387. ISSN: 2321-2705.

Starck, N. M., Cai, Z., and Carll, C. 2010. Wood handbook, chapter 11: Wood-based composite materials-panel products-glued laminated timber, structural composite lumber, and wood non-wood composite materials. *General Technical Report FPL-GTR-190*. Madison, WI: U.S. Department of Agriculture, Forest Service, Forest Products Laboratory.

Tanrattanakul, V., and Chumeka, W. 2010. Effect of potassium persulfate on graft copolymerization and mechanical properties of cassava starch/natural rubber foams. *Journal of Applied Polymer Science*, 116(1): 93–105.

Tay, C. C., Hamdan, S., and Bin Osman, M. S. 2016. Properties of sago particleboards resonated with UF and PF resin. *Advances in Materials Science and Engineering*. doi:10.1155/2016/5323890.

Walker, J. C. F. 2006. *Primary Wood Processing: Principles and Practice* (2nd edition). Netherlands: Springer.

Wang, Y. 2006. *Adhesive Performance of Soy Protein Isolate Enhanced by Chemical Modification and Physical Treatment*. Dissertation Doctor of Philosophy. Manhattan: Kansas State University.

Wang, Z., Gu, Z., Li, Z., Hong, Y., and Cheng, L. 2013. Effects of urea on freeze–thaw stability of starch-based wood adhesive. *Carbohydrate Polymers*, 95(1): 397–403.

Wool, R., and Sun, X. S. 2011. *Bio-Based Polymer and Composites*. Cambridge: Academic Press.

Wu, Y. B., Lv, C. F., and Han, M. N. 2009. Synthesis and performance study of polybasic starch graft copolymerization function materials. *Advanced in Material Research*, 79–82: 43–46.

Yu, H. W., Cao, Y., Fang, Q., and Liu, Z. 2015. Effects of treatment temperature on properties of starch-based adhesives. *BioResources*, 10(2): 3520–3530.

Zhu, L., Zhang, Y., Ding, L., Gu, J., and Tan, H. 2015. Preparation and properties of a starch-based wood adhesive with high bonding strength and water resistance. *Carbohydrate Polymers*, 115: 32–37.

12 Laminated Veneer Lumber from Acacia Hybrid

Grace Singan and Kang Chiang Liew

CONTENTS

12.1 INTRODUCTION

The increasing demand for solid wood material nowadays has highlighted the potential for timber production from fast-growing timber species. *Acacia* spp. is a fast-growing plantation wood that is appealing for its timber and pulp in the tropics (Tuong and Li, 2011) and grows extensively, especially in Southeast Asia, where over 10 million hectares have been planted (Muhammad et al., 2017). Both *Acacia mangium* and *Acacia auriculiformis* are leguminous trees from the Mimosoideae sub-family, with *Acacia mangium* being categorized as pea flowering tree family species known as *Fabaceae* (Bakri et al., 2018). The hybrid between these two Acacia species produces Acacia hybrid, which is more appealing than its parent trees in terms of its strength, durability, physical properties and adaptabilities.

In Malaysia, the first Acacia hybrid was discovered in a natural state in Sabah near Ulu Kukut Plantation back in 1971 by Hepburn and Shim (Pinso and Nasi, 1991; Rufelds, 1987). Acacia hybrid may occur either naturally or through biclonal orchard production or controlled crosses (Rusli et al., 2013). Where natural regeneration took place, both *Acacia mangium* and *Acacia auriculiformis* had been planted in close proximity, and the seedlings of the hybrid were subsequently collected from planted area and brought to other localities, giving rise to Acacia hybrid plantations

(Paiman et al., 2018). Acacia hybrid is an upgraded version of its predecessors as it is denser, has better stem straightness and circulatory system, self-pruning ability, susceptible to heart rot disease, lighter branches, rapid growth, and less tapering (Bakri et al., 2018; Mohd. Hamami and Semsolbahri, 2003; Paiman et al., 2018; Rusli et al., 2013; Yahya et al., 2010).

Meanwhile, in Sarawak, Acacia hybrid was introduced along with the clone of second-generation *Acacia mangium* known as Acacia super bulk (Bakri et al., 2018). By 2020, there will be approximately one million hectares of land reserve in Sarawak for commercialization and development programs such as reforestation and replantation of Acacia and other fast-growing timber trees. This is an effort by the state to meet the demand for raw materials from the timber industries as well as to maintain forest conservation (PERKASA, 2009). The rise in Acacia hybrid exploitation in the near future could mean potential alternative raw materials for use in various wood-based industries. If plantation productivity is sustained and increased, it could also provide increased wood supply for local processing industries, strengthen the rural economy and improve living standards for small growers and local workers (Nambiar, 2015).

12.2   ACACIA HYBRID AS WOOD PRODUCT

The aim of establishing a forest plantation is to create a sustainable timber resource to supply various wood-based industries. Raw materials such as timber trees are declining as the result of high demand for tropical trees on both the domestic and international markets. Acacia hybrid is a well-known fast-growing timber that could meet current demand for its product. As a renewable natural resource, it is important to understand its characteristics, such as the wood's properties and growth rate (Paiman et al., 2018). For use in furniture making, its machining properties are very important to specify its surface quality, which in turn determines the finishing and adhesive strength properties (Sofuoglu, 2017). Wood products that have good machining properties are useful for high value-added products in industrial applications (Ruiz-Aquino et al., 2018).

12.3   PROPERTIES OF ACACIA HYBRID

Acacia mangium is a leguminous tree that comes from the Mimosoideae family species. It is vastly planted which is useful for commercialization of wood products. However, *Acacia mangium* poses certain drawbacks that make it susceptible and lacking in terms of its log condition and susceptibility to heart rot damage (Weinland and Yahya, 1991). This led to the introduction of Acacia hybrid, a mixture of *Acacia mangium* and *Acacia auriculiformis* that has a faster growth rate, high resistance to heart rot, straighter bole structure and reduced tapering (Bakri et al., 2018). In terms of its mechanical properties, Acacia hybrid has much lower Modulus of Rupture (MOR) and Modulus of Elasticity (MOE) compared to *Acacia mangium* (Jusoh et al., 2014). According to Sunarti et al. (2013), Acacia hybrid has a better tolerance to pests and diseases. The hybrid is currently developed through natural hybrid selection from trees grown in plantation.

12.4   PRODUCTION OF LAMINATED VENEER LUMBER (LVL)

Acacia hybrid logs were harvested from SAFODA located at Ulu Kukut, Sabah, taken from a stand of approximately 3–4 years old and a diameter of breast height (DBH) between 20 and 25 cm. The experimental design is shown in Figure 12.1.

The logs are peeled to a thickness of 2 mm using a hydraulic spindleless peeling machine, with moisture content maintained at 10%–11%. The LVL were then produced by 300 × 300 mm veneers that were glued using urea formaldehyde resin with the same grain direction arrangement. A urea formaldehyde resin was mixed with ammonium chloride as a hardener to hasten its cure time. For each layer of veneer,

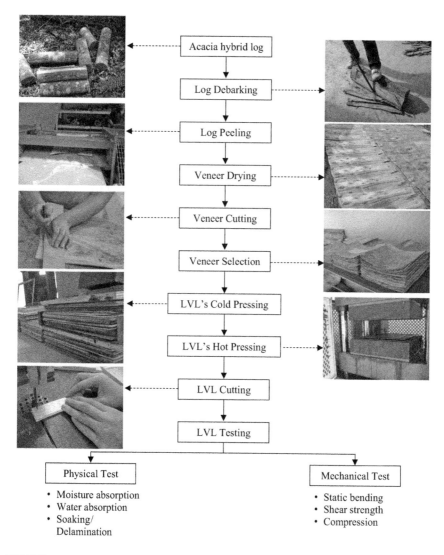

FIGURE 12.1   The experimental design for producing LVL from Acacia hybrid.

FIGURE 12.2   LVL made from Acacia hybrid.

the resin was applied at 148 g/m². The LVL production was produced at 3-ply, 5-ply and 7-ply. The total replicates for each layer of LVL was ten, and so altogether 30 LVL were produced. After the resin spreading process, the LVL underwent cold pressing for 2–3 hours before being hot pressed. For the hot press process, the LVL required 20 kg/cm² pressure at 115°C for nearly 5 minutes. The LVLs (Figure 12.2) were put to rest at room temperature before sample cutting.

One way analysis of variance (ANOVA) was performed at 5% of significant difference in order to determine the physical characteristics of LVL, mainly for moisture absorption and soaking among 3-ply, 5-ply and 7-ply Acacia hybrid LVL. An LSD test was used to further analyze the difference among the means. All analyses were computed using IBM SPSS Statistics version 24 for Windows.

12.5   THE PHYSICAL CHARACTERISTICS OF LAMINATED VENEER LUMBER FROM ACACIA HYBRID

The mean value of moisture absorption and soaking of Acacia hybrid LVL are shown in Table 12.1. LVL of 7-ply recorded the highest value for both moisture absorption and soaking, while the LVL of 3-ply shows the lowest value. For the moisture absorption test, there are significant differences at $p \leq 0.05$ for all types of LVL layers, while for the soaking test, there are no significant differences among them. The results show that there are no significant differences for the soaking test, probably because of the same raw materials used and manufacturing conditions for 3-ply, 5-ply and 7-ply LVL (Tenorio et al., 2011). The increasing value for both moisture absorption and soaking test of three types of LVL layers indicate that the more veneer layers in the LVL production, the more moisture it can absorb, hence affecting the soaking characteristics. There are several factors that affect the moisture absorption and soaking of LVL, such as the type of raw materials it is made of, glue strength and the initial moisture of the veneer (Abdul et al., 2010). Since Acacia

---

TABLE 12.1

The Moisture Absorption and Soaking Properties of LVL Made from Acacia Hybrid for 3-ply, 5-ply and 7-ply

LVL Layers	Moisture Absorption	Soaking
3-ply	10.75[a]	22.5[a]
	(± 0.48)	(± 22.67)
5-ply	11.52[b]	26.25[a]
	(± 0.59)	(± 14.67)
7-ply	13.19[c]	27.5[a]
	(± 0.71)	(± 18.13)

Note: Mean value followed by a letter (a, b or c) within a column represents a significant difference at $p \leq 0.05$ @ 5%. Values in parentheses represent standard deviation.

---

hybrid is a fast-growing wood species, its density is less than the hardwood timber species, enhancing its ability to absorb more moisture.

For the water absorption test, it took two time periods to determine the total water absorption of LVL from Acacia hybrid: 2 hours and 24 hours. Figure 12.3 shows that there are inclining values from 2 hours of water absorption to 24 hours. For 3-ply LVL, the water absorption value increased from 27.88% to 53.42%, and for 5-ply LVL, it increased from 21.98% to 40.67%. However for 7-ply LVL, the water absorption value increases from 25.04% to 42.90%, slightly higher than 5-ply value. The reason for this is that, since wood is a hygroscopic material itself, at some point the panels become less susceptible to dimensional change. Tenorio et al. (2011) suggest that even though the different panels are fabricated from the same raw material, the difference between them can only be distinguished through the veneers' orientation. The pattern of water absorption shown in Figure 12.3 was nearly the same and it proves this statement.

12.6 THE MECHANICAL STRENGTH OF LAMINATED VENEER LUMBER FROM ACACIA HYBRID

The mechanical strength properties of LVL were investigated in this study to determine the structural characteristics of Acacia hybrid wood. The wood's MOR and MOE is an important indicator of the materials' properties required for use in building construction (Bal and Bektas, 2012).

The MOR and MOE were determined using Universal Testing Machine GOTECH A1-7000L-10, by adopting ASTM D5456 standard. From the result shown in Figure 12.4, both the MOR and MOE values decline with the increasing number of LVL layers. The 3-ply LVL shows higher MOR and MOE values, while the 7-ply LVL shows lower MOR and MOE values. Since urea formaldehyde resin is used in the production of LVL, it can be a factor in the bending strength result, since Bal and Bektas (2012) had mentioned that formaldehyde-based adhesives could affect its flexural

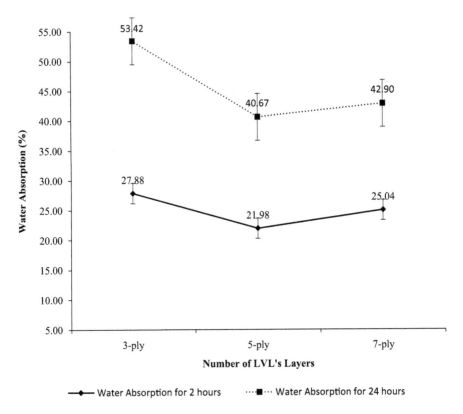

FIGURE 12.3   The water absorption of LVL 3-ply, 5-ply and 7-ply from Acacia hybrid over 2 hours and 24 hours respectively.

properties, especially the internal bond strength. However, the shear strength result shown in Figure 12.5 gave a different answer. It shows that 3-ply LVL has the higher shear strength at 1.10 N/mm² while 5-ply LVL has the lowest value at 0.58 N/mm². According to the JAS standard, the passing value is 0.7 N/mm², which means that the 3-ply and 7-ply LVL achieve the passing standard and are eligible to be used in structural and construction works. The bonding of LVL was much affected by its wettability, as shown in Figure 12.3. According to Alamsyah et al. (2007), the higher wettability resulted in better adhesive spread and more contact between wood surface and adhesive. Wettability in wood usually refers to the rate at which a liquid can wet and spread on its surface (Wellons, 1980). It is a quick way to predict the gluability of wood-based products and can be measured by determining the contact angle between the solid–liquid interface and the liquid–air surface (Bodig, 1962).

Meanwhile, a compression test was done to determine the compressibility and elasticity of LVL, especially the one that produced from plantation species. From Figure 12.6, the maximum compression strength of materials against the 500 kgf pressure are evaluated. It was observed that 3-ply LVL has a lower compression value and this declines until 7-ly LVL. Since compression test is mainly been undergone to evaluate its stiffness, physical strength, hardness and abrasion resistance,

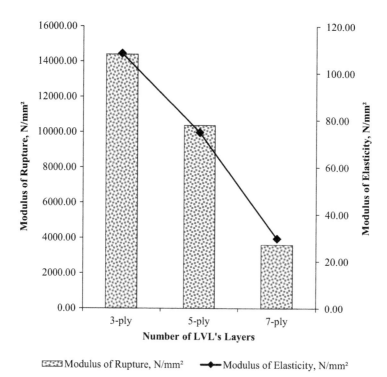

FIGURE 12.4 The bending strength of Acacia hybrid LVL 3-ply, 5-ply and 7-ply respectively that shows declined value.

7-ply LVL had shown a favorable result that is 3063.30 N/mm² (Inoue et al., 2008). When in contact with moisture and heat, the compressed LVL recovered its initial state, given the same mechanical strength as sawn timber (Inoue et al., 2008). So, the thicker the LVL, the similar the characteristics of LVL made from sawn timber in terms of physical strength.

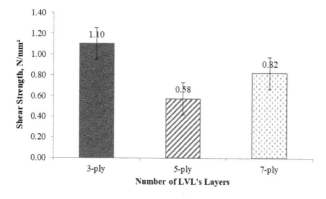

FIGURE 12.5 The shear strength of Acacia hybrid LVL for 3-ply, 5-ply and 7-ply.

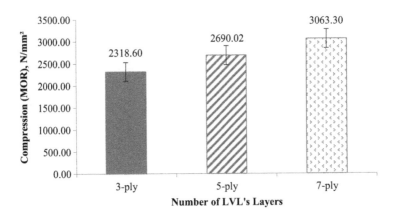

FIGURE 12.6 The compression test of Acacia hybrid LVL for 3-ply, 5-ply and 7-ply, showing the highest value of 7-ply LVL.

12.7 CONCLUSION

The study shows that LVL made from Acacia hybrid has a low moisture content despite its poor water holding capacities, which was proven by the soaking test. The reason is that the LVL posses high hygroscopic characteristics; the water absorption test shows that it can absorb more water over a 24-hour period. In terms of mechanical strength, the 3-ply LVL has a high rupture rate compared to thicker LVL. It shows that the thinner the LVL layer, the more the elasticity of Acacia hybrid wood products declines, making it easier to break. The thicker LVL layers manage to retain high strength due to their lower rupture properties. However, the 3-ply LVL has high shear strength because of its capacity to hold more glue due to its thinner veneer layer.

ACKNOWLEDGMENTS

The authors extend special thanks to the following individuals for providing the wood materials for the experiments in this chapter: SAFODA Sabah and both Mr Azli Sulid and Mr Rizan Gulam Hussein of Universiti Malaysia Sabah.

REFERENCES

Abdul, H.P.S., Nurul, M.R., Bhat, A.H., Jawaid, M., and Nik, N.A. 2010. Development and material of new hybrid plywood from oil palm biomass. *Materials and Design*, 31: 417–424.

Alamsyah, E.M., Liu, C.N., Yamada, M., Taki, K., and Yoshida, H. 2007. Bondability of tropical fast-growing tree species I: Indonesian wood species. *Journal of Wood Science*, 53(1): 40–46.

Bakri, M.K., Jayamani, E., Hamdan, S., Rahman, M.R., and Kakar, A. 2018. Potential of Borneo Acacia wood in fully biodegradable bio-composites' commercial production and application. *Polymer Bulletin*, 75(11): 5333–5354.

Bal, B.C., and Bektas, I. 2012. The effects of wood species, load direction, and adhesives on bending properties of laminated veneer lumber. *BioResources*, 7(3): 3104–3112.

Bodig, J. 1962. Wettability related to gluabilities of five Philippine Mahoganies. *Forest Product Journal*, 12: 265–270.

Inoue, M., Kawai, S., Walinder, M., and Rowell, R.M. 2008. Dimensional stabilization of compressed laminated veneer lumber by hot pressing in an airtight frame. *Wood Material Science and Engineering*, 3–4(3–4): 119–125.

Jusoh, I., Abu Zaharin, F., and Adam, N.N. 2014. Wood quality of acacia hybrid and second generation acacia Mangium. *BioResources*, 9(1): 150–160.

Mohd Hamami, S., and Semsolbahri, B. 2003. Wood structures and wood properties relationship in planted acacias: Malaysian examples. Proceedings of International Symposium on Sustainable Utilization of Acacia Mangium, Kyoto, October 24–34.

Muhammad, A.J., Ong, S.S., and Ratnam, W. 2017. Characterization of mean stem density, fibre length and lignin from two Acacia species and their hybrid. *Journal of Forestry Resources*, 29(2): 549–555.

Nambiar, E.K.S. 2015. Forestry for rural development, poverty reduction and climate change mitigation: We can help more with wood. *Australian Forestry*, 78(2): 55–64.

Paiman, B., Lee, S.H., and Zaidon, A. 2018. Machining properties of natural regeneration and planted *Acacia mangium* × *A. Auriculiformis* hybrid. *Journal of Tropical Forest Science*, 30(1): 135–142.

PERKASA 2009. Sarawak timber Ind. Seminar on Viability Assessment of Indigenous Tree Species and Propagation Techniques for Planted Forest Development in Sarawak. *Dev. Corp. Newsletter*, 5(6): 6–8.

Pinso, C., and Nasi, R. 1991. The potential use of *Acacia mangium* and *A. Auriculiformis* hybrid in Sabah. *In*: Carron, L. and Aked, K. (Eds.) *Breeding Technologies for Tropical Acacias*. ACIAR, Canberra, Pp. 17–21.

Rufelds, C.W. 1987. Quantitative Comparison of Acacia *mangium* Willd. versus Hybrid Acacia *auriculiformis*. FRC Publication No. 40. Forest Research Centre, Sepilok.

Ruiz-Aquino, F., Gonzalez-Pena, M.M., Valdez-Hernandez, J.I., Romera-Manzanares, A., and Fuentes-Salinas, M. 2018. Mechanical properties of wood of two Mexican oaks: Relationship to selected physical properties. *European Journal of Wood and Wood Products*, 76(1): 69–77.

Rusli, R., Samsi, H.W., Kadir, R., Ujang, S., Jalaludin, Z., and Misran, S. 2013. Properties of small diameter acacia hybrid logs for biocomposites production. *Borneo Science*, 33: 9–15.

Sofuoglu, S.D. 2017. Determination of optimal machining parameters of massive wooden edge glued panels which is made of Scots pine (*Pinus sylvestris* L.) using Taguchi design method. *European Journal of Wood and Wood Products*, 75(1): 33–42.

Sunarti, S., Na'iem, M., Hardiyanto, E.B., and Indrioko, S. 2013. Breeding strategy of acacia hybrid (Acacia Mangium x A. auriculiformis) to increase forest plantation productivity in Indonesia. *Journal of Tropical Forest Management*, 19(2): 128–137.

Tenorio, C., Moya, R., and Munoz, F. 2011. Comparative study on physical and mechanical properties of laminated veneer lumber and plywood panels made of wood from fast-growing Gmelina arborea trees. *The Japan Wood Research Society*, 57: 134–139.

Tuong, V.M., and Li, J. 2011. Changes caused by heat treatment in chemical composition and some physical properties of acacia hybrid sapwood. *Holzforschung: International Journal of the Biology, Chemistry, Physics, and Technology of Wood*, 65(1): 67–72.

Weinland, G., and Yahya, A.A. 1991. *Management of Acaia, Amgium Stands: Trending Issues*. Resource Institute Malaysia, Forestry, Pp. 41–53.

Wellons, J.D. 1980. Wettability and gluability of Douglas-fir veneer. *Forest Product Journal*, 30: 53–55.

Yahya, R., Sugiyama, J., and Silsia, D., and Gril, J. 2010. Some anatomical features of an acacia hybrid, *A. Mangium* and *A. Auriculiformis* grown in Indonesia with regard to pulp yield and paper strength. *Journal of Tropical Forest Science*, 22(3): 343–351.

Index

Printed and bound by CPI Group (UK) Ltd, Croydon, CR0 4YY

17/10/2024

01775709-0009